Isaac Norman Broomell

Anatomy and Histology of the Mouth and Teeth

Isaac Norman Broomell

Anatomy and Histology of the Mouth and Teeth

ISBN/EAN: 9783337405281

Printed in Europe, USA, Canada, Australia, Japan

Cover: Foto ©berggeist007 / pixelio.de

More available books at **www.hansebooks.com**

Isaac Norman Broomell

Anatomy and Histology of the Mouth and Teeth

ISBN/EAN: 9783337405281

Printed in Europe, USA, Canada, Australia, Japan

Cover: Foto ©berggeist007 / pixelio.de

More available books at **www.hansebooks.com**

ANATOMY AND HISTOLOGY

OF THE

MOUTH AND TEETH

BY

I. NORMAN BROOMELL, D.D.S.

PROFESSOR OF DENTAL ANATOMY, DENTAL HISTOLOGY, AND PROSTHETIC TECHNICS IN THE
PENNSYLVANIA COLLEGE OF DENTAL SURGERY, PHILADELPHIA

With 284 Illustrations

PHILADELPHIA
P BLAKISTON'S SON & CO.
1012 WALNUT STREET
1899

TO

C. N. PEIRCE, D.D.S.

AS A SOUVENIR OF A LONG AND VALUED FRIENDSHIP AND A

TESTIMONY OF ESTEEM FOR HIS PROFESSIONAL

AND PRIVATE WORTH

This Volume is Respectfully Dedicated

BY THE AUTHOR

PREFACE.

In the preparation of this work it has been the aim of the author to systematically describe those parts of human anatomy which come directly under the care of the stomatologist. In the earlier chapters, which are devoted to a gross description of the mouth and those tissues which enter into its construction, there has been no attempt at originality other than in the arrangement, which includes a complete description of one part before another is taken up.

In the writing and classification of the succeeding chapters the writer has attempted what others, though wiser and better qualified, appeared unwilling to undertake, and it is from the works of such as these that the foundation for the present work has been derived.

Within the last few years the progress in nearly every branch of dental education has made a work of this character an imperative want. Dental therapeutics and dental chemistry have been well-nigh reconstructed, while the investigations of the microscopist and physiologist have brought forth many valuable revelations. Next in importance has been the advance in, or rather the introduction of, technic teaching. Considerable space has, therefore, been devoted to the surface anatomy of the individual teeth, with a hope that it may be of value in dental anatomy technic.

While in one or two instances the writer has departed from the field assigned as a text, the parts thus included are so closely associated with the mouth, both in a constructive and in a functional manner, that the work would be lacking in completeness if they were omitted.

The illustrations are, with but few exceptions, the original work of the author, being reproduced by photograph from the actual subject. In many instances dissections were required to reveal the parts, this being particularly true of those illustrations included in the chapter on the Development of The Teeth, about one hundred dissections being required to accomplish the purpose. In preparing the illustrations descriptive of the various surfaces of the individual teeth, the progress of the work was materially interfered with by the difficulty experienced in securing normal teeth out of the mouth; may their number ever grow less.

The author desires to thus publicly acknowledge obligations to the works of Tomes, Black, Morris, Stöhr, Klein, and Stricker. He is also indebted to Prof. A. P. Brubaker and to Dr. C. P. Shoemaker for valuable assistance rendered, and to P. Blakiston's Son & Co. for their many courtesies during the preparation of the volume.

That there is a place for such a work as this purports to be the writer has but little doubt; that the following pages will fill that demand is his earnest desire, and it remains for the reader to ascertain how far these demands have been met in the direction of its aim and endeavor.

302 North Fortieth St., Oct. 20, 1898.

TABLE OF CONTENTS.

PART I.—ANATOMY.

ANATOMY AND HISTOLOGY

OF THE

MOUTH AND TEETH.

PART I.—ANATOMY.

CHAPTER I.

GENERAL DESCRIPTION OF THE MOUTH.—THE BUCCAL ORIFICE (THE LIPS).—THE LATERAL WALLS OF THE MOUTH (THE CHEEKS). —THE HARD PALATE OR DOME OF THE MOUTH.—THE SOFT PALATE AND FAUCES.—THE FLOOR OF THE MOUTH.—THE TONGUE AND ITS ATTACHED MUSCLES.

The mouth (Fig. 1) (*stoma*, pl. *stomata*) is the entrance or gateway to the alimentary canal, and is situated between the superior and inferior maxillary bones and their attached tissues. It contains the *active* organs of mastication, *the teeth*, the organs of taste, of which the tongue is chief, together with a greater part of the organs of speech. Anatomists usually divide this cavity into two compartments, the teeth serving to separate one from the other, the inner space being called *the mouth*, while that between the teeth and lips or cheeks is known as the *vestibule of the mouth*. In this description all that space bounded anteriorly by the lips, posteriorly by the fauces, and laterally by the cheeks, will be considered as a single cavity, and the organs and structures contained therein, together with all parts directly interested in its formation, will constitute a text for this work. The *entrance to the* cavity of the *mouth* is formed and controlled by a freely movable transverse orifice or slit, the *buccal orifice*, while it communicates with the pharynx posteriorly through the fauces. Entering into the construction of the mouth and assisting in the performance of its functions are bones, ligaments, muscles, blood-vessels, nerves, glands, ducts, etc., each of which will be described in turn.

THE BUCCAL ORIFICE,

or entrance to the cavity of the mouth, is a transverse opening somewhat variable in extent, the extremities of which are known as the *corners* or *angles of the mouth*. The orifice is bounded by two fleshy folds, the *upper* and *lower lips* (*labia*), the former

FIG. I.—A GENERAL VIEW OF THE MOUTH.

usually being in the form of a double curve, coming together in the center and forming a small teat or tubercle, while the latter is made up of a single curve extending from angle to angle. While this general description applies to the labial forms most frequently met with, it must not be mistaken for a constant condition. In some instances the lips are thin, with straight parallel margins, firmly set against the teeth, and seldom separated from each other when at rest. In another class they are thick, full, and prominent, with their margins strongly curved, resting lightly against the teeth, and more or less separated from each other during rest. Accompanying the extreme as well as the intervening conditions are various other peculiarities, such as the color, the rigidity or flexibility of the muscular structure, etc. The upper lip generally overhangs the lower, but in some instances the lower lip is the most prominent. Externally the lips are covered by the common integument, internally and over their contiguous surfaces by a continuation of the integument, the mucous membrane. Between the external and internal coverings and forming the substance of these fleshy folds are muscular fibers in which are imbedded numerous blood-vessels, nerves, and glands (*labial glands*). By the various muscles which enter into their construction the lips are attached to the surfaces of the maxillary bones.

The **integument**, or **external covering** of the lips, is similar to the skin covering other parts of the body. In the male it is subject to a peculiar change and modification of its outer layer, resulting in the production of a hairy growth.

The **mucous membrane**, or **internal covering** of the lips, the beginning of which is strongly manifest by its bright-red color, is without moisture on the contiguous surfaces, is extremely sensitive, and contains a number of vascular papillæ, many of which are accompanied by nerve terminals. Mucous membranes are described as lining certain cavities or tracts, as the digestive tract, the respiratory tract, and the genito-urinary tract, and it is upon the contiguous surfaces of the lips that the digestive tract begins. The line of junction between the integument and the mucous membrane, is quite variable in form, but usually corresponds with the general curvature of the lips. Internally at the median line each lip is provided with a pronounced fold of mucous membrane, which is attached to the basal portion of the gum, the *frenum of the lip* (frænum labium superioris and inferioris), which, in a measure, check the movements of the lips.

Muscles of the Lips.

The muscular fibers within the substance of the lips are principally those of a single muscle, the *orbicularis oris*, but associated with it is a portion of the elevator and depressor muscles
of the lips, the *levator labii superioris alæque nasi, levator labii
superioris, depressor labii inferioris* or *quadratus menti*, and the
zygomaticus minor.

Orbicularis Oris.—This is the sphincter muscle which surrounds and controls the buccal aperture. In form it is an oval
sheet with the long axis placed transversely, the fibers being
continued from one lip to the other by passing around the
angles of the mouth. It is divided into an internal or labial
portion, and an external or facial portion. The labial portion
forms the red part of the lips, and has no bony attachment
except through the medium of the adjacent muscles. The
external or facial portion forms the deeper layer and blends
with the surrounding muscles, works in conjunction with them,
and is provided with the following small bony *attachments:*
The nasolabial slips are attached to the septum of the nose,
other fibers are attached to the incisive fossa of the superior
maxilla over the position of the lateral incisor tooth, and to the
incisive fossa of the inferior maxilla near the socket of the lateral
incisor or cuspid tooth.

Structure.—The muscle consists of three sets of fibers, one of
which runs transversely, one in a vertical, and one in an anteroposterior direction. The transverse set is continuous with the
fibers of the buccinator or cheek muscle, and form the greater
part of the muscle. The red or labial portion of the muscle is
also formed from the same fibers, while the vertical fibers form
the superficial part of the facial portion and are continuous with
the fibers of the levator and depressor muscles. Some of these
latter fibers pass around the corners of the mouth, thus becoming transverse, those from above passing to the lower lip, while
those from below pass to the upper lip. The anteroposterior
fibers pass from before backward between the transverse fibers,
and unite the mucous membrane to the skin. These are chiefly
found in the labial portion of the muscle.

Relations.—The inner margin of the superficial surface is closely connected with the integument, while superimposed between this and the outer portion is a layer of fatty tissue. Upon the deep surface lies the mucous membrane of the mouth, separated from the muscular tissue by blood-vessels, mucous glands, and small salivary glands.

Action.—To bring the lips together, to draw the upper lip downward, and the lower lip upward; to draw together the corners of the mouth; to throw both lips outward; to draw them back against the teeth, and to oppose the action of all other muscles that blend into it and inclining to draw it in various directions.

Levator Labii Superioris Alæque Nasi.—As its name implies, this muscle is an elevator of the upper lip and the wing of the nose. It is one of the superficial facial muscles, is thin and triangular, and is situated by the side of the nose, extending from the infra-orbital ridge to the upper lip.

Origin.—From the nasal process of the superior maxilla near its orbital margin.

Insertion.—From its origin it passes almost directly downward, dividing into two portions, the smaller of which is inserted into the nasal wing, while the larger portion is prolonged downward, blending into the orbicularis oris and levator labii superioris, and forming a part of the substance of the upper lip.

Relations.—Superficially, by the integument; deeply, by the levator anguli oris and compressor narium.

Action.—By its smaller and shorter portion to raise the wing of the nose and to dilate the nostril; by its larger and longer portion to elevate the inner half of the upper lip.

Levator Labii Superioris.—This muscle belongs to the superficial layer, and derives its name from its action.

Origin.—From the facial surface of the superior maxilla, at a point between the orbital cavity and the infra-orbital foramen. Also by the attachment of a few fibers to the malar bone.

Insertion.—Passing downward and inward, it is inserted into the orbicularis oris and the integument of the upper lip. Near its lower third it joins the levator labii superioris alæque nasi, and acts in conjunction with it. Occasionally it is reinforced by fibers from the orbicularis palpebrarum, which it receives at its outer border.

Relations.—Superficially, by the orbicularis palpebrarum and the integument; deeply, by the levator anguli oris; the compressor nasi at its origin, and by the infra-orbital vessels and nerves.

Action.—To elevate the upper lip.

Depressor Labii Inferioris, or Quadratus Menti.—The name of this muscle is derived from its form and action. It belongs to the second layer of facial muscles, is quadrilateral in shape, and consists of parallel fibers which meet above in the median line.

Origin.—At the outer aspect of the lower border of the inferior maxilla, from a point near the symphysis to the space beneath the first bicuspid tooth.

Insertion.—Its fibers pass upward and inward, and after uniting with its fellow of the opposite side, blend into the body of the orbicularis oris of the lower lip.

Relations.—By its superficial surface with the integument and a portion of the depressor anguli oris; deeply, with the mental nerve and vessels, a portion of the orbicularis oris, the mucous membrane lining the lower lip, and the labial glands.

Action.—To draw down and somewhat evert the lower lip.

Zygomaticus Minor.—An extremely slender muscle belonging to the superficial set of facial muscles. It is closely associated with a larger muscle, the zygomaticus major, belonging to the angular series, to be described in connection with the muscles of the cheek.

Origin.—From the anterior inferior part of one of the facial bones,—the malar,—close to its junction with the superior maxilla.

Insertion.—It passes downward and forward, its fibers becoming lost in the special elevator muscle of the upper lip about midway between the median line and the angle of the mouth.

Relations.—Superficially, by the integument, by its deep surface with the levator anguli oris, facial portion of the orbicularis oris, and the infra-orbital branch of the facial nerve.

Action.—To elevate and somewhat evert the upper part of the lip.

The Blood-supply to the Lips.

The blood-supply to the lips is principally through the *superior* and *inferior coronary* arteries, both of which are branches of the *facial* artery. In addition to these the *inferior labial* artery, also a branch of the *facial*, and the *mental branch* of the *inferior dental* artery supply a part of the lower lip.

The superior coronary artery courses along the inferior margin of the upper lip, between the mucous membrane and the fibers of the orbicularis oris muscle. At the median line it anastomoses with its fellow of the opposite side.

The inferior coronary artery, somewhat smaller than the superior, supplies the lower lip by coursing through its substance in a manner similar to the superior coronary and also anastomoses with its fellow of the opposite side at the median line.

Course of the Blood From the Heart to the Lips.—From the heart to the aorta, to the common carotid, to the external carotid, to the facial, to the superior and inferior coronary and the inferior labial arteries. After passing through the labial capillaries the blood is returned to the heart through the *superior* and *inferior coronary veins*, and the larger veins of which they are branches.

Nerves of the Lips.

The general nerve-supply to the lips is principally by small branches of the *infra-orbital* nerve for the upper lip, and by branches of the *mental* nerve for the lower lip. The *buccal* and *superior maxillary* branches of the lower division of the facial nerve supply the orbicularis oris muscle; the upper division of the facial nerve sends branches which supply the levator labii superioris alæque nasi, as well as the levator labii superioris and the zygomaticus minor, while the *superior maxillary* branch of the lower division of the facial supplies the depressor labii inferioris.

THE LATERAL WALLS OF THE MOUTH.
The Cheeks (*buccæ*).

The cheeks are continuous with and similar in structure to the lips, being covered internally by mucous membrane and ex-

ternally by the common integument. Immediately beneath the
mucous membrane are a number of transverse muscular fibers,
covered externally by a layer of subcutaneous fat, and lying
between this and the integument other muscular tissue, the
fibers of which radiate in various directions, according to the
action of the muscle to which they belong. Besides muscular
and fatty tissue, there are imbedded within the substance of the
cheek blood-vessels, nerves, and glands. The fatty tissue
spoken of as intervening between the muscular fibers gives to
the cheek the fullness and rotundity desired by so many but
possessed by so few.

The integument, or external covering of the cheek, is similar
in structure to the skin covering other parts of the body, and,
like the lips in the male, is productive of a hairy growth.

The mucous membrane, or internal covering of the cheek,
is similar to that of the lips, containing numerous glands (*buccal
glands*) which are almost identical to, but smaller than, the labial
glands. In addition to the buccal glands which are distributed
over the entire membrane, there are about five of larger size,
which open into the mouth in the region of the molar teeth, and
are called *molar glands* (see Glands of the Mouth).

The Muscles of the Cheeks.

The transverse muscular fibers referred to as being immediately beneath the mucous membrane are those of the *buccinator*,
a muscle named from its action, that of being the chief muscle
employed by the trumpeter. External to the buccinator is the
masseter, one of the muscles of mastication, the elevator and
depressor muscles of the angle of the mouth, the *levator anguli
oris*, and the *depressor anguli oris*, and the dermal muscles,
zygomaticus major and *zygomaticus minor*, and the *risorius*.

Buccinator.—This muscle forms the greater portion of the
lateral wall of the mouth. It is deep-seated in the cheek, being
one of the third stratum of facial muscles.

Origin.—The fibers are distinct in their origin from a part of
the alveolar process of the superior maxillary bone, at a point
immediately over the second and third molar teeth, from the
anterior border of the pterygomaxillary ligament, a narrow

band of tendinous fibers or raphe extending from the pterygoid plate of the sphenoid bone to the mylohyoid ridge of the inferior maxilla near the position of the third molar tooth. Some of its fibers also arise from the outer wall of the alveolar process of the inferior maxilla below the second and third molars.

Insertion.—The fibers pass forward and converge as they reach the lateral margins of the orbicularis oris ; here the fibers of the upper portion pass downward and blend into the muscles of the lower lip, while the lower fibers pass upward and blend into those of the upper lip. Those fibers which arise from the inferior maxilla pass forward and also blend into the lower lip.

Relations.—Superficially, by the skin and subcutaneous fat, the duct of Steno, the masseter muscle, a portion of the angular group, and the facial artery and vein. Passing over it are branches of the facial and buccal nerves, also a layer of deep fascia continuous with that which covers the upper part of the pharynx. By its deep surface it is in relation with the mucous membrane and buccal glands.

Action.—To draw outward or backward the angles of the mouth, thus enlarging the buccal orifice and pressing the lips tightly against the teeth ; to force the food between the occlusal surfaces of the molar and bicuspid teeth during mastication; to diminish the concavity of the cheek, compressing the air contained therein and forcing it forward.

Masseter.—This muscle is placed immediately external to the buccinator, and is one of the principal muscles of mastication. It is short, thick, and somewhat quadrate in form, and is composed of two sets of fibers, superficial and deep. The fibers of the former are directed obliquely downward and backward ; those of the latter, which are much shorter, pass almost vertically downward.

Origin.—The superficial layer, from the malar process of the superior maxilla, and from the anterior portion of the zygomatic arch of the malar bone. The deep layer from the posterior third of the zygomatic arch, as well as from the greater part of its inner surface.

Insertion.—The superficial fibers are inserted into the ramus and angle of the inferior maxilla, and the deep fibers into the upper half of the outer surface of the ramus.

Relations.—By its external surface with the zygomaticus major, risorius, and platysma myoides muscles, the parotid gland and its duct; by the transverse facial artery, the facial vein, and facial nerve, and by the integument. By its internal surface with the ramus of the inferior maxilla, a mass of fat which separates it from the buccinator, and with the temporal muscle. Its posterior margin is in relation with the parotid gland, and its anterior with the facial artery and vein.

Action.—The principal action of this muscle is to close the jaw and to draw it slightly forward. (For further description, see Muscles of Mastication, part I, chap. III.)

The Angular Series.—The remaining muscles of the cheek are those of the angular series, or those muscles which are inserted into the angle of the mouth, two coming obliquely from above,—the *levator anguli oris* and the *zygomaticus major*,—one running almost horizontally forward,—the *risorius*,—and one ascending from below,—the *depressor anguli oris*.

Levator Anguli Oris.—This muscle, which receives its name from its action, belongs to the second layer of facial muscles. It is formed in the shape of a triangular sheet.

Origin.—From the canine fossa of the superior maxilla, immediately below the infra-orbital foramen.

Insertion.—Passing downward and outward it is inserted into the angle of the mouth, its fibers blending with those of the orbicularis oris and the other angular muscles.

Relations.—Superficially, with the levator labii superioris, the zygomaticus minor, and the infra-orbital vessels and nerves; deeply, with the facial portion of the orbicularis oris and buccinator muscles, and the mucous membrane of the mouth.

Action.—Especially to elevate the angle of the mouth, and to assist in drawing these angles inward, decreasing the size of the buccal orifice.

Zygomaticus Major.—This muscle, the companion of which has been described in connection with the muscles of the lips, belongs to the first facial layer. It is composed of a long, fleshy band of muscular fibers, which run direct from their point of origin to their point of insertion.

Origin.—From the malar bone, in close proximity to the zygomatic suture.

Insertion.—From its origin it passes obliquely downward to the angle of the mouth, and blends into the fibers of the orbicularis oris and depressor anguli oris.

Relations.—Superficially, with the skin and subcutaneous fat ; deeply, with the malar bone, the masseter and buccinator muscles, the facial and transverse facial arteries, the facial vein, and branches of the facial nerve.

Action.—To draw upward and outward the angles of the mouth, as in smiling or laughing. By contracting, it throws into prominence the cheek tissues in front of the malar bone, and forces the lower eyelid upward. When acting simultaneously with its fellow of the opposite side, the buccal aperture is widened, and the upper lip is elevated, exposing the superior teeth.

Risorius.—One of the superficial set of facial muscles, receiving its name from its supposed action in laughter (*ridere*, to laugh). It is flat and ribbon-shaped, and is frequently very small and poorly developed.

Origin.—From the deep fascia covering the masseter muscle and parotid gland, some of its fibers occasionally arising from the mastoid process of the temporal bone.

Insertion.—Passing transversely forward and inward to the angle of the mouth, its fibers blend with those of the orbicularis oris, and the depressor anguli oris.

Relations.—Superficially, with the integument and subcutaneous fat ; deeply, with the masseter and buccinator muscles, the facial artery and vein, and branches of the facial nerve.

Action.—To draw the angles of the mouth directly outward, thereby increasing the width of the buccal orifice.

Depressor Anguli Oris.—Also one of the superficial layer of facial muscles, deriving its name in accordance with its action upon the angle of the mouth. It is a triangular-shaped muscle with its base below, becoming narrow as it ascends.

Origin.—From the lower border of the inferior maxilla, and from its external oblique line below the cuspid, bicuspid, and first molar teeth.

Insertion.—Passing upward and inward it is inserted into the integument at the angle of the mouth, its fibers blending into those of the muscles previously described as coming together at this point.

Relations.—Externally, with the integument; deeply or internally, with the depressor labii inferioris, the buccinator, and the inferior coronary artery.

Action.—To draw down the angle of the mouth and to slightly extend it.

Blood-supply to the Cheeks.

The blood-supply to the cheeks is principally through the *facial* artery and its direct branches, the superior and inferior *coronary*, the *transverse facial,* and branches from the internal maxillary.

The Facial Artery and Branches.—The facial artery, also called the *external maxillary,* enters the cheek after passing over the body of the inferior maxilla at the anterior edge of the masseter muscle. It courses obliquely forward and upward through the substance of the cheek, until it reaches the inner angle or canthus of the eye, where it joins the nasal branch of the ophthalmic artery, and is called the *angular artery.* Near the center of the cheek the *inferior labial artery* is given off, which passes forward and downward to the lower lip, but supplies a portion of the cheek in so doing. Midway between the center of the cheek and the angle of the mouth the *superior and inferior coronary arteries* are given off, supplying that part of the cheek immediately adjacent to the angle of the mouth, after which they pass on to supply the upper and lower lips. The *masseteric branch* is given off in the immediate center of the cheek, at a point immediately below the inferior labial, passes directly upward over the masseter muscle, and anastomoses with branches of the internal maxillary and transverse facial. There are also given off from the main trunk near the center of the cheek the *buccal* branches, which pass upward over the buccinator muscle, and also anastomose with branches of the internal maxillary and transverse facial arteries.

The Transverse Facial Artery.—This is the largest branch of the temporal artery. It is at first deeply-seated in the sub-

stance of the parotid gland, after leaving which it courses transversely over and supplies the masseter muscle, sends off small
branches which supply the integument of the cheek, and anastomoses with the buccal, infra-orbital, and the facial arteries.
Besides the arteries already named, the deeper portions of the
cheek receive blood from two branches of the internal maxillary
artery, the *masseteric branch* and the *buccal branch*. The
former supplies the masseter muscle and anastomoses with the
masseteric branch of the facial, while the latter supplies the
buccinator muscle and anastomoses with the buccal branches
of the facial.

Course of the Blood From the Heart to the Cheeks.—
From the heart to the aorta, to the common carotid, to the
external carotid, to the facial and its direct branches, or from the
external carotid to the temporal, to the transverse facial and
branches.

From the cheeks the blood is returned to the heart principally through the *facial vein*, a division of the *anterior superficial vein*. It enters the cheek at a point midway between the
lower eyelid and the wing of the nose, passes obliquely downward, being in close contact with the anterior edge of the masseter muscle over the body of the lower jaw, joining the internal jugular vein in the neck. The *transverse facial vein* which
follows the course of the transverse facial artery, and the *superior* and *inferior coronary veins* also collect and convey a portion of the blood from the cheeks to the larger veins and thence
to the heart.

Nerves of the Cheeks.

The nerve-supply to the cheeks is principally from the
seventh or *facial nerve* and its branches, the *buccal* branch supplying a greater part of their substance. There are also a few
fibers of the *infra-orbital* branch of the seventh nerve distributed
to the labiobuccal region. The *buccal branch* of the lower
division of the facial, also the *buccal branch* of the inferior maxillary division of the fifth nerve, supplies the buccinator muscle.
The *infra-orbital branch* of the upper division of the facial nerve
supplies the zygomaticus major and the levator anguli oris ; the

buccal branch supplies the risorius, and the *supramaxillary branch* of the lower division of the facial nerve supplies the depressor anguli oris.

THE INTERIOR OF THE MOUTH.

For convenience of description the mouth may be divided into two parts—a *superior portion* and an *inferior portion.* In dis-

Fig. 2.—The Superior Portion or Roof of the Mouth.

section this division may be accomplished by an incision beginning at the angles of the mouth and carried backward and slightly upward through the substance of the cheeks until the temporomaxillary articulation is reached. After disarticulating this joint another incision is made, beginning at the joint on either side, carried downward and forward, then obliquely across the throat, until the two come together at the median line. This latter incision must be deep enough to completely sever the tissues of the throat. The superior portion of the mouth contains the *hard palate*, or *roof of the mouth*, the *soft palate*, and the sixteen *superior teeth*, firmly set in the bone and surrounded by a dense fibrous tissue—*the gums*. The inferior portion contains the tongue and its attached muscles, forming the *floor of the mouth*, the sixteen *inferior* teeth and the *gums* surrounding them.

THE SUPERIOR PORTION OF THE MOUTH (FIG. 2).

The osseous framework, or base upon which this half of the mouth is constructed, is composed of a part of four bones— the two superior maxillary, or upper jaw bones, and the two palate bones (see Bones of the Mouth, p. 50).

The Hard Palate, or Roof of the Mouth (Fig. 2).

This is formed by the union of the palatal processes of the superior maxillary bones and the horizontal plates of the two palate bones at the median line. It is limited in front and laterally by the margins of the alveolar process, or that portion of the bone which gives support to the teeth, and ends posteriorly in an irregular border, to which is attached a muscular, membrane-like curtain—the soft palate. The hard palate is covered throughout by a thick and firm mucous membrane, seldom so highly colored as the membrane lining the lips and cheeks. The mucous membrane is closely adherent to the bone through its covering, the periosteum. In the center of the hard palate is a ridge or fold of the mucous membrane, which follows the median line from before backward; this is called the *palatal raphe*. Anteriorly, the raphe ends in a small papilla, which marks the opening of a canal in the bone—the anterior palatal canal. Posteriorly, the raphe usually diminishes, but occasionally is well marked through the whole extent of the hard palate. Near the center of the hard palate it frequently separates into

two or more smaller ridges, which are proportionately diminished in size, and are continued backward side by side. On
either side of this central ridge, anteriorly, the mucous membrane presents a number of fantastically arranged folds, the
palatal rugæ (wrinkles). These folds are usually quite
numerous and prominent, but are occasionally almost absent.
The nature of these wrinkles is strongly indicative of the character or temperament of the individual; thus, in the four basal
temperaments they may be divided as follows: in the bilious,
heavy and strong, composed of angles rather than curves; in

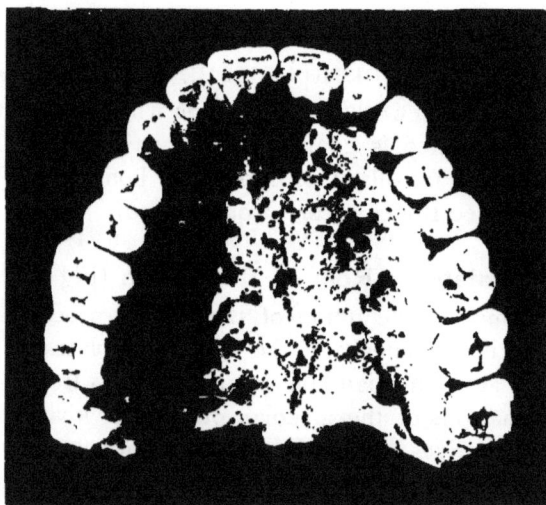

FIG. 3.—THE HARD PALATE, OR ROOF OF THE MOUTH, WITH ITS MEMBRANOUS
COVERING REMOVED.

the nervous, few in number, close together, not prominent, and
composed of long curves; in the sanguine, quite numerous,
fairly prominent, well rounded and graceful in outline; in the
lymphatic, few in number, flat, widely separated, and but little
curved. Accompanying these varying conditions in the rugæ
will be found a corresponding variation in the raphe. The anterior and lateral margins of the mucous membrane covering
the hard palate form the palatal portion of the gums, which is
called the *gingiva* (the gum) or gingival border.

In figure 3 the hard palate is shown with its membranous covering removed. It will be observed that the bony plates are perforated by numerous small foramina, through which the body of the bone receives its nourishment, broken by depressions for the accommodation of the various mucous glands, and traversed by longitudinal grooves which give lodgment to blood-vessels and nerves.

The arch formed by the hard palate from side to side varies greatly in form, imparting much knowledge in regard to the temperament of the individual, and in a measure controlling the quality of the voice. Thus, in the sanguine temperament the roof of the mouth presents almost a perfect oval. In the bilious type it is comparatively high and flat, extending from the base of one alveolar process to another, from which point it descends abruptly to the necks of the teeth. In the nervous subject the roof is high and semi-elliptical or parabolic in shape, and in the lymphatic it is low and segmental in form.

In the same illustration the union or suture between the two bones may be observed at the median line. Near the anterior third of this central suture is the opening of the incisive or anterior palatal canal, the *anterior palatal foramen*, the location of which has been referred to in the description of the mucous membrane. Near the posterior border, and situated within the suture which unites the superior maxillary bones with the palate bones (the palatomaxillary suture), are two other foramina, the posterior palatal; and immediately behind these, and separated by a thin ridge of bone, are the accessory palatal foramina, these being in the tuberosity of the palate bones. (For further description of these foramina, see Bones of the Mouth, p. 50.) By the vessels and nerves which enter the hard palate through these various foramina, the mucous membrane and glands receive their blood- and nerve-supply.

Blood-supply to the Hard Palate.—This is principally derived from the *posterior* or *descending palatal* branch of the *internal maxillary* or *deep facial artery*, which passes downward in the posterior palatal canal and emerges through the posterior palatal foramen. Immediately on reaching the palate it divides

3

into an anterior and a posterior branch, the former passing forward in a groove provided for it to the anterior palatal foramen, where it anastomoses with the nasopalatal artery. The groove in which the artery lies in its passage forward is usually at the base of the alveolar process, and in some instances is converted into a canal for a part of its length. The posterior branches pass backward and downward to supply the soft palate. In connection with supplying the hard palate proper, this artery carries blood to the palatal alveolar walls, to the mucous glands, the mucous membrane, and the gums.

Course of the Blood from the Heart to the Hard Palate.—From the heart to the aorta, to the common carotid, to the external carotid, to the internal maxillary, to the posterior or descending palatal branch of the latter. From the hard palate the blood is returned to the heart by the *superior palatal* and *inferior* or *descending palatal veins*, the former following the course of the superior palatal artery, while the latter originates at a point near the junction of the hard and soft palates, passes downward, and joins the facial vein below the body of the inferior maxilla.

Nerves of the Hard Palate.—The nerves of the hard palate are the *anterior* or *large palatal* and branches from the nasopalatal, both of which are branches of the sphenopalatal (Meckel's) ganglion. The anterior palatal nerve arises from the inferior angle of the ganglion, passes downward, accompanied by the descending palatal artery, through the posterior palatal canal, from which it emerges at the posterior palatal foramen. From this point it passes forward in a groove of the hard palate, and joins the nasopalatal nerve as it emerges from the anterior palatal foramen. Accompanying this nerve in its course through the posterior palatal canal are other branches of the sphenopalatal ganglion, which pass to the soft palate, and will be described in that connection. The nerves of the soft palate are all sensory, and filaments are distributed to the mucous membrane and glands and to the palatal portion of the gums.

THE SOFT PALATE (Fig. 2).

The soft palate is attached to the posterior border of the hard palate, from which it is continued as a backward prolongation. Hanging downward, with its free borders inclining backward, it may be considered as forming a part of the posterior boundary of the mouth. It partially separates the mouth from the nasal cavity and from the pharynx. It is attached laterally to the walls of the pharynx, while its lower border is free. The *substance* of the soft palate is composed of a number of thin, but dense, muscular fibers, blood-vessels, nerves, and mucous glands, the latter being similar to those of the hard palate. The anterior surface of this muscular curtain is concave, directed forward and downward, and is traversed by a median raphe. The posterior surface is convex, directed backward and upward, and is continuous with the nasal cavity. Suspended from the center of its free border is a small rounded or conic membranous appendix, the *uvula*, and passing outward from the base of this at each side are two curved folds of mucous membrane, which extend outward and downward, and are known as the *pillars of the fauces*. From the position which these folds occupy they are divided into the *anterior* and the *posterior* pillars of the fauces. The anterior pillar is formed from muscular fibers which extend from the soft palate to the side and base of the tongue (palatoglossus muscle), and is somewhat prominent as it passes downward, outward, and forward. The posterior pillar approaches more closely to its fellow of the opposite side than does the anterior. It is formed of muscular fibers which extend from the soft palate above to the pharynx below (palatopharyngeus muscle). It is somewhat concave in its downward and backward course, and while closely united to the anterior pillar above, is separated from it below, leaving a triangular interval or niche in which is lodged a small, almond-shaped body, the *tonsil*, the space being known as the *tonsillar recess*. The intervening space—bounded by the margins of the soft palate above, the root of the tongue below, and the pillars laterally—is called the *isthmus of the fauces*, and establishes the communication between the mouth and the pharynx. The free margins of the soft palate, assisted by the fauces, mark the *posterior boundary of the mouth.* The entire surface of the soft palate and its prolongations, the pillars of the fauces, is covered with mucous membrane, being continuous with that of the mouth on its anterior surface, and with that of the nasal cavity on its posterior surface.

Muscles of the Soft Palate.

On each side the muscles of the soft palate are the *palato-glossus, palatopharyngeus, levator palati,* and *tensor palati,* together with the *azygos uvulæ.*

Palatoglossus.—A small fasciculus of fibers, somewhat cylindric in form, expanding at either end into a thin sheet. It is named from its attachment to the soft palate and to the tongue. It is the prominence of this muscle, together with its covering of mucous membrane, that forms the anterior pillar of the fauces.

Origin.—By a thin muscular sheet from the under surface of the aponeurosis of the soft palate near the median line, its fibers uniting with those of the opposite side. It passes downward in front of the tonsil and against the pharyngeal wall.

Insertion.—Into the side and base of the tongue.

Relations.—Superficially, it is covered by the mucous membrane of the soft palate and tongue ; deeply, in contact with the aponeurosis of the soft palate, the superior constrictor muscle of the pharynx, and one of the muscles of the tongue—the hyoglossus.

Action.—The lateral walls of the soft palate are drawn down, and the sides of the tongue are drawn upward and slightly backward. Acting in conjunction with the palatopharyngeus, the opening of the fauces is constricted.

Palatopharyngeus.—This muscle—also named from its attachments—is broad above, where it forms the greater part of the lower half of the soft palate. Near the median line a few of its fibers blend with those of its fellow of the opposite side.

Origin.—By two heads from a point near the raphe or median line of the soft palate, passing downward and slightly backward, forming, with its covering of mucous membrane, the posterior pillar of the fauces.

Insertion.—Into the posterior border of the thyreoid cartilage, and to the inner surface of the lower part of the pharynx.

Relations.—In the soft palate, superficially, with the mucous-membrane, both anteriorly and posteriorly; above, with the levator palati; and beneath, by the mucous glands. In the posterior pillar it is surrounded with mucous membrane, and in the pharynx by the constrictor muscles of the pharynx and the mucous membrane.

Action.—To constrict the opening of the fauces, by bringing together the posterior pillars, thus depressing the soft palate and elevating the pharynx.

Levator Palati.—This is a moderately thick muscle, and derives its name from its action upon the soft palate.

Origin.—By a short tendon from the under surface of the petrous portion of the temporal bone, and from the posterior and inferior aspect of the cartilage of the Eustachian tube.

Insertion.—After passing downward by the side of the posterior nares it is inserted into the median line of the soft palate, where its fibers unite with those of its fellow of the opposite side.

Relations.—Externally, with the tensor palati and superior constrictor muscles; internally and posteriorly, with the mucous membrane.

Action.—To raise the soft palate, bringing it against the posterior wall of the pharynx.

Tensor Palati.—This is a slender and flattened muscular sheet, and receives its name from its action upon the soft palate.

Origin.—From the scaphoid fossa at the base of the internal pterygoid plate and the spinous process of the sphenoid bone; also from the outer side of the anterior aspect of the Eustachian tube.

Insertion.—After descending between the internal pterygoid muscle and the internal pterygoid plate, and winding around the hamular process of the latter, it is inserted into the transverse ridge on the horizontal portion of the palate bone, and at the median line of the soft palate, where it is continuous with the aponeurosis of the opposite side.

Action.—To tighten or spread the soft palate laterally, forming a septum between the posterior nares and the pharynx. It also opens the Eustachian tube during deglutition.

The azygos uvulæ (so named because it was at one time supposed to be a single muscle) is composed of a pair of small muscles which *originate* from the aponeurosis of the soft palate and the nasal spine of the palate bone. They pass downward and form or are *inserted* into the uvula.

Relations.—Anteriorly, with the levator palati, palatoglossi, and a part of the palatopharyngei; posteriorly, to the mucous membrane.

Action.—To shorten or draw up the uvula.

Blood-supply to the Soft Palate.

The *posterior* or *descending palatal branch* of the *deep facial artery*, after emerging from the posterior palatal canal, sends its posterior division backward and downward to the soft palate, in the substance of which they anastomose with the ascending palatal artery. After passing over the superior border of the pharynx, the *ascending pharyngeal artery* sends off branches which are distributed to the soft palate. A few branches of the superior palatal branch of the *internal maxillary* and a few twigs from the *lingual artery* also convey blood to the parts.

Course of the Blood from the Heart to the Soft Palate.—From the heart to the aorta, to the common carotid, to the external carotid, to the facial or lingual, to the various branches named above, to the soft palate. The return of the blood to the heart is principally through the *superior* and *inferior* or *descending palatal veins*, both of which closely follow the course of the arteries of the same name.

Nerves of the Soft Palate.—The *small*, or *posterior*, *palatal*, the *external palatal* (both of which are branches of Meckel's ganglion), branches of the *glosso-pharyngeal* nerve, and the following nerves which supply the various muscles: filaments from the pharyngeal plexus to the palatoglossus and palatopharyngeus, branches of the Vidian to the levator palati and azygos uvulæ, and from the mandibular division of the fifth nerve to the tensor palati.*

THE INFERIOR PORTION OR FLOOR OF THE MOUTH (FIG. 4).

This half of the cavity of the mouth contains the *tongue* and its attached muscles, the sixteen inferior *teeth* firmly implanted

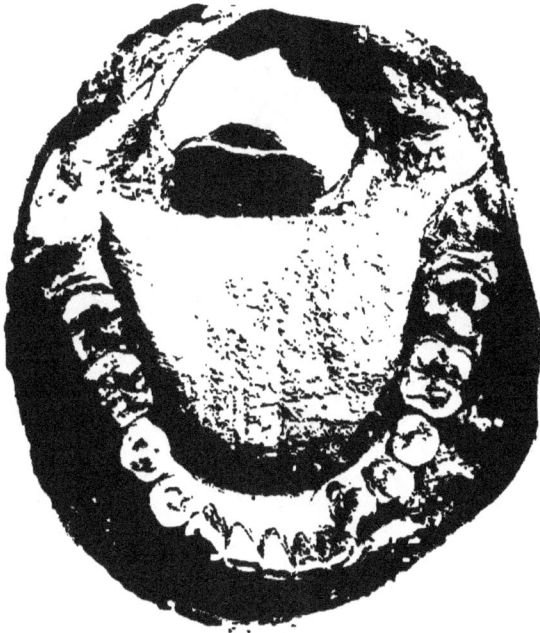

FIG. 4.—THE INFERIOR PORTION OR FLOOR OF THE MOUTH.

in the jaw, and the gums covering the alveolar walls. The base or osseous framework upon which this portion of the mouth is constructed is made up of a single bone, the inferior maxillary, mandible, or lower jaw bone (see Bones of the Mouth, p. 50). The floor of the mouth is bounded anteriorly by the lower lip,

* A description of the superior teeth will be found in another chapter.

laterally by the cheeks, and below by the muscles attached to the external and internal oblique lines of the mandible.

THE TONGUE (Fig. 5).

The tongue is a freely movable, highly sensitive, muscular organ. Its special function is that of speech, participating,

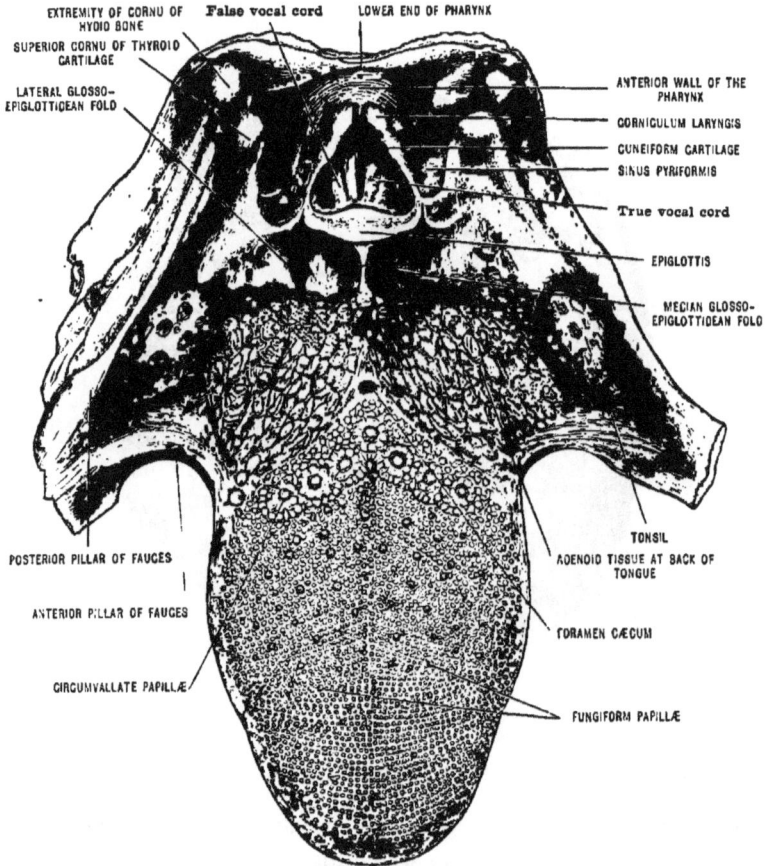

FIG. 5.—DORSUM OF THE TONGUE.—(*Morris.*)

moreover, in the special sense of taste, and in mastication and deglutition. The organ is attached posteriorly to a U-shaped bone, the *hyoid bone*, which of itself is movable, and is placed in the neck between the angles of the lower jaw and the thyreoid cartilage.

The tongue is suspended and kept in its position in the mouth by numerous muscles, some of which are attached to the base of the skull, and others to the lower jaw and hyoid bone.

The size of the tongue bears little or no relation to the size of the individual, but is proportioned to the capacity of the alveolar arch, which space it completely fills when at rest. The shape of the tongue is controlled by the shape of the alveolar arch; thus, when the arch is contracted, narrow, and pointed, the margins of the tongue, when at rest, will assume that form (*a*, Fig. 6) ; but when the arch is broad and rounded anteriorly, the margins of the tongue will also be broad and rounded (*b*, Fig. 6).

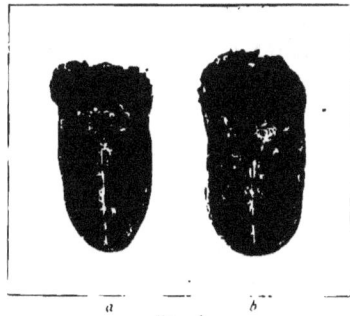

FIG. 6.

The substance of the tongue is chiefly composed of muscular fibers, which are arranged in a complicated manner, crossing one another at various angles, thus making the movements of the organ exceedingly varied and extensive. Fibrous, areolar, and fatty tissues enter into its structure, and it is freely supplied with blood-vessels and nerves. Its free surface is covered by sensitive mucous membrane, and over its entire surface are numerous mucous follicles and glands.

Before continuing the description of the tongue, it will be necessary to name its parts. The upper surface, or that facing the roof of the mouth, is the *dorsum ;* those portions directed toward the cheeks are known as the *sides* of the tongue ; while the hem which unites these two borders at the median line, and extending a short distance backward on either side, is the *tip.* That portion between the frenum and extending back to the pillars of the fauces is the *base,* while that part of the dorsum immediately posterior to the tip is the *post-tip,* that region which lies between the post-tip and the base being the *prebase.* The

dorsum, sides, and tip are free, while the base is attached by
muscles to the lower jaw and hyoid bone.

From the base to the epiglottis is a fold which serves to limit
the movement of the latter organ, and from the sides of the
base the pillars of the fauces are given off. Under the anterior
free extremity in the median line is a fibrous muscular lamina
or ligament, the *frenum*, which connects this part of the organ
with the lower jaw and marks the anterior border of the base of
the tongue. The tongue is divided through its anterior two-
thirds by a slight longitudinal furrow, the *median raphe*, which
ends posteriorly near a small foramen, not constant in the adult
tongue, but plainly observed in the fetus or infant, the *foramen
cæcum*. This foramen represents the upper termination of the
thyreoglossus duct.

Papillæ of the Tongue.—Over the anterior two-thirds of
the dorsum and the sides and tip of the tongue are a number
of small, soft, conic eminences, which are known as the papillæ
of the tongue. These are most numerous over the anterior part
of the dorsum, and at the back they are covered and partly
hidden by an epithelial coating. In general, the papillæ are
quite similar to those of the integument, not being compound
organs in their vascular and nervous supply. In consequence
of their variation in form, the papillæ are variously named. The
largest papillæ, being arranged like the letter V, are called the
circumvallate or *calciform ;* those of medium size, the *fungiform*,
are so named from their resemblance to a young mushroom ;
and the smallest and most numerous are known as the *conic*
or *filiform* papillæ. Each papilla presents a broad, free end,
and is attached by a constricted base, which rests in a small, cup-
like concavity, about the margins of which is a well-formed
circular rim. Beneath the thick epithelium of these parts are
numerous secondary papillæ, and about the base of each papilla
are the openings of one or more glands.

The *circumvallate* or *calciform* papillæ (Fig. 5) form a V-shaped
line at the posterior boundary of the dorsum. They are few in
number (varying from six to twelve), but are largest in size, not
infrequently measuring $\frac{1}{5}$ of an inch in diameter. These
papillæ are generally regarded as being gustatory, or directly

interested in the sense of taste. Each papilla is capped with a small secondary papilla.

The *fungiform* papillæ (Fig. 5), of medium size, varying from $\frac{1}{50}$ to $\frac{1}{30}$ of an inch in diameter, are scattered over the dorsum, sides, and tip of the tongue at irregular intervals, and are much more highly colored than the smaller papillæ which surround them. They vary greatly in number, and being principally gustatory, account in a great measure for the diversity in the acuteness of the sense of taste in different individuals. These papillæ, like the circumvallate, are capped with smaller secondary papillæ.

The *conic* or *filiform* papillæ (Fig. 5) are the smallest and most numerous, and are thickly scattered over the entire surface of the dorsum in front of the circumvallate as well as over the sides and tip of the tongue. They are placed closely together, and with such regularity that they fairly ridge the tongue with delicate lines, which run parallel with the circumvallate in that region, but as the tip is approached they become transversely inclined. These papillæ are generally regarded as being tactile or directly interested in the sense of touch, and are concerned in directing the movements of the food during mastication. They also possess secondary papillæ upon their surfaces.

Immediately posterior to the circumvallate papillæ are two shallow grooves which follow the V-shaped line of the papillæ and unite at the foramen cæcum. These grooves serve to indicate the line of junction between the anterior and posterior portions of the tongue. The latter not being within the cavity of the mouth will not be described.

Muscles of the Tongue (Fig. 7).

The muscles of the tongue are both *extrinsic-outward* or external, and *intrinsic-inherent*, inward, or special.

The extrinsic muscles include those which have their origin from the base of the skull, the hyoid bone, or the mandible, and are the *hyoglossus, geniohyoglossus, styloglossus, palatoglossus,* and a few fibers of the superior constrictor of the pharynx. The intrinsic muscles which make up the bulk of the tongue are two in number, the *superior lingualis* and *inferior lingualis*.

Hyoglossus.—As its name implies, this muscle extends from the hyoid bone to the tongue. Its fibers are so arranged that they form a thin square sheet.

Origin.—It arises from the whole length of the upper border of the great cornu, from the body, and by a few fibers from the lesser cornu of the hyoid bone. At their point of origin the

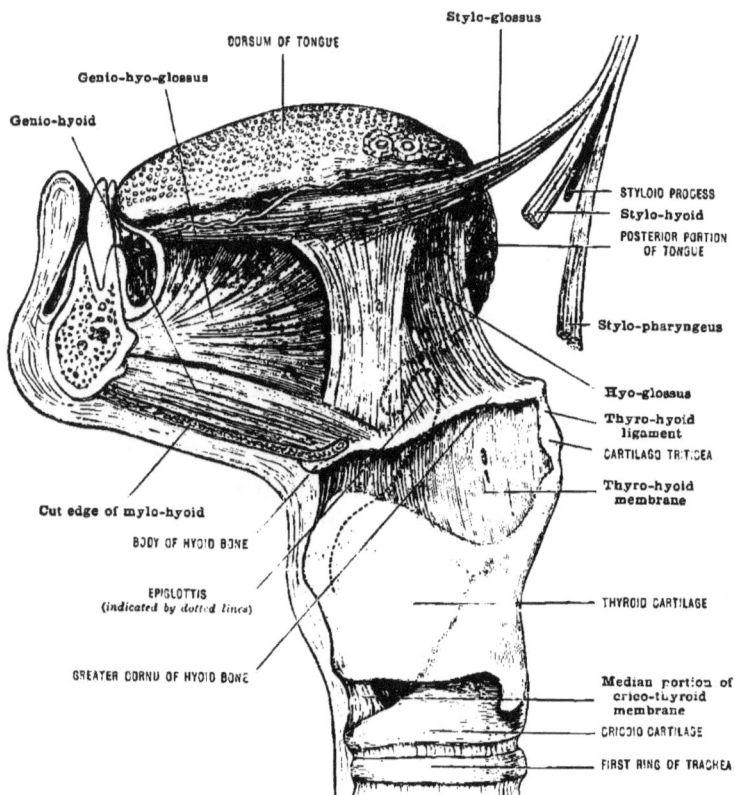

DORSUM OF TONGUE

Stylo-glossus

Genio-hyo-glossus

Genio-hyoid

STYLOID PROCESS

Stylo-hyoid

POSTERIOR PORTION OF TONGUE

Stylo-pharyngeus

Hyo-glossus

Thyro-hyoid ligament

CARTILAGO TRITICEA

Thyro-hyoid membrane

Cut edge of mylo-hyoid

BODY OF HYOID BONE

EPIGLOTTIS (indicated by dotted lines)

GREATER CORNU OF HYOID BONE

THYROID CARTILAGE

Median portion of crico-thyroid membrane

CRICOID CARTILAGE

FIRST RING OF TRACHEA

FIG. 7.—SIDE VIEW OF THE TONGUE, WITH ITS MUSCLES.—(*Morris*.)

fibers are in the form of a thin sheet, and ascend toward the tongue almost parallel to one another, but before reaching the tongue the anterior fibers pass slightly forward, and at the upper margin of the side of the tongue bend inward and join the fibers of the superior lingualis. In their distribution to this part of the tongue they form a kind of submucous covering to the organ.

Insertion.—Into the posterior half of the side of the tongue, between the styloglossus and superior lingualis muscles.

Relations.—Externally, with the digastricus, styloglossus, stylohyoid, and mylohyoid muscles, the lingual and hypoglossal nerves, Wharton's duct, and the sublingual gland. Internally, with the lingualis, geniohyoglossus, and middle constrictor of the pharynx muscles, the lingual artery, and the glossopharyngeal nerve.

Action.—To extend the tongue and to draw it backward, also to draw downward the sides of the tongue, making its dorsum more convex transversely.

Geniohyoglossus (Fig. 7).—This muscle also receives its name from its three points of attachment, the chin internally, the hyoid bone, and to the tongue. It is a triangular-shaped muscle, narrow and pointed at its attachment to the mandible, and broad and fan-shaped on approaching the tongue. Being near the median line, it is separated from its fellow of the opposite side by a thin layer of connective tissue, the *septum* of the tongue.

Origin.—It arises by a short tendon from the upper genial tubercle of the lower jaw, from which point its fleshy fibers diverge fan-like to its extensive insertion.

Insertion.—To the whole length of the tongue from base to apex immediately external to the median line; into the body of the hyoid bone, and by a few fibers into the side of the pharynx.

Relations.—By its inner surface, with the septum of the tongue and its fellow of the opposite side ; by its outer surface, with the hyoglossus, mylohyoides, styloglossus, and lingualis muscles, sublingual gland, lingual artery, and hypoglossal nerve. Superiorly, with the mucous membrane of the floor of the mouth ; inferiorly, with the geniohyoid muscle.

Action.—Its anterior fibers assist in drawing back the tip of the tongue, its posterior fibers throwing forward and protruding the tongue. This muscle also depresses the center of the dorsum longitudinally, making it concave transversely, and some of the lower fibers which are attached to the hyoid bone elevate the bone and assist in raising the tongue.

Styloglossus.—Also named from its attachment, is a long fan-shaped muscle, somewhat compressed laterally.

Origin.—From its point of origin at the tip of the styloid process of the temporal bone, and from a portion of the stylo-maxillary ligament, it passes with a long curve, forward, slightly downward, then upward and inward to its place of insertion at the side of the tongue.

Insertion.—Upon reaching the side of the tongue it divides into two portions, the fibers of one portion passing transversely inward, while the others pass longitudinally along the side of the tongue.

Relations.—Externally, with the internal pterygoid muscle, parotid and sublingual glands, lingual nerve, and the mucous membrane of the mouth ; internally, with the superior constrictor and hyoglossus muscles, and with the tonsil.

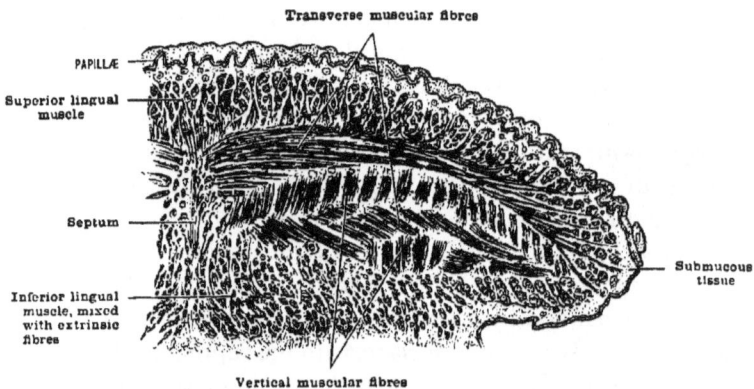

Transverse muscular fibres

PAPILLÆ

Superior lingual muscle

Septum

Submucous tissue

Inferior lingual muscle, mixed with extrinsic fibres

Vertical muscular fibres

FIG. 8.—TRANSVERSE SECTION THROUGH THE LEFT HALF OF THE TONGUE (Magnified).—(*Morris.*)

Action.—To draw the tongue backward and to produce a transverse concavity to its upper surface by elevating its sides.

Superior Lingualis.—This is one of the intrinsic muscles, and is situated immediately beneath the mucous membrane, extending from the base to the tip of the organ.

Inferior Lingualis (Fig. 9).—This muscle is placed near the under surface of the tongue, and is composed of two bands which extend from base to apex, some of its fibers being attached posteriorly to the hyoid bone, and in passing forward are placed between the hyoglossus and geniohyoglossus. Anteriorly, its fibers blend with those of the styloglossus.

Many of the fibers of this muscle run transversely and are placed between the two former intrinsic muscles. These, together with some fatty tissue, compose the greater part of the substance of the tongue. The fibers are attached at the median line to the fibrocartilaginous septum of the tongue, and laterally

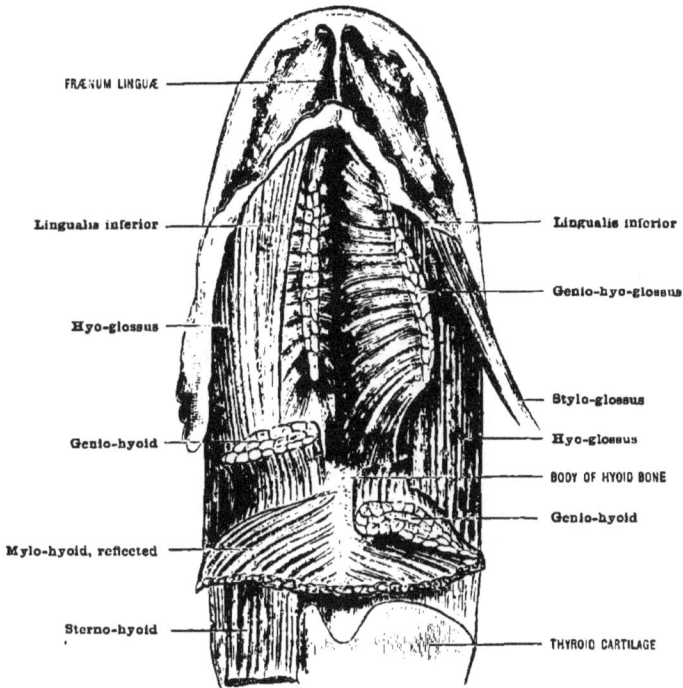

FRÆNUM LINGUÆ

Lingualis inferior

Lingualis inferior

Genio-hyo-glossus

Hyo-glossus

Stylo-glossus

Hyo-glossus

BODY OF HYOID BONE

Genio-hyoid

Genio-hyoid

Mylo-hyoid, reflected

Sterno-hyoid

THYROID CARTILAGE

FIG. 9.—UNDER SURFACE OF THE TONGUE WITH MUSCLES.—(*Morris.*)

to the mucous membrane. In connection with the transverse fibers there are a few placed vertically, which pass by long curves from the dorsum to the under surface of the tongue.

Blood-vessels of the Tongue (Fig. 10).

This organ receives its blood principally through the *lingual*, *facial*, and *ascending pharyngeal* arteries. The *lingual artery* arises from the front of the external carotid near the facial, and often as a common trunk with it. From its point of origin to the tongue it is divided into three portions, the first or oblique, the second or horizontal, and the third or ascending, and it is this latter portion which directly supplies the tongue. Ascending tortuously beneath the hyoglossus muscle, it reaches the under surface of the tongue, and, lying between the lingualis and hyoglossus muscles, it is continued to the under surface of the tip of the tongue at which point it is called the *ranine artery*. At a point about corresponding with the posterior margin of the hyoglossus muscle a branch is given off (the *dorsalis linguæ*), which

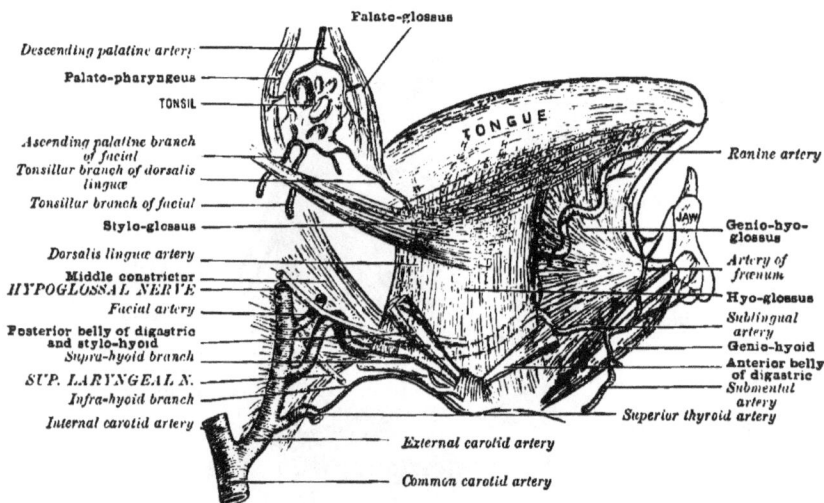

FIG. 10.—SCHEME OF THE RIGHT LINGUAL ARTERY.—(*Walsham.*)

passes almost directly upward, and, after dividing into two or more small branches, supplies the back part of the dorsum of the tongue and the mucous membrane about the circumvallate papillæ. At the anterior border of the hyoglossus muscle another branch is given off (the *sublingual artery*) supplying the anterior muscular structure of the floor of the mouth. The facial artery by one of its muscular branches supplies the styloglossus muscle.

Course of the Blood From the Heart to the Tongue.—
From the heart to the aorta, to the common carotid, to the ex-

ternal carotid, to the lingual artery, and its smaller branches to the tongue. From the tongue the blood is returned to the heart principally through the *lingual* vein, which begins at the ranine vein beneath the tip of the tongue, passes backward under cover of the mucous membrane, following the course of the lingual artery until the hyoglossus muscle is reached, beyond which point the fibers of the muscle separate the artery from the vein. After receiving the sublingual and dorsalis linguæ veins, the course of which corresponds to the arteries of the same name, the vein passes backward and downward and empties into the internal jugular.

Nerves of the Tongue.

The *mandibular division* of the *fifth* nerve by its lingual branch supplies the papillæ of the anterior portion and sides of the tongue, while the *lingual branch* of the *glossopharyngeal* supplies the circumvallate papillæ, the base, and posterior sides. A few branches of the superior laryngeal are distributed to the back part of the root of the organ. The motor nerve of the tongue is the *hypoglossal* or *ninth*, supplying both the extrinsic and intrinsic muscles.*

---

* A description of the inferior teeth will be found in another chapter.

THE BONES OF THE MOUTH: THE SUPERIOR MAXILLÆ, THE PALATE BONES, THE INFERIOR MAXILLA OR MANDIBLE.

SUPERIOR MAXILLARY BONES.

The *superior maxillary bones*, two in number, one on each side of the median line or center of the face, are irregular in shape, and may be classed as the largest bones of the face, with the probable exception of the mandible or inferior maxilla. From the central position which they occupy they contribute largely to the bony framework of this portion of the skull. They are not only instrumental in forming the major portion of the roof of the mouth or hard palate, but assist in the formation of the floor of the orbit, and the sides and base of the nasal chamber. They furnish a solid and firm foundation for the sixteen superior teeth, and by their variety in form contribute much to the character and quality of the voice. The outer or facial surface of these bones provides attachment for numerous muscles. Each superior maxillary bone presents for examination a *body*, *four surfaces*, and *four processes*. The body may be described as forming an irregular triangle, its general contour depending much upon the temperament of the subject and consequent character of the teeth. Within the body of the bone is an irregular cavity, the maxillary sinus or antrum of Highmore.

The four surfaces of the body of the bone are the superior or orbital, the lateral or facial, the proximal or nasal, and the posterior or zygomatic.

The orbital surface, which assists in forming the greater portion of the floor of the orbit, is slightly concave over its anterior two-thirds, and somewhat convex over the remaining or posterior third. The three borders of this surface form almost an equilateral triangle, and are named, as indicated by their location, the anterior, the posterior, and the mesial or proximal. The *anterior border* is convex from before backward and slightly concave throughout its length. That portion which forms a part of the lower border of the completed orbit is smooth, while the remaining portion is roughened to form an articulation with the malar bone. The *posterior border* extends from the center of the malar process backward and inward to the orbital process of the palate bone, which articulates with the superior maxillary at this point. A portion of this border, together with the orbital process of the palate bone, is instrumental in forming the anterior boundary of the sphenomaxillary fissure.

The mesial or proximal border is marked by an irregular thin edge, which articulates with a portion of two bones, the lacrymal anteriorly, and the os planum of the ethmoid bone posteriorly. Only the posterior two-thirds of this border presents an articulating edge, the remaining or anterior third being smooth and forming the commencement of the lacrymal groove, which in the articulated skull becomes a canal, passing downward and backward to communicate with the inferior meatus of the nose. Beginning at the posterior border of this surface and running forward will be found a deep groove—the *infra-orbital groove*. When near the center of the surface, this groove dips down and is covered by a layer of bone, from which point it passes forward as a canal,—the *infra-orbital canal,*—making its exit at a point about ½ of an inch below the border of the orbit, near the center of the facial surface of the bone, the foramen thus formed being the *infra-orbital foramen*. Near the root of the nasal process, and immediately within the anterior border of this surface, is a small depression which marks the origin of the inferior oblique muscle of the eyeball.

The Lateral or Facial Surface (Fig. 11).—This surface is made up of the anterior part of the bone ; it is irregularly concave, and presents a greater variety in form than any other part of the bone, with the single exception of the palatal process. It is bounded above by the infra-orbital ridge, and the roughened surface of the malar process which articulates with the malar bone ; below, by the border of the alveolar process ; anteriorly, by the frail concave border of the opening into the nasal cavity, the anterior nasal spine, and the perpendicular margins of the bone beneath. Posteriorly, this surface is separated from the posterior or zygomatic surface by a strong projecting eminence, the malar process.

The canine fossa is a deep depression, situated almost in the center of this surface, the bone at this point being extremely thin and closely related to the floor of the antrum. The concave floor of this fossa is frequently traversed by one or two smaller convex ridges, corresponding to the roots of the bicuspid teeth.

The canine eminence is a prominent ridge running vertical to

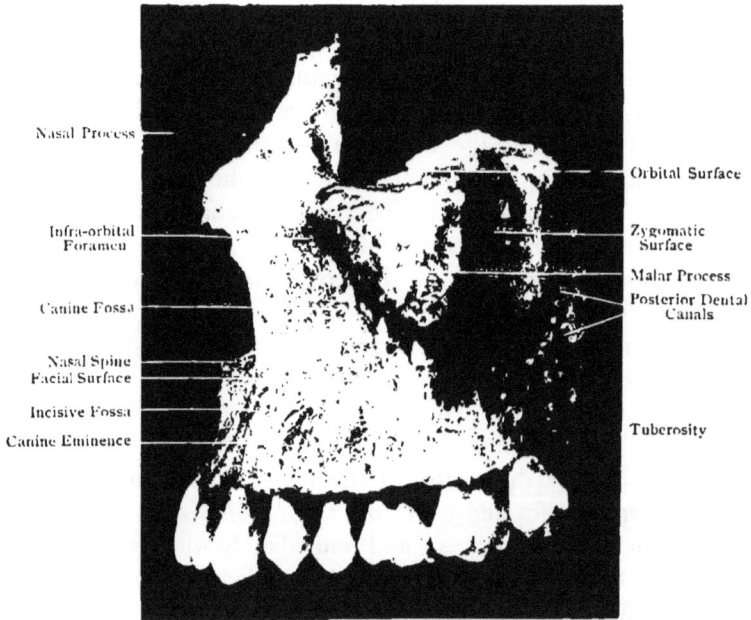

FIG. 11.—LEFT SUPERIOR MAXILLA, OUTER OR FACIAL SURFACE.

the body of the bone immediately anterior to the canine fossa, and corresponding in position to the root of the canine tooth, the size and type of the tooth having much to do with its extent and prominence. This ridge gives origin to one of the depressor muscles of the upper lip, and also to one of the depressor muscles of the wing of the nose. The *incisive* or *myrtiform fossa* is a depression found between the canine eminence and the inner margin of the bone. The depth of this fossa is in a measure controlled by the position and size of the teeth, and by the amount of prominence in the canine eminence.

The infra-orbital foramen, which transmits the infra-orbital nerves and blood-vessels, is immediately below the center of the infra-orbital ridge, and near the upper margin of the canine fossa. It is oval in form and faces almost directly toward the median line. Between this foramen and the infra-orbital ridge is the point of origin for the principal levator muscle of the upper lip, the levator labii superioris proprius. The whole extent of the facial surface may present a number of vertical ridges, or the same space may be regular and smooth, the condition being controlled by the size and shape of the tooth-roots and the thickness of the bone covering them. One of the levator muscles of the angle of the mouth, the levator anguli oris, is attached to this surface near the upper border of the canine fossa.

The Proximal or Nasal Surface (Fig. 12).—Above, this surface presents a large, irregular opening into the maxillary sinus, this opening being nearly closed in the articulated skull by neighboring bones. In front of the opening into the sinus, and standing perpendicular from the body of the bone, is the strong ascending plate of the nasal process, marked near its lower extremity by a rough, horizontal ridge, the *inferior turbinated crest*, which gives attachment to the inferior turbinated bone. The smooth, concave surface immediately above this ridge corresponds to the middle meatus of the nose, and forms the external wall of that passage. Below the opening into the sinus and the nasal process, and occupying the anterior two-thirds of the middle of this surface, is a large semicircular space, forming the outer wall of the inferior meatus of the nose. Below this space, and projecting inward from the body of the bone, is *the palatal process*, which articulates with the corresponding process of the opposite bone. At the anterior superior angle of the nasal surface, and passing downward just behind the nasal process, is the lacrymal groove. In the articulated skull this groove becomes a canal, the *lacrymal canal*, the ethmoid and the inferior turbinated bones assisting in its formation. The canal passes downward and slightly backward, and opens into the inferior meatus of the nose. It is about ½ of an inch in length, and gives passage to the lacrymonasal duct.

The lacrymal tubercle is a small prominence of bone formed at the junction of the anterior border of this surface, with the external surface of the nasal spine. The extended portion of the lacrymal duct, the lacrymal sac, finds lodgment at this point.

The posterior palatine or palatomaxillary canal commences

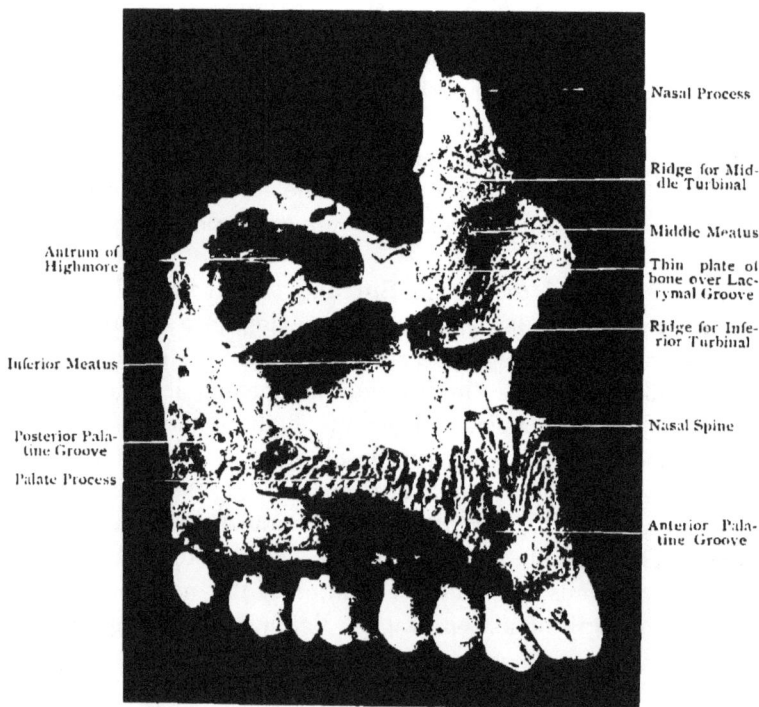

Nasal Process

Ridge for Middle Turbinal

Middle Meatus

Thin plate of bone over Lacrymal Groove

Ridge for Inferior Turbinal

Nasal Spine

Anterior Palatine Groove

Antrum of Highmore

Inferior Meatus

Posterior Palatine Groove

Palate Process

Fig. 12.—Left Superior Maxilla, Internal, Proximal, or Nasal Surface.

near the middle of the posterior border of this surface, appearing in the disarticulated bone as a groove, and, passing downward and forward, gives passage to the posterior palatine vessels and anterior palatine nerves. The canal is made complete by the articulation of the superior maxillary with the vertical plate of the palate bone. On the posterior portion of the nasal surface, extending from the irregular opening into the antrum downward to a point opposite the palatal process, is a roughened surface about ½ of an inch in width, which marks the extent of articulation with the palate bone.

The proximal or nasal surface presents four borders—superior, inferior, anterior, and posterior. The superior border is irregular, and articulates with the lacrymal and ethmoid bones. The inferior border projects inward, and forms a strong horizontal plate—the palatal process. This process defines the border from before backward to the posterior third, at which point it is marked by the lower border of the roughened surface which articulates with the palate bone. The anterior border is sharp, frail, and irregular in outline, and forms the free margin of the opening into the nasal cavity. The posterior border is marked by the inner margin of the zygomatic surface, being smooth upon its upper half, and roughened upon its lower half, at which point it articulates with the palate bone.

The Posterior or Zygomatic Surface (Fig. 11).—This surface is partly convex and partly concave, and is bounded above by a well-defined margin, which serves as the dividing-line between this and the superior or orbital surface. This border is also marked by a roughened margin on the posterior portion of the malar process, the orbital portion of the palate bone articulating at this point. The major portion of this border is smooth and rounded, forming the lower border of the sphenomaxillary fissure, and marked by a notch, the commencement of the infraorbital groove. The outer border of the surface is formed by the malar process, and by a line drawn from this point directly downward to the alveolar process. The inner border is smooth and somewhat irregular above, while below it is roughened for articulation with the palate bone.

The tuberosity, which also forms a portion of the inferior border of this surface, is a roughened and rounded eminence of bone, and is penetrated by a number of nutrient vessels, which enter the many small foramina at this point. Between the tuberosity and the body of the zygomatic surface are several large apertures leading into canals, which pass into, and give nourishment to, the substance of the bone. These canals transmit the posterior dental blood-vessels and nerves, one of which, after passing over the outer wall of the maxillary sinus, unites with the anterior dental canal. The tuberosity is posterior to, and above, the third molar tooth, in some instances extending directly backward from this tooth for the distance of half an inch or more, but usually the tooth penetrates the base of the tuberosity, leaving but a thin layer of bone posterior to it.

The inferior border of the posterior or zygomatic surface is formed by that portion of the alveolar process which supports the second and third molar teeth.

The bone presents four processes for examination—the nasal, the malar, the palatal, and the alveolar.

The nasal process is a strong, irregular piece of bone, standing vertically above the body of the bone proper, and forming the lateral boundary of the nose. This process is greatly increased in strength by the infra-orbital ridge joining it at or near its base, and ascending its external anterior surface to some extent. That portion of the process posterior to its junction with the infra-orbital ridge assists in forming the inner wall of the orbit.

The external or anterior surface of the nasal process is marked by a number of shallow grooves, traces of the development of the bone. Scattered over this surface are a number of small foramina, the entrances to minute canals transmitting nutrient vessels to the body of the bone. This surface gives origin to one of the lip muscles, the levator labii superioris alæque nasi.

The internal surface of the nasal process is usually described as including all that portion between the superior border and the floor of the anterior nares. The surface is marked by two concave portions and two ridges. The two ridges divide the surface into three parts—the superior meatus, the middle meatus, and the inferior meatus of the nose. The *superior meatus* is the smallest of the three, and occupies the slightly concave space above the superior ridge. The *middle meatus*, partly concave and partly convex, includes the space between the superior and the inferior ridges, and extends from the free margin of the bone in front to the lacrymal groove behind. The *inferior meatus*, which is much the largest, occupies all that concave surface between the inferior ridge above and the palatal process below, and extends from the anterior margin of the bone backward to the point of articulation with the palate bone. The two ridges previously referred to are known as the *superior turbinated crest*, which articulates with the middle turbinated bone, and the *inferior turbinated crest*, which articulates with the inferior turbinated bone.

The malar process is a large, irregular portion of bone situated at the angle of separation between the facial and zygomatic surfaces, and presents a triangular, roughened surface for articulation with the malar bone. The superior boundary of this process is formed by the orbital surface and the outer end of the infra-orbital ridge ; the inferior boundary may be marked by an irregular imaginary line running from the upper margin of the canine fossa to a point between the first and second molar teeth, while the posterior inferior boundary may be traced from the outer superior angle of the zygomatic surface downward and forward to the point above referred to. This process, as well as the nasal process, is subject to much variety in form and general outline. The malar process, assisting as it does in forming what is commonly called the cheek bone, is particularly variable in size, and in certain types and races it is so prominent as to become a controlling feature in the facial form. One of the muscles of mastication—the masseter—has a portion of its origin from the malar process.

The palatal process is more directly interested in the formation of the cavity of the mouth than any other portion of the superior maxillary bone. By articulating with its fellow of the opposite side, it forms about three-fourths of the hard palate or roof of the mouth, the remaining fourth being formed by a portion of the palate bones. It is thick and strong, and projects horizontally inward from the inner surface of the body of the bone. It presents two surfaces for examination,—a superior or nasal surface, and an inferior or oral surface. The *superior* or *nasal surface* is smooth and more or less concave, and forms the floor of the nares. The *inferior* or *oral surface* is also concave, but is much roughened by numerous small projections, between which are lodged the mucous glands. Upon the anterior portion of this surface are a number of small foramina, which mark the entrance to numerous small canals giving passage to nutrient vessels to supply the body of the bone. Near the center of the posterior third are the anteroposterior grooves, which accommodate the posterior palatine nerves and blood-vessels. This process also presents for examination three borders and various other points of interest. The three borders are the anterior, posterior, and mesial. The *anterior* border is thick and somewhat irregular; the *posterior* border is thin and frail, and articulates with a portion of the palate bone. The *mesial* border presents a wide articulating surface in front, behind it is narrow, the whole extent of this border articulating with the corresponding process of the opposite bone.

The Nasal Spine.—At the anterior superior angle of the palatal process is a well-defined spine,—the nasal spine,—being formed by a prolongation of the process beyond the level of the facial surface of the bone. This process, when articulated with its fellow of the opposite side, forms the base of the nose.

The Nasal Crest.—Beginning at the base of the nasal spine, and extending backward along the median border of the bone, is a sharp, irregular ledge of bone, the nasal crest. This portion of the process articulates with the vomer.

The incisor crest is a continuation of the nasal crest anteriorly, projecting beyond the nasal spine in the form of a sharp, spear-like point.

The incisive foramen, or *foramen of Stenson,* is situated immediately back of the incisor crest, and leads downward and forward from the nasal chamber toward the mouth, entering that cavity just back of the central incisor tooth. This passage in the single bone is a simple groove, but in the articulated skull it becomes the anterior palatine canal, which, after passing downward, opens on the nasal surface of the palatal process by four foramina—the incisive foramina, and the foramina of Scarpa, or the naso-palatine foramina. These foramina transmit the naso-palatine nerves.

The palatal process of the superior maxilla is subject to a greater variety in form than any other portion of the bone, this variation in the articulated skull being the cause of the many different curves assumed by the roof or dome of the mouth.

The Alveolar Process.—This process forms the lower margin of the bone, and extends from the base of the tuberosity behind to the median line in front, at which point it articulates with the same process of the opposite bone. It has an outer and an inner margin corresponding to the buccal and palatal surfaces of the roots of the teeth, which are firmly imbedded in it. Its general form from before backward is that of a gradual curve, somewhat variable in different bones, the extent of this variation depending on the type or race to which the bone belongs. The body of the process is made up of an outer and an inner plate, which are connected by numerous septa of cancellated bone. The *outer plate* of the process is continuous with the facial and zygomatic surface of the body of the bone, and assists in forming these surfaces. It is quite thin and frail, and the position of the alveoli beneath are well shown by the numerous vertical ridges upon it. The inferior margin of the outer plate is reinforced by an additional thickness of bone, forming the border of the alveolar sockets. That portion of the plate supporting the molar teeth is heavier in general, and the position of the sockets are not so plainly outlined. The *inner plate* of the alveolar process is much heavier and stronger than the outer plate, and extends from the margins of the alveoli below to the palatal process above. The inferior margin of this plate is not reinforced except in the region of the molars. The construction of the inner plate is, to a great degree, controlled by the shape and position of the palatal process. In the lymphatic temperament this process, when articulated with its fellow, forms a flat or shallow dome to the oral cavity, and in so doing gradually curves into the alveolar process, giving additional thickness to it. The depth of the process in this type is not great, and the roots of the teeth are short and heavy in proportion. In the bilious temperament the inner alveolar plate is deep and abrupt, extending from the inferior margin upward in almost a perpendicular direction to the palatal process, which joins it almost at right angles. The alveolar process gives origin to one of the cheek muscles,—the buccinator, —which is attached to the outer plate near its upper margin, and directly over the space occupied by the second bicuspid and first molar teeth.

The Alveoli or Tooth Sockets.—These cavities, which are variable in number, are formed by the outer and inner plate of the alveolar process, and by numerous connecting septa of bone placed between the two plates. The shape and depth of these cavities is regulated by the form and length of the roots of the teeth which they support. The first socket, or that next to the mesial surface of the bone, gives support to the central incisor tooth. It forms almost a perfect cone, and has an average depth of half an inch. Its lower border is circular, and the anterior or labial portion describes a larger circle than the posterior or palatal half. The mesial and distal walls are somewhat flattened. The second cavity, proceeding backward from the median line, supports the lateral incisor tooth. It is also conic, but much smaller than the preceding. It is seldom over $\frac{3}{8}$ to $\frac{7}{16}$ of an inch in depth. It is much flattened on its mesial and distal walls, giving the appearance of an oblong, rather than a round, cavity. This socket, as well as that for the central incisor, occupies an almost vertical position in the process. Very frequently the socket for the lateral incisor presents a slight distal curve at its upper extremity. The third socket, or that giving support to the cuspid tooth, is much larger and deeper than either of those previously described. It extends upward, inward, and backward to the average depth of $\frac{5}{8}$ to $\frac{3}{4}$ of an inch.

In transverse section, its labial wall presents a much larger circle than its palatal margin. The labial and distal walls are much flattened and somewhat convex. The general direction of this socket is slightly to the distal. The socket which supports the first bicuspid is usually divided from mesial to distal by a thin septum of bone, thus forming an outer or buccal socket, and an inner or palatal socket. This division seldom exists to the full depth of the cavity, but usually begins about midway of its length. The lower margin of this socket is oblong or egg-shaped, its outer or buccal portion forming a larger curve than its palatal. The lateral walls are slightly concave, or flattened, until the point of separation is reached, when they become more circular, the alveoli above this point becoming cone-shaped. It is not uncommon for this socket to be a single cavity, and when thus formed it resembles a flattened cone, with the buccal and palatal margins rounded. The next socket gives support to the second bicuspid tooth, in most instances being a single cavity, but in rare instances it is divided near its upper extremity. In general outline it resembles the socket for the first bicuspid.

The socket for the first molar is much larger than any of those previously described; its inferior margin presents a circular outline on its buccal and palatal portions, the former curve being larger than the latter. The mesial and distal walls are flattened and slightly concave. The upper three-fourths of this socket is divided into three separate compartments, being so arranged that two are upon the buccal and one upon the palatal side. The septa separating the two buccal cavities from the palatal cavity are heavy and strong, while that placed between the two buccal sockets is thin and frail. The two buccal cavities are usually flattened upon their mesial and distal sides. The palatal socket is larger and somewhat deeper than the buccal, the average depth of all being about ½ of an inch. The socket for the second molar is similar in most respects to that for the first molar, except that it is somewhat smaller. The same description might answer for the third molar socket, which in general is similar to the alveoli for the other molars. It is smaller than the second molar socket, and may be a single cavity, or it may be divided into three or more compartments.

Articulations.—The superior maxillary bone articulates with its fellow of the opposite side, with the frontal, lacrymal, ethmoid, palate, vomer, malar, and inferior turbinated bones. Occasionally it articulates with the sphenoid bone.

Attachment of Muscles.—The muscles attached to this bone are eleven in number, and are as follows:

Compressor nares,	Internal pterygoid,
Orbicularis oris,	Orbicularis palpebrarum,
Levator labii superioris alæque nasi,	Levator labii superioris proprius,
	Inferior oblique,
Levator anguli oris,	Buccinator,
Depressor alæ nasi,	Masseter.

Blood-supply.—The maxilla receives its vascular supply from numerous large arteries. They are derived from the alveolar, infra-orbital, nasopalatal, descending palatal, ethmoidal, nasal, frontal, and facial branches.

Development.—The superior maxilla arises from four points of ossification, which are deposited in membrane. These four centers make their appearance as early as the eighth fetal week, this early beginning making it somewhat difficult to accurately follow its growth. The four centers are named, as located, pre-

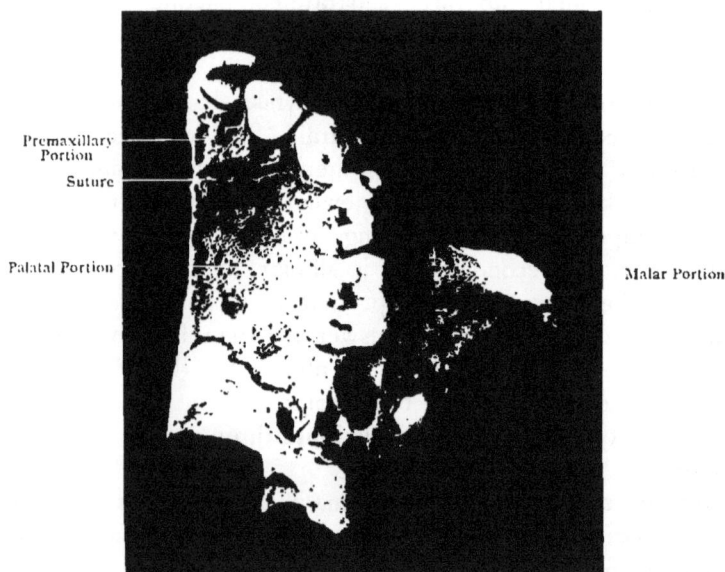

FIG. 13.—LEFT SUPERIOR MAXILLARY, ABOUT THE SECOND YEAR, ENLARGED.

maxillary, maxillary, malar, and prepalatal. The premaxillary nucleus gives rise to the incisive portion of the bone, or that part supporting the incisor teeth. During early life this division of the bone is separated from the body of the bone, and is known as the premaxillary portion (Fig. 13). Union between the premaxillary portion and the maxilla proper takes place about birth, and the suture thus formed is visible on the facial surface until the sixth or seventh year, and on the palatine surface until the adult period. The palatal suture extends as far back as the posterior border of the anterior palatal canal. This nucleus also sends a narrow process upward which forms part of the outer boundary of the anterior narial aperture. On the palatal aspect it furnishes a speculum which surrounds the anterior and mesial walls of Stenson's canal. The posterior limit of the premaxillary portion is indicated by the suture on the palatal surface. The maxillary nucleus forms the greater portion of the body of the true maxilla and the nasal process. The malar center gives origin to the malar process, and all that portion external to the infra-orbital groove. The prepalatine center

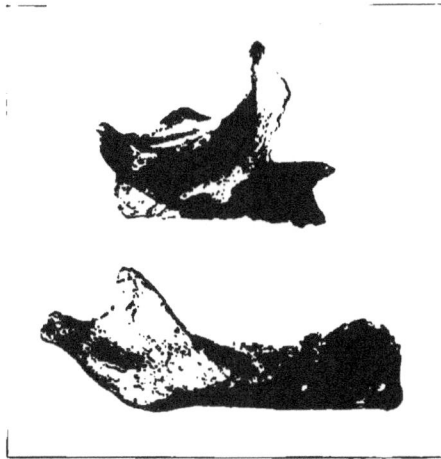

FIG. 14.—DEVELOPING MAXILLARY BONES ABOUT THE FIFTH MONTH AFTER BIRTH.

gives rise to the nasal surface of the bone and that portion of the palatal process posterior to Stenson's canal.

Development of the Alveolar Process.—This process is represented at birth by the walls of a deep groove, in which are lodged the partly calcified, deciduous teeth and the germs of most of the permanent teeth (see Development of the Teeth).

The growth of the process continues with the growth of the teeth until, finally, at about the seventh month after birth, the dental organs are completely encased within its walls. With the decalcification of the roots of the deciduous teeth comes the loss of the process surrounding them, and, as the permanent teeth advance to take their place in the arch, the process is again built up about their roots.

The Maxillary Sinus, or Antrum of Highmore[*] (Fig. 12).—This is a large cavity situated within the body of the maxilla. Its general shape is that of a pyramid, with its base directed toward the median line, or nasal surface, its apex pointing toward and extending into the malar process, and, in some instances, penetrating the malar bone. The size of the cavity varies in different subjects and in the opposite bone of the same subject. The average capacity is about three fluid-rams, but this may be increased to six or eight fluidrams. The size of the bone and the prominence of the malar process control, in a measure, the size of the cavity; but not infrequently the largest bone will present the smallest sinus. Sex also appears to exert a controlling influence over the capacity of the cavity, it being greater in the male than in the female. In youth the cavity is quite small, the walls being much thicker proportionately than in the adult. The walls of the sinus in the matured subject are quite thin and frail, and are four in number. The superior wall is formed by a thin plate of bone, the floor of the orbit. This surface is almost flat, and serves as a roof to the cavity. Near the anterior margin of this surface is a thick rib of bone which marks the course, and forms one of the walls of the infra-orbital canal.

The inner wall, or that looking toward the nasal surface, is formed by the thin bony layer separating this cavity from that of the nares. The outer or lateral surface, formed by the facial and zygomatic surfaces of the bone, is smooth, and convex from before backward. Near the center of this surface the cavity may penetrate the malar process, and in the disarticulated skull would present an opening at this point. The inferior wall is formed by the alveolar process, and is marked by a number of irregular eminences corresponding to the roots of the neighboring teeth. The teeth referred to are generally the first and

---

[*] Described separately, in preference to including in general description of the bone.

second molars, and occasionally the second bicuspid. It is not unusual for the roots of one or more of these teeth to penetrate the floor of the sinus, in consequence of which the lining membrane of the cavity may suffer disease generated in the teeth.

The inferior wall is much the strongest of the four, and, besides the unevenness of the surface produced by the tooth-roots, it frequently supports a number of thin, bony partitions, which may completely or partly divide the floor of the cavity into numerous small compartments.

The posterior portion of the lateral wall is marked by the posterior dental canals, which give passage to the posterior nerves and blood-vessels. In like manner the anterior portion of the lateral wall is grooved for the reception of the anterior dental nerves and blood-vessels. Upon the inner wall, or that forming the base of the pyramid, is an opening which communicates with the middle meatus of the nose. In the articulated skull this opening is quite small, being from $\frac{1}{5}$ to $\frac{1}{4}$ of an inch in diameter. The correct idea of this opening can not be obtained by studying the individual bone, as the numerous perforations then to be observed are closed or partly closed by articulation with adjacent bones. The mucous membrane lining the nasal cavity enters the sinus through the small aperture above referred to, and forms a continuous lining over its entire surface. The recent work of Prof. M. H. Cryer ("Studies of the Maxillary Bones" *) has thrown much light upon the relations of the maxillary sinus to the mouth and teeth, and he has demonstrated beyond a doubt that the relationship existing between the parts is susceptible to extensive variation. He has shown that in some instances the cavity upon one side will be large, with its floor broken by the tooth-roots, while that upon the opposite side will be extremely small and far removed from the root apices. In fact, the researches of this gentleman have so revolutionized the subject under consideration that the foregoing description is only reliable in so far as it treats of the conditions most frequently met with.

---

* " Dental Cosmos," vol. XXXVIII.

THE PALATE BONE.

The *palate bones* (Figs. 15 and 16), two in number, are situated immediately posterior to the two maxillæ, and with them complete the hard palate. They also assist in forming the boundaries of the orbital and nasal cavities, the sphenomaxillary,

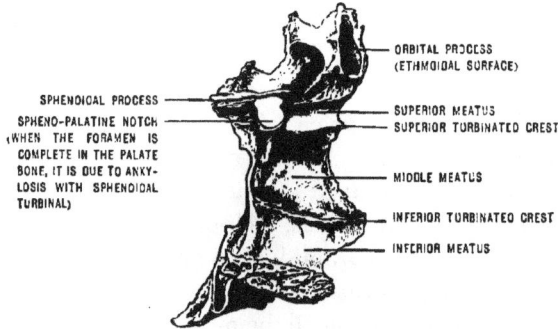

ORBITAL PROCESS
(ETHMOIDAL SURFACE)

SPHENOIDAL PROCESS

SPHENO-PALATINE NOTCH
(WHEN THE FORAMEN IS
COMPLETE IN THE PALATE
BONE, IT IS DUE TO ANKY-
LOSIS WITH SPHENOIDAL
TURBINAL)

SUPERIOR MEATUS
SUPERIOR TURBINATED CREST

MIDDLE MEATUS

INFERIOR TURBINATED CREST

INFERIOR MEATUS

Fig. 15.—Palate (Left) Bone (Inner View).—(*Morris.*)

the sphenopalatine, and the pterygoid fossa, the sphenomaxillary fissure, the posterior ethmoidal cells, and the maxillary sinus. When in position in the skull, these bones are wedged between

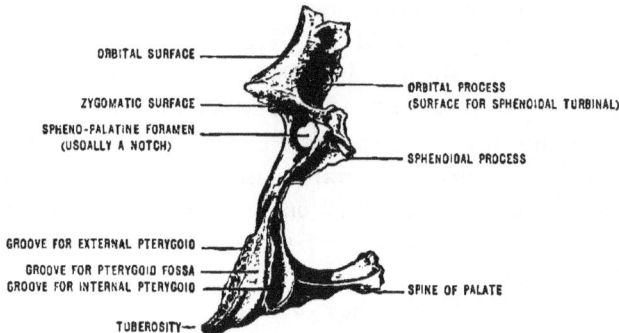

ORBITAL SURFACE

ZYGOMATIC SURFACE

SPHENO-PALATINE FORAMEN
(USUALLY A NOTCH)

ORBITAL PROCESS
(SURFACE FOR SPHENOIDAL TURBINAL)

SPHENOIDAL PROCESS

GROOVE FOR EXTERNAL PTERYGOID
GROOVE FOR PTERYGOID FOSSA
GROOVE FOR INTERNAL PTERYGOID

SPINE OF PALATE

TUBEROSITY

Fig. 16.—Palate Bone (Posterior View).—(*Morris.*)

the maxillæ and the sphenoid bone. They are rectangular in outline, and each bone presents for examination a horizontal and a vertical plate, a tuberosity, and two processes, the orbital and the sphenoid.

The horizontal plate, smaller than the vertical, assists in forming the hard palate, and corresponds to the palatal process of the maxilla. In entering into the construction of the hard palate the form of this plate varies to the same degree as the palatal plate of the maxilla. In general, it is described as quadrilateral in shape, having two surfaces and four borders. The superior surface, which is concave from side to side, forms the posterior floor of the nasal chamber. The inferior surface completes the hard palate posteriorly, and presents, near its posterior border, a transverse ridge for the attachment of one of the muscles of the soft palate (the tensor palati) ; the anterior border is serrated for articulation with the palatal process of the maxilla. The posterior border is free, curved, and sharp, and marks the posterior boundary of the hard palate. At the median line this border terminates in a sharp point, which, when articulated with the corresponding bone of the opposite side, forms the posterior nasal spine ; to this point the azygos uvulæ muscle is attached. The external border is situated just below the junction of the horizontal and vertical plates. In this portion is a groove which assists in forming a portion of the posterior palatal canal. The internal border is broad and serrated for articulation with its fellow of the opposite side. When the palate bones are in position in the skull, these borders form a ridge, continuing the crest formed by the palatal process of the maxilla, this crest receiving the inferior border of the vomer.

The vertical plate is thin and frail, and extends from the floor of the nasal chamber below to the upper extremity of the sphenopalatine notch above. It has two surfaces and four borders.

The external surface is roughened for articulation with the maxilla, excepting a small triangular surface near the upper extremity, which forms a portion of the sphenomaxillary fossa, and a small portion near the middle of the surface close to the anterior border, which forms a portion of the wall of the maxillary sinus. Near the posterior boundary of this surface is a vertical groove, which forms, when articulated with the maxilla, the posterior palatal canal, transmitting the descending palatal nerves and vessels.

The internal surface is divided into three shallow depressions by two transverse ridges—the superior and inferior turbinated crests. The lower depression thus formed assists in the construction of a portion of the inferior meatus of the nose. The crest immediately above this depression articulates with the inferior turbinated bone. The central depression, the largest of the three, forms a portion of the middle meatus of the nose, the crest above articulating with the middle turbinated bone. The superior depression—much smaller but deeper than either of those previously described—forms a large part of the superior meatus. The anterior border of the vertical plate is thin and sharp, the inferior turbinated crest protruding near the center of the border, and forming the maxillary process. This process assists in closing the maxillary sinus by being received into the maxillary fissure of the maxilla. At the upper extremity of this border is the orbital process, which presents for examination five surfaces, three of which are articular. The anterior or maxillary surface is directed outward, upward, and downward. It is oblong in form and articulates with the posterior superior angle of the inner surface of the maxilla. The posterior or sphenoidal surface is directed backward, upward, and inward, and articulates with the vertical plate of the ethmoid bone. The superior or orbital surface is triangular in form, extending upward and outward, forming the posterior angle of the floor of the orbit. The external or zygomatic surface is smooth, oblong, and directed outward, backward, and downward, forming a portion of the sphenomaxillary fossa.

The posterior border of the vertical plate is irregular and serrated, and comes into relation with the internal pterygoid process, terminating below in a prominent tuberosity. This presents three grooves or flutes. The inner receives the internal pterygoid, the outer the external pterygoid process, while the middle groove completes the pterygoid fossa, and gives attachment to a portion of the internal pterygoid muscle. This process also gives rise to the superior constrictor of the pharynx. Passing through the tuberosity are a number of small canals, those on the nasal side being the accessory palatal canals. Near the junction of the tuberosity with the horizontal plate is the opening of the posterior palatal canal, and beyond this the small external palatal canals.

The sphenoidal process is at the superior end of the posterior border. It is variable in shape and curves upward, backward, and inward. It presents a superior, an external and an internal surface, and two borders—an anterior and a posterior.

The superior surface, the smallest of the three, is marked by a groove, which assists in forming the sphenopalatine canal. This surface articulates with the horizontal ·portion of the sphenoidal turbinated bone. The external surface assists in forming the sphenomaxillary fossa by its anterior portion, while the posterior portion is rough for articulation with the pterygoid plate of the ethmoid bone. The internal surface is instrumental in forming a portion of the outer wall of the posterior nares, and for this purpose is smooth and concave. The anterior border forms the posterior margin of the sphenopalatine notch. The posterior border is serrated, and articulates with the inner surface of the pterygoid process.

The superior border of the vertical plate is divided by a deep notch or foramen, which divides the orbital from the sphenoidal process. This opening is the sphenopalatine notch or foramen, and transmits the sphenopalatine vessels and nerves from the sphenopalatine fossa to the nasal chamber.

The inferior border of the vertical plate joins the external border of the horizontal plate. Extending downward and backward from the inferior and posterior borders is the pyramidal process, the borders of which are serrated for articulation with both pterygoid plates of the sphenoid bone.

Articulations.—The palate bone articulates with the sphenoid, superior maxilla, sphenoidal turbinated, inferior turbinated, ethmoid, and with its fellow of the opposite side.

Attachment of Muscles.—The following muscles are attached to the palate bone :

Tensor palati,	Internal pterygoid,
Azygos uvulæ,	Superior constrictor of pharynx.

Blood-supply.—The arteries which supply this bone are derived from branches of the descending palatine, the sphenopalatine, and pterygopalatine.

Development.—The palate bone is developed from a single
center deposited in membrane. This center makes its appearance about the eighth or ninth fetal week, near the line of
junction between the horizontal and vertical plates. At birth
these plates are about the same length, but soon after this
period, when the nasal sinuses increase in height, the vertical

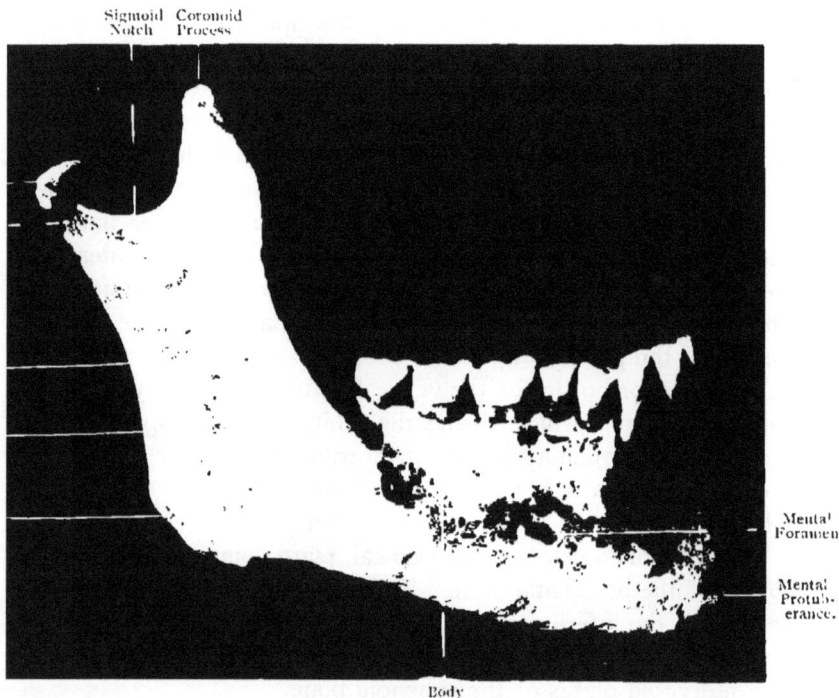

FIG. 17.—THE MANDIBLE OR INFERIOR MAXILLA. RIGHT SIDE, EXTERNAL OR FACIAL
SURFACE.

plate begins to lengthen, and continues to do so until it becomes nearly double the length of the horizontal plate.

INFERIOR MAXILLARY BONE.

The Inferior Maxillary, Mandible, or Lower Jaw Bone (Fig.
17).—This bone, having no osseous union with the skull proper,
may be considered as one of its appendicular elements. It is

the heaviest and strongest bone of the head, gives support to the sixteen inferior teeth, and serves as a framework for the lower half or floor of the mouth. It is situated at the lower extremity of the face, and immediately below the superior maxillary and malar bones, while its posterior extremity rests against the glenoid fossa of the temporal bone, forming a movable articulation with this cavity. In general, the bone is symmetric in outline, and presents for examination a horizontal portion, or *body*, and two vertical portions, or *rami*, which in the adult are almost perpendicular to, or at right angles with, the body of the bone.

The body, or horizontal portion, consists of two identical halves, which meet at the median line and form a slight vertical ridge, the *symphysis*. This line indicates the point of union between the two lateral halves, which at birth are usually separated, but soon after this period become firmly united. Each lateral half of the body presents two surfaces—an external and an internal ; and two borders—a superior and an inferior.

The external or facial surface (Fig. 17) is smooth and convex, and furnishes a number of points for examination. Beginning at the median line, the symphysis ends inferiorly in a prominent triangular surface—the *mental protuberance*, or chin.

The Incisive Fossa.—Passing backward from the symphysis, and immediately above the triangular ridge which forms the mental process, is a decided but shallow depression—the incisive fossa. This fossa gives origin to one of the elevator muscles of the chin—the levator menti. Slightly posterior to and below this fossa, on a line corresponding to the position of the cuspid tooth, is an oblong depression for the origin of the depressor muscle of the lower lip—depressor labii inferioris.

The External Oblique Line.—Extending obliquely across the facial surface from the mental process to the base of the vertical portion of the bone, and continuous with its anterior margin, is a well-defined ridge—the external oblique line. Near the center of this ridge, or below the position occupied by the bicuspid and first molar teeth, is the point of attachment of the depressor muscle of the angle of the mouth—the depressor anguli oris. Somewhat anterior to and above this point is the origin of the depressor muscle of the lower lip—the depressor labii inferioris. Between the line of origin of the depressor anguli oris and the inferior border of the bone is a roughened surface for the attachment of the platysma myoides muscle. This roughened surface divides the body of the bone into an upper and a lower portion. That portion above is known as the alveolar or mucous portion, while that below is called the basalar or non-mucous portion. The attachment of the platysma myoides muscle at this point marks the lower boundary or floor of the mouth. The superior or alveolar portion of the bone is within the cavity of the mouth, and is covered with mucous membrane and mucoperiosteum ; while the inferior or basalar portion is outside and below the cavity, and is covered with periosteum similar to other bones.

The Mental or Anterior Dental Foramen.—Midway between the superior and inferior border of the body, and usually below the second bicuspid tooth, is a large foramen,—the mental or anterior dental foramen,—giving passage to the mental branches of the inferior dental nerve and accompanying blood-vessels. The position of this foramen is not constant, but, as previously stated, it is usually below the second bicuspid, or between this point and the first bicuspid. The buccinator muscle, which forms a large portion of the lateral wall of the mouth, has its origin from the facial surface of the mandible, being attached to the alveolar portion immediately below the molar teeth.

The Internal Surface of the Body of the Bone (Fig. 18). —The median line is marked by a slight vertical depression, representing the line of union, and corresponding to the symphysis externally.

The Mylohyoid, or Internal Oblique Ridge.—The internal surface is divided into two portions by a well-defined ridge—the mylohyoid, or internal oblique ridge. It occupies a position closely corresponding to the external oblique ridge on the facial surface. Beginning near the base of the bone at the median line, it passes backward and upward, increasing in prominence until the base of the vertical portion of the bone is reached, into which it gradually disappears. This ridge gives origin to the mylohyoid muscle, which forms the central portion of the floor of the mouth. In correspondence to the facial surface of the bone, the attachment of the mylohyoideus muscle forms the dividing line between the mucous membrane and mucoperiosteum covering the upper portion of the body of the bone, and the periosteum covering the inferior portion.

The Genial Tubercles.—Near the lower third, at the median line, is a roughened, eminence—the genial tubercles. Taken collectively, these are in two pairs—a superior and an inferior. The superior pair (usually the largest) give origin to the geniohyoglossus muscle, and the lower pair to the geniohyoid muscle.

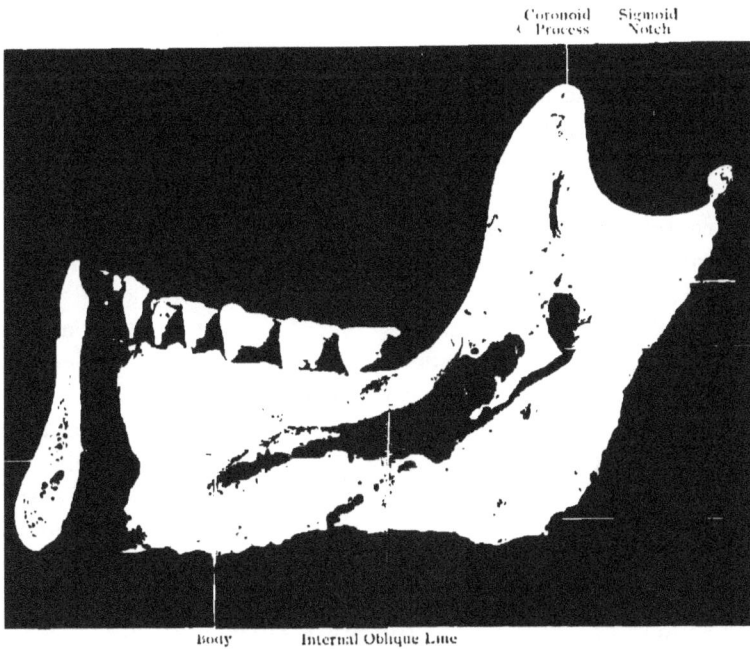

FIG. 18.—THE MANDIBLE OR INFERIOR MAXILLA. RIGHT SIDE, INTERNAL SURFACE.

The Sublingual Fossa.—By the side of the genial tubercles, and above the mylohyoid ridge, is a shallow, smooth depression —the sublingual fossa. One of the salivary glands—the sublingual—is partially supported in this fossa.

The Digastric Fossa.—Below the mylohyoid ridge, and near the median line, is a slight depression,—the digastric fossa,— which affords attachment for the digastric muscle.

The Submaxillary Fossa.—In the center of the internal surface, extending from before backward, between the mylohyoid ridge and the lower border of the bone, is an oblong depression —the submaxillary fossa. In this fossa rests another of the salivary glands—the maxillary.

The Superior or Alveolar Border.—This border extends from the junction of the body, with the vertical plate on one side, to the corresponding point on the other. The construction of this border is similar to the alveolar border of the superior maxilla. At the anterior portion it is narrow, but gradually increases in width as it proceeds backward—in some instances following the line of the body of the bone; in others, inclining inward, or to the lingual. Each lateral half is marked by eight sockets, for the accommodation of the sixteen inferior teeth. They are smaller in proportion than the alveolar sockets in the superior maxilla. The socket nearest the median line receives the central incisor tooth, and is the smallest of the number. It has an average depth of $\frac{7}{16}$ of an inch, is conic from above downward, oblong in transverse section, with its lateral walls flattened. The second socket gives support to the lateral incisor tooth; it is a trifle larger than the central incisor socket, but in other respects is quite similar. The socket for the cuspid tooth is situated at the anterior angle of this border, and is much larger and deeper than either of the incisor sockets. It has an average depth of $\frac{9}{16}$ of an inch; its lateral walls are compressed, and sometimes slightly concave. In transverse section the labial wall forms a larger curve than the internal or lingual wall. Passing backward, the next two sockets are for the support of the bicuspids; they are circular in outline, with an average depth of $\frac{1}{2}$ of an inch. The cavity for the first bicuspid is usually a little larger than that for the second. In rare instances one or the other of these sockets will be divided for the accommodation of two roots. The sockets for the first and second molars present a circular outline upon their free margins, but below they divide into two flattened, cone-shaped cavities—one anterior and one posterior. The flattened sides of these cavities are concave in the center, and at their lower third curve backward. The average depth of these sockets is

½ of an inch. The socket for the third molar, like its superior fellow, is variable both in form and position, frequently being crowded inside or outside of the tooth-line. In some instances it is divided into two or more compartments. The average depth is not over ⅜ of an inch.

The alveolar process, which composes the superior border of the body of the mandible, differs from the same process in the superior maxilla in one very important particular: instead of the outer plate being thin and frail, it is equally as heavy as the inner or lingual plate. When the tooth-line is inclined inward from the body of the bone, the posterior outer wall is much heavier than the interior.

The Inferior Border of the Body of the Bone.—This border extends from a slight depression, to be observed at the point of union between the body and ramus, to the corresponding point upon the opposite side. It is strong, rounded, and compact, and gives to the bone the greatest portion of its strength. Near its junction with the ramus is the facial notch, so named from the facial artery passing over this point.

The Ramus, or Vertical Portion of the Bone.—This vertical plate is quadrilateral in outline, and presents two surfaces—external and internal; four borders—superior, inferior, anterior, and posterior; and two processes—the condyloid and the coronoid.

The external surface is flat and smooth. Near the center it is slightly concave and roughened for the attachment of one of the muscles of mastication—the masseter.

The internal surface presents near the center an oblong opening—the *inferior dental* or *mandibular foramen*—leading into the inferior dental or mandibular canal. Surrounding this foramen, on its posterior internal margin, is the mandibular spine, to which is attached the sphenomandibular ligament. Running obliquely downward from the base of the foramen, and beneath the spine, is a decided groove,—the mylohyoid groove,—which accommodates the mylohyoid nerve, artery, and vein, which pass forward to supply the floor of the mouth. Below and behind this groove the surface is roughened for the attachment of another muscle of mastication—the internal pterygoid.

The Inferior Dental or Mandibular Canal.—Beginning at the foramen of the same name, this canal enters the body of the bone, passes downward and forward horizontally, until it finds an exit at the mental foramen. This canal lies immediately below the alveolar sockets, and from it are given off smaller canals which open into the tooth-sockets through minute foramina. Near the mental foramen the canal divides into a number of smaller ones, which pass forward through the substance of the bone to the sockets of the cuspid and incisor teeth.

The Superior Border of the Ramus.—This border is crescent-shaped, and is otherwise known as the sigmoid notch. Arising from its anterior portion is a flattened, cone-shaped process—the coronoid process. On its posterior portion is a rounded or oblong eminence—the condyloid process. The concave or crescent-shaped margin of this border is thin and smooth in front, becoming wider and heavier as it approaches the condyle.

The Coronoid Process.—The anterior margin of this process, being a continuation of the external oblique line, is heavier at the base than at the apex. The outer surface is smooth, and affords attachment to the masseter, and a few fibers of the temporal muscle. The internal surface is marked by a vertical ridge, which passes downward, increasing in size, and finally joining the internal oblique line at a point posterior to the third molar. The surface anterior to this ridge is grooved, and gives attachment to a part of the temporal muscle above, and the buccinator muscle below. The surface posterior to this ridge affords attachment for the greater part of the temporal muscle. The posterior border of this surface is thin, and forms the anterior margin of the sigmoid notch.

The Condyloid Process.—This may be described as the expanded extremity of the posterior border of the ramus, and is quite variable in form (see Occlusion of the Teeth). It is divided into a superior or articular portion, and an inferior portion, or neck.

The articular portion of the condyle is more or less oblong, and is convex above, fitting into the glenoid fossa of the temporal bone, and forming, with the interarticular cartilage which lies between the two surfaces, the temporomaxillary articulation.

The neck is that constricted portion immediately below the articular surface. It is flattened in front and presents a pit,—the pterygoid fossa,—to which a portion of the pterygoid muscle is attached. Immediately below the point of junction between the neck and the articular surface externally is the condyloid tubercle, to which is attached the external lateral ligament.

The Inferior Border of the Ramus.—This border is thick, rounded, and continuous with the lower border of the body of the bone. At the point of junction between this and the posterior border is the angle of the jaw. The angle has a slight outward inclination, and is roughened for the attachment of a part of the superficial portion of the masseter muscle.

The anterior border has been described in connection with the coronoid process.

The posterior border is smooth and rounded on its upper half, the lower half being roughened for the attachment of the stylomaxillary ligament.

Attachment of Muscles. — The following muscles are attached to the mandible :

Buccinator,	Superior constrictor of pharynx,
Depressor labii inferioris,	Masseter,
Depressor anguli inferioris,	Orbicularis oris,
Levator menti,	Internal and external pterygoid,
Geniohyoglossus,	Geniohyoid,
Platysma myoides,	Mylohyoid,
Digastric,	Temporal.

Development.*—On account of its early functional activity, the mandible is among the first bones to be ossified. Development takes place from six centers for each lateral half, the nuclei being deposited as early as the sixth or eighth fetal week, and after their establishment the developmental process takes place very rapidly. The six centers of ossification are principally named according to their position. The early preparation for the development of the bone is found in the appearance of what is known as the mandibular plates, which are thrown out from the sides of the cranial base, and finally unite at

---

* See " Development of the Teeth."

the median line. Not long after this period a cartilaginous band
—Meckel's cartilage—is developed in the substance of the mandibular plates, and it is about this cartilaginous framework that
ossification first takes place. The various centers are distributed
along the line of Meckel's cartilage, and are named as follows :
Mental, dentinary, coronoid, condyloid, angular, and splenic.

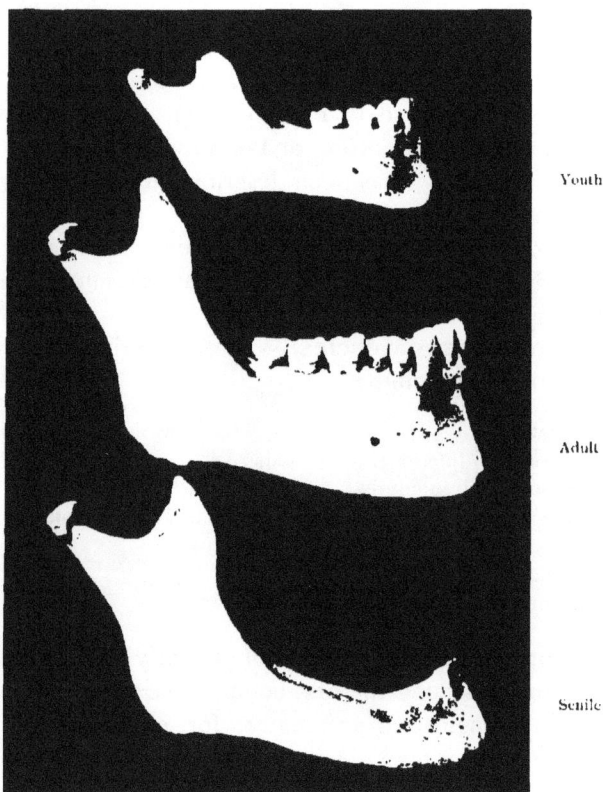

Youth

Adult

Senile

FIG. 19. —CHART SHOWING THE EVOLUTION AND DEGENERACY OF THE MANDIBLE.

The *mental* center provides for the development of that portion
of the bone between the median line and the mental foramen.
The *dentinary* center forms the lower border and outer plate,
and provides for the establishment of the crypts inclosing the
developing tooth-follicles.

The *coronoid* and *condyloid* centers are both instrumental in constructing these processes, and the *angular* center provides for the angle of the bone. The *splenic* center is somewhat later in making its appearance, and from it the inner plate of the mandible is formed, the line of union between it and the dentinary center being indicated by the mylohyoid groove. While, as above stated, most of the centers of development are along or near the line of Meckel's cartilage, the condyloid and coronoid processes are developed from other cartilage. Soon after birth the two lateral halves of the mandible begin to coalesce at the median line, this union taking place from below upward; and by the eighth or tenth month union is complete and the individual bone is established. The inferior maxilla is subject to a continuous change in form, not only in regard to its general contour, but also accommodating itself to the movements and growth of the teeth, the former taking place at or about the angle, while the latter occurs in the alveolar portion of the bone.

Figure 19 represents the changes which take place in the angle of the mandible from youth to old age. It will be observed that the angle formed in the adult bone, with the teeth in position, is almost a right angle ; and that in youth, with the deciduous teeth in the alveoli, the angle is much more obtuse, which condition is again approached in old age.

CHAPTER III

THE TEMPOROMANDIBULAR ARTICULATION. — THE MUSCLES OF
MASTICATION.

TEMPOROMANDIBULAR ARTICULATION.

Although external to the cavity of the mouth, this articulation is so closely associated with the masticatory function that it seems important that a brief description of its construction

FIG. 20.—TEMPOROMANDIBULAR ARTICULATION.

and action should be given. It receives its name from the two bones which enter into its formation—the temporal and the mandible, or inferior maxillary.

This joint is the seat of motion in the mandible, and entering into its construction are bones, ligaments, cartilage, and synovial membrane, these being the tissues essential to all diarthrodial or movable articulations. The various movable joints of the body are classified according to the nature of the movement, and correspond to the mechanical actions known as hinge joint, ball-and-socket joint, gliding joint, pulley joint, etc. The temporomandibular joint is of the diarthrodial class, and the movements which it possesses are a combination of the gliding movement (arthrodia) and of the hinge movement (ginglymus). The osseous parts entering into the formation of the joint are the anterior portion of the glenoid fossa of the temporal bone and the condyloid process of the mandible (Fig. 20).

The glenoid fossa may be described as an oblong cavity, with its base directed upward, being bounded anteriorly by a heavy bony ridge (the anterior root of the zygoma), posteriorly by an irregular, flattened portion of the bone (the tympanic plate of the petrous portion), internally by a union of the anterior and posterior boundaries, and externally by the middle root of the zygoma. The floor of the fossa is traversed by a well-marked fissure—the glenoid fissure (fissure of Glaserius)—which divides the fossa into two portions, an anterior and a posterior. The anterior half is deeper and more concave than the posterior, and is the articulating portion, being occupied by the condyle, while the posterior half gives lodgment to the parotid gland.

The condyloid process of the mandible having been described with that bone, in this connection reference will be made to the variety of forms which it presents, and the influence which it exerts over the nature of the tooth occlusion. This process, when narrow and oblong (Fig. 20), closely resembles the ginglymus, or hinge joint, and will be accompanied by teeth presenting deep, penetrating cusps, forming a positive and well-locked occlusion (Fig. 21, *B*), with little or no lateral motion. If the condyle presents the appearance shown in figure 22, which resembles the enarthrodia, or ball-and-socket joint (although it can not be considered as such), the teeth associated with such a formation will be provided with short, rounding cusps, and the occlusion will be loose and wandering (Fig. 21, *A*). This difference in the form of the condyle will be accompanied by a corresponding variation in the concavity of the glenoid fossa. Not

6

only does the osseous structure in the joint partake of individual characteristics, but likewise the muscles and ligaments; their functions being to operate the articulation, they are developed

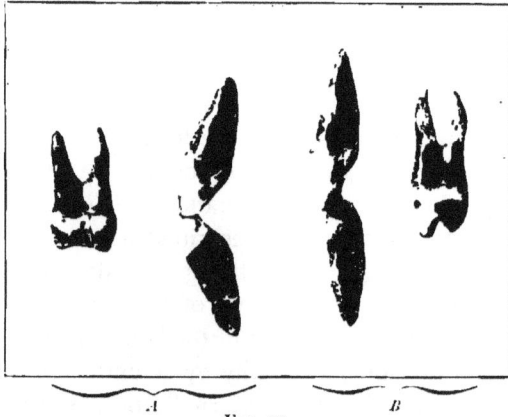

A   B
FIG. 21.

FIG. 22.

in accordance with the action required of them, which action is, in a measure, dependent upon the conditions existing in the mouth.

Both the condyle and the glenoid fossa are covered with articular cartilage. In the latter this membrane extends over its anterior border, to facilitate the play of the joint. The condyle is held in position in the fossa by three ligaments—the capsular, the sphenomaxillary, and stylomaxillary. The *capsular ligament* is divided into four portions—anterior and posterior, external and internal. The anterior portion consists of a few fibers connected with the anterior margin of the fibrocartilage, attached below to the anterior margin of the condyle and above to the front of the glenoid ridge. The posterior portion is attached above just in front of the glenoid fissure, and is in-

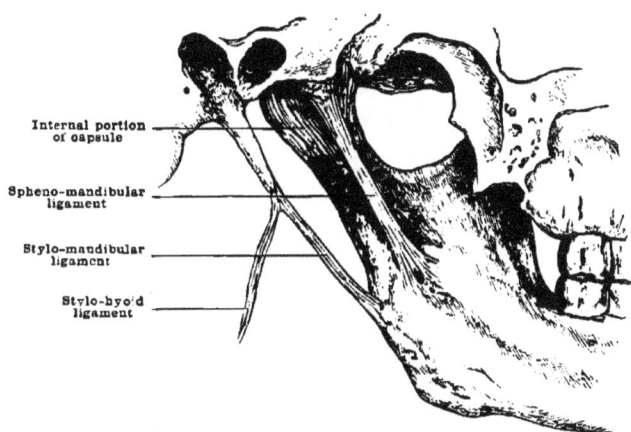

Internal portion of capsule

Spheno-mandibular ligament

Stylo-mandibular ligament

Stylo-hyo'd ligament

FIG. 23.—INTERNAL VIEW OF TEMPOROMANDIBULAR JOINT.—(*Morris.*)

serted into the posterior margin of the ramus of the maxilla just below the neck of the condyle. The external portion, otherwise known as the external ligament, is the strongest portion of the capsular ligament. It has a broad attachment above to the zygoma, from which point it passes downward and backward, and is inserted into the outer side of the neck of the condyle. The internal portion, or short internal lateral ligament, is composed of well-defined fibers, having a broad attachment above to the inner edge of the glenoid fossa and to the alar spine of the sphenoid bone ; below it is inserted into the inner side of the neck of the condyle.

The *sphenomaxillary*, or long internal lateral *ligament*, is a thin, loose band, situated some distance from the joint proper, and, as its name implies, has its attachment above to the alar spine of the sphenoid bone, and also to that portion of the temporal bone contiguous to it. It passes downward and forward, and is inserted into the mandibular spine of the maxilla.

The *stylomaxillary ligament* extends, from the styloid process of the temporal bone, downward and forward, to be inserted into the posterior border of the ramus of the inferior maxilla, at a point between the masseter and internal pterygoid muscles.

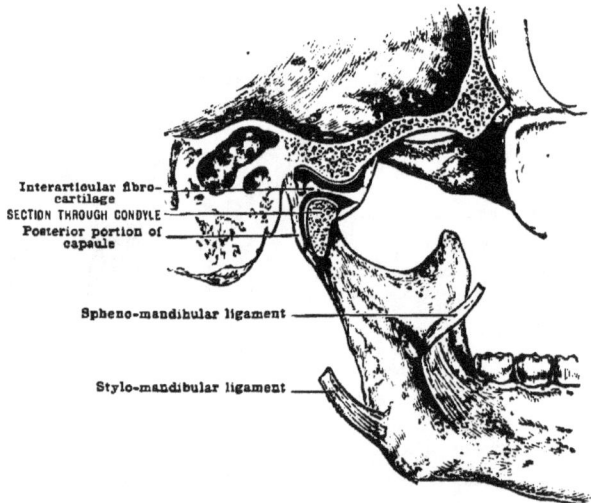

FIG. 24.—VERTICAL SECTION THROUGH THE CONDYLE OF JAW TO SHOW THE TWO SYNOVIAL SACS AND THE INTERARTICULAR FIBROCARTILAGE.—(*Morris.*)

The *interarticular fibrocartilage* is an oval sheet placed between the two articulating surfaces. It is thinnest at the center and becomes thicker as the margins of the fossa are approached, at which point it is connected with the fibers of the capsular ligament. Being placed immediately between the two articular surfaces it divides the joint into two separate synovial cavities. Each of these synovial cavities is occupied by a synovial membrane, that occupying the upper compartment being the largest, and passes from the margins of the glenoid fossa above to the

upper surface of the interarticular cartilage below. The mem
brane which occupies the lower cavity is smaller, and passes
from the under surface of the interarticular cartilage above to
the margins of the condyle below. The *blood-supply* to this
articulation is derived from the temporal, middle meningeal, and
ascending pharyngeal arteries.

The *nerves* are derived from the masseteric and auriculo-
temporal.

The movements of this articulation present a greater range
than any other joint in the human body. While the chief move-

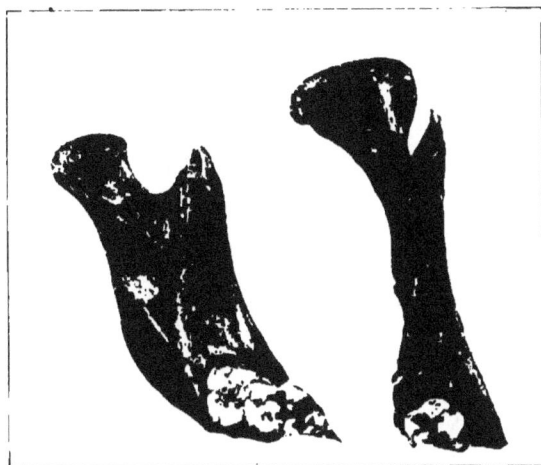

FIG. 25.—SHOWING VARIATION IN THE SHAPE OF THE CONDYLES.

ment is of the ginglymoid or hinge character, brought into play
in simple depression and elevation of the mandible, it also has
the power of extension and retraction, may be rotated from side
to side, together with all the motions intermediate between
these. When the mandible is depressed, the condyle moves on
the fibrocartilage, and at the same time glides forward and
slightly downward until it rests on the anterior border of the
glenoid fossa ;—this movement does not extend sufficiently to
allow the condyle to rest upon the extreme summit of the
border, except in cases of excessive movement, as in yawning,
when the condyle may glide over the summit and the joint

become disarticulated. When the mandible is elevated, the condyle slides backward and upward, and at the same time the fibrocartilage, which has extended with it, also retracts until the condyle is settled in the fossa. The movement of extension and retraction is by a horizontal gliding action, by which the mandible is thrust forward and drawn back again. In this movement, as well as in the one previously described, both condyles are similarly and simultaneously engaged. The lateral or triturating movement is made in an oblique direction. This consists in a

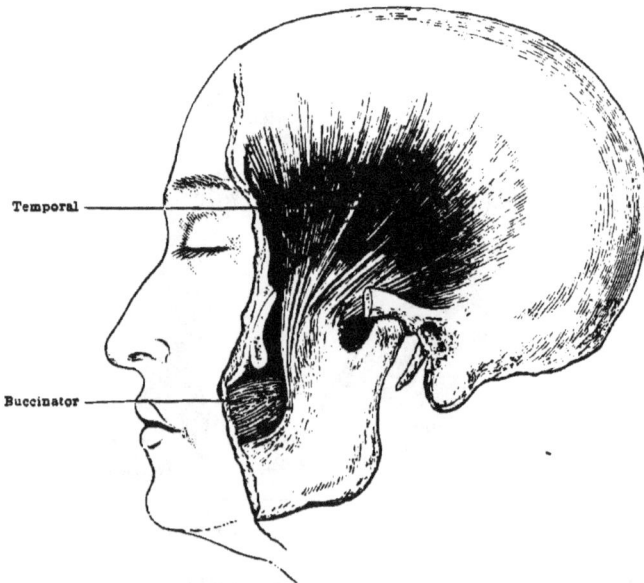

FIG. 26.—THE TEMPORAL MUSCLE.—(*Morris.*)

rotation of the condyles within the fossæ, the cartilage gliding obliquely forward and outward on one side, and backward and inward on the other, this action taking place alternately. This movement is more or less developed in accordance with the nature of the occlusion, being favored by those teeth possessing but little cusp formation, with a consequent loose and wandering occlusion; while in that type of tooth associated with a long overbite and deep penetrating cusps, forming a firm and well-locked occlusion, this movement will be but little developed. If this movement be employed to throw the symphysis to one side and back again, and not from side to side, the condyle of that side rotates in the glenoid fossa, while the condyle of the opposite side is drawn forward and inward.*

The Muscles of Mastication (Figs. 26 and 27).

The muscles actively engaged in the process of mastication are four in number: The masseter, the temporal, the internal pterygoid, and the external pterygoid.

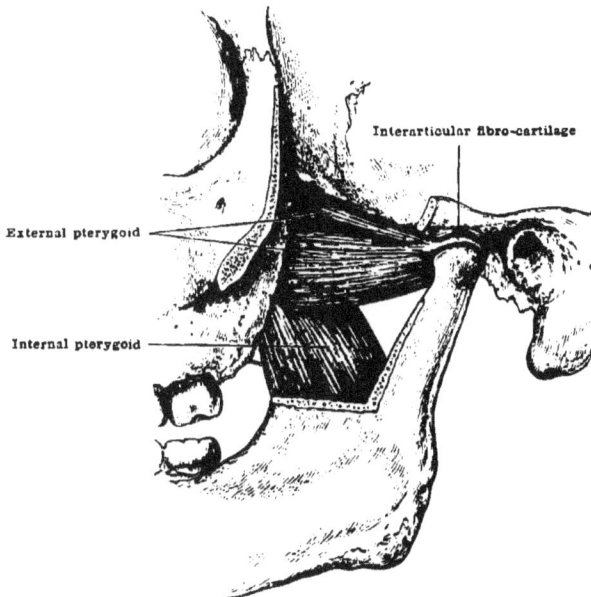

FIG. 27.—THE PTERYGOID MUSCLES.—(*Morris.*)

The masseter muscle, composed of two portions, superficial and deep, is a thick, short muscle, quadrilateral in outline. The superficial portion, the largest, arises from the lower border of the anterior part of the zygomatic arch and the inferior margin of the malar bone; it passes downward and backward, and is inserted into the outer and lower half of the angle of the

---

* The reader may obtain much valuable information upon this subject by a careful study of the writings of Dr. W. G. A. Bonwill.

mandible. The deep portion arises from the internal surface of the zygomatic arch, near its lower border, passes downward and somewhat forward, and is inserted into a portion of the ramus and external surface of the coronoid process of the mandible.

The temporal muscle is also composed of a superficial and a deep portion. It is an extensive, fan-shaped muscle, occupying the temporal fossa, and, passing downward, is inserted into the coronoid process of the mandible.

The internal and external pterygoid muscles are two short, thick muscles, arising from the pterygoid process of the sphenoid bone and from a part of the tuberosity of the palate bone, the former being inserted into the inner surface of the angle of the jaw, while the latter is inserted into the condyle of the mandible and into the interarticular cartilage of the temporomaxillary articulation.

Actions.—Collectively, the muscles of mastication act as elevators of the mandible, the mouth being opened by the relaxation of their fibers, and the force of gravity, and by the assistance of some of the muscles of the neck.

CHAPTER IV.

A GENERAL DESCRIPTION OF THE TEETH.—THE PERMANENT TEETH:
CLASSIFICATION, SURFACES, ETC.—THE ROOTS OF THE TEETH.
—THE DENTAL ARCH.

A GENERAL DESCRIPTION OF THE TEETH.

A Tooth (Fig 28).—One of the thirty-two specialized organs for the seizure and mastication of food, placed at the entrance to the alimentary canal (the mouth). The typical form of a tooth is a modified cone or combination of cones, and is composed of two fundamental parts—the *crown* and the *root* or *roots*. The crown is that part which projects beyond the gum and is visible in the mouth; while the root is that part which is implanted in the bone and covered by the gum. Intervening between these two extremities, and usually occupying a portion of each, is a third division—the *neck*.

Completely covering the crown of a tooth is a hard, vitreous-like substance, *enamel ;*[*] the root is covered by a hard, bone-like substance, *cementum ;*[*] while the interior or body of the organ is composed of a hard substance closely resembling bone, the *dentin*.[*] The neck of a tooth, which serves to unite the crown to the root, and which is usually formed at the expense of each, is covered partly by enamel and partly by cementum. This brief description in the singular number can best be continued in the plural. *Teeth* are

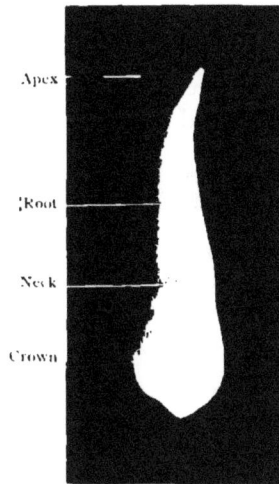

Fig. 28.

---

[*] See Tissues of the Teeth, part II.

classified according to their form, which is always in accordance with their function, into *simple* and *complex*. In the simple class (Fig. 28) the single modified cone is the predominating form, the free extremity of the crown serving as the base of the cone, while the apex is formed by the free end of the root. Included in this same classification are those teeth which are made up of a double cone, or a simple cone and an inverted cone attached to each other at a common base. The purposes for which such teeth are adapted are those of grasping, incising, and tearing, and they are usually so arranged that the free extremities of their crowns interlock or overhang each other.

In the complex class (Fig. 29) the external form of the tooth is produced by a combination of cones, some of which are simple,

others inverted, but all uniting at a common base—the neck of the tooth. In this class the simple cones form the roots of the teeth, while the crowns are made up of a number of smaller cones, much modified. These teeth are adapted to crushing and grinding, and are less inclined to interlock during active service.

The teeth are divided into two grand divisions—those of infancy and childhood, called *deciduous* or *temporary teeth*, and those of the adult period, known as *permanent teeth*. The latter class being most important, will first receive consideration.

Fig. 29.

The Permanent Teeth (Fig. 30).—The permanent teeth, thirty-two in number, are divided into those of the superior portion of the mouth, *superior teeth*, and those of the inferior portion, the *inferior teeth*. In number they are equally divided, each jaw giving support to sixteen. They are firmly imbedded in the alveolar sockets of three of the bones of the mouth, the superior sixteen being attached to the two superior maxillary or upper jaw-bones, and the inferior sixteen to the mandible, inferior maxillary, or lower jaw-bone. As above referred to, the attachment of the teeth to the bones is by implantation in sockets, the alveoli (see description, "Bones of the Mouth").

In this attachment there is a special development of bone, closely modeled to the roots of the teeth, and which is subservient to the ever-varying changes which take place during the development of the organs. The joint thus formed between the roots of the teeth and the alveoli is of the immovable or synarthrodial class, and is styled *gomphosis*. Intervening between the roots of the teeth and the walls of the alveoli is a delicate membrane—the alveolodental periosteum, or peridental membrane.

Before continuing the description of the teeth, a further clas-

Palatal Surfaces   Labial and Buccal Surfaces

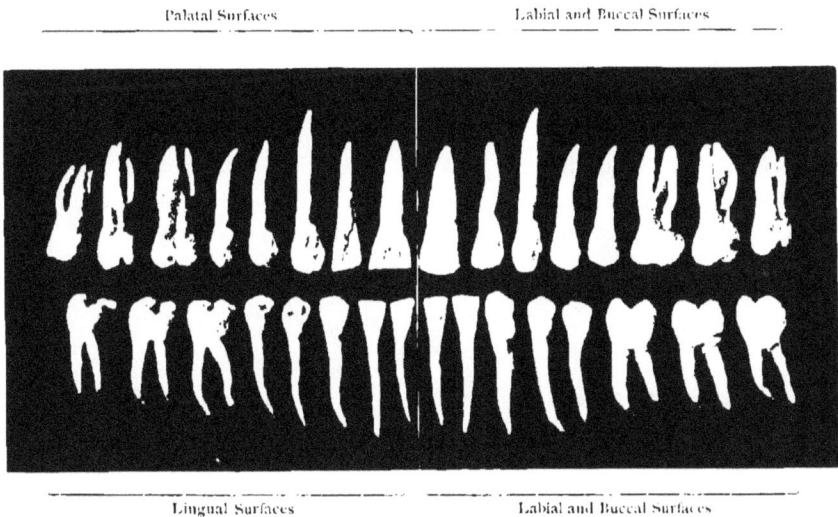

Lingual Surfaces   Labial and Buccal Surfaces

FIG. 30.—THE PERMANENT TEETH.

sification, which refers alike to the superior and inferior organs, must be presented. This classification is derived from the function and form of the teeth. Figure 30 shows the thirty-two teeth removed from the jaws and placed side by side in two straight lines. In the center is a perpendicular line, which corresponds to the median line or center of the mouth, the teeth at either extremity being those which occupy the back part o the mouth. Without confining the description to either the superior or inferior teeth, it will be observed that the first two teeth upon either side of the median line are similarly formed.

and all four are called *incisors* (*incidere*, to cut) ; the two largest incisors, being nearest the median line or center, are called *central incisors;* while the two smallest, being placed at the side of the centrals, are known as *lateral incisors* (*lateralis*, the side). The third tooth from the median line upon either side is the *cuspid* (*cuspis*, a point), so named from possessing a single cusp or point. Passing to the right or left on the chart, or backward in the mouth, the fourth and fifth teeth from the median line are the *bicuspids* (*bi*, two ; *cuspis*, a point), having two points or cusps. The bicuspid nearest the median line is the *first bicuspid;* that most distant from the median line is the *second bicuspid.* The sixth, seventh, and eighth teeth from the median line upon either side are those of another class, the *molars* (*mola*, a millstone), being named according to their function, that of crushing or grinding the food. Proceeding from before backward, the molars are denominated *first molar, second molar,* and *third molar.* To sum up, the names and number of the permanent teeth may be given by the **dental formula,** as follows :

Incisors, ... $\frac{4}{4}$		Bicuspids, ... $\frac{4}{4}$	
Cuspids, ... $\frac{2}{2}$		Molars, ... $\frac{6}{6}$	

The Surfaces of the Teeth.—The crown of each tooth presents five surfaces, which are variously named, in accordance with the duty which they are called upon to perform or suggestive of their location. The outer surface of the incisors and cuspids, or that contiguous to the lips (labia), is called the *labial surface ;* the corresponding surface of the bicuspids and molars, or that contiguous to the cheeks (buccæ), is the *buccal surface.* Those surfaces of the superior teeth facing the palate are known as the *palatal surfaces,* while the corresponding surfaces of the inferior teeth, or those facing the tongue (lingua), are named *lingual surfaces.*

The proximate surfaces of the teeth are named with regard to their relation to the median line, those surfaces nearest to this point being called *mesial,* those most distant, *distal.* In addition to these four surfaces, which represent what might be termed the sides of the teeth, a fifth surface is present, that which occludes with the teeth of the opposite jaw, and called the *occlusal surface.*

In the incisors and cuspids this surface is formed by the converging of the labial and palatal or lingual surfaces, forming an edge, to the free extremity of the crown, named, from its action in mastication, the *incisive* or *cutting-edge.* In the bicuspids and molars the various sides of the crowns remain nearly parallel to each other throughout their extent, thus providing a surface nearly equal to, or greater than, any of the others, and one well adapted to the purposes for which it is intended—that of grinding or crushing the food.

The Roots of the Teeth.—The superior incisors and cuspids are each provided with one root ; the superior first bicuspid may have one or two roots, most frequently the latter ; while in the second bicuspid a single root is usually present. The superior first and second molars are each supported in the jaw by three roots, and, while in the superior third molar three roots are most common, the number is quite variable, ranging from a single cone-shaped root to three, four, five, or even six smaller branches given off from a common base.

In the inferior incisors, cuspids, and bicuspids, a single root is most frequently met with, although the latter class, in rare instances, may be provided with two. The inferior first and second molars are each provided with two roots, but in the third molar, like its superior fellow, the number may be diminished or increased. In the superior molars, two of the three roots are placed above the buccal half of the crown, and are called *buccal roots ;* the remaining root is placed above the palatal half of the crown, and is designated as the *palatal root.* In the inferior molars, one of the two roots is placed below the anterior or mesial half of the crown, and is named the *mesial root,* and the other below the posterior or distal half, and is known as the *distal root.* In those teeth with a complicated root formation, it would seem to be a question whether they are possessed of a single root, with two or more branches, or separate and distinct roots throughout. To determine this, some account must be taken of the point at which the bifurcation or trifurcation takes place. If this separation be in close proximity to the crown, the tooth should be considered as having more than one root (Fig. 31, A) ; but, on the other hand, if the point of separation be some distance from the crown, with a solid mass of root substance intervening, the tooth may be said to possess a single root, with two or more branches (Fig. 31, B). In the latter instance, that part of the tooth between the point of separation and the crown is called the root or root base ; while the prolongations beyond the point of separation are known as the branches of the root. The roots of the teeth are not only variable in number, but are also subject to much variety in form. In the *anterior teeth* (the incisors and cuspids) the roots are inclined to the form of the simple cone, which form, however, is frequently more or less broken by a slight curvature near their extremities, or by a slight compression of their lateral walls. In the *posterior teeth* (the bicuspids and molars) the roots, root bases, or root branches are all inclined to the conical form, but do not approach so nearly the perfect cone as those of the anterior teeth. These roots are also more or less crooked and flattened laterally. The free extremities of the roots of the teeth, forming as they do the apex of these cone-like prolongations of the crowns, are known as the apices or *apical extremities*.

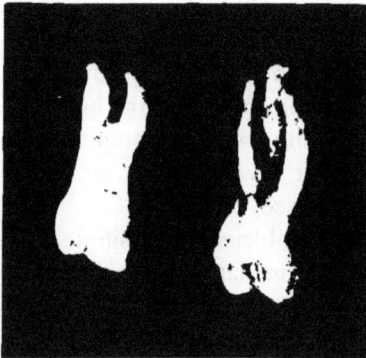

B   A

FIG. 31.

The extent of the enamel covering to the crowns of the teeth is marked by a well-defined line, which completely encircles the neck of the tooth, the *cervical line*.

The Dental Arch (Fig. 32).—The teeth are arranged in the jaws in the form of two parabolic curves, the superior arch describing the segment of a larger circle than the inferior, as a result of which the superior teeth slightly overhang the inferior. Figure 32 represents the sixteen superior teeth in position in the bone,

presenting their occlusal surfaces, a part of their palatal surfaces being also visible. Viewed in this direction, the gradual
change in the crowns of the teeth from the simple incisors to
the complex molars may be observed. An examination of the
central incisors will show how perfectly they are adapted to the
process of cutting or incising the food, the cutting-edge being
sharp and the palatal surface comparatively smooth and
unbroken. In the lateral incisors the cutting feature predominates, but the palatal surface is broken near the neck of the

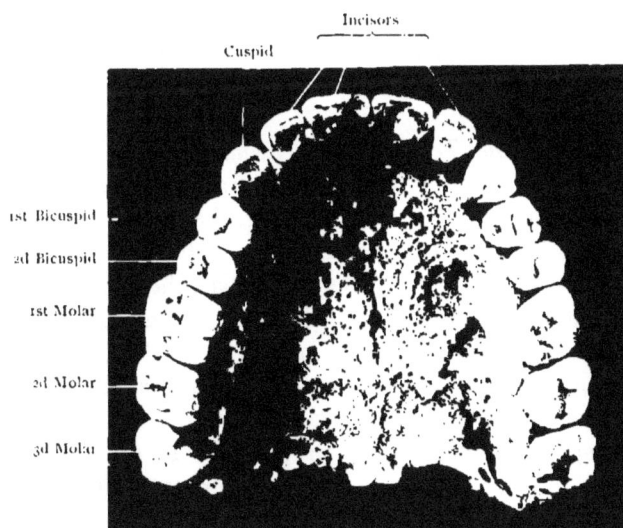

FIG. 32.—THE DENTAL ARCH.

tooth by a slight depression, surmounted by a more or less pronounced fold of enamel, in many instances resembling a small
cusp. The crown of the cuspid tooth furnishes the intermediate
form between the simple and the complex. This tooth, instead
of being provided with a straight cutting-edge, is surmounted
at the center of its occlusal surface with a well-defined point or
cusp, descending from the summit of which are two cutting-
edges, one passing to the mesial and one to the distal. The
palatal surface of this tooth presents a marked contrast to the

corresponding surface of the incisors, being broad and full, and frequently provided with a prominent ridge of enamel in the region of the neck, showing a rapid approach to the complex form. In the bicuspids the buccal half of the crown is quite similar to the crown of the cuspid, but in the palatal half a complete revolution has taken place. The enamel fold—but slightly apparent in the incisors, and somewhat increased in the cuspids—has now become a fully developed cusp, resulting in the production of an occlusal surface adapted to crushing or grinding, instead of incising or tearing. In the molars, the increase in the size of the tooth-crown is accompanied with an occlusal surface much more complex than any of the teeth previously described, and one well adapted to its function—that of crushing and grinding the food.

Arrangement of the Teeth in the Dental Arch (Fig. 32).—Beginning with the superior teeth, the central incisors are found occupying the center of the arch, and are, therefore, slightly in advance of the laterals. These teeth are so implanted in the alveoli that their crowns are not perpendicular, the cutting-edge being slightly more prominent than the neck of the tooth. The roots are also somewhat inclined from the median line, and as a result the crowns have a slight mesial inclination, the mesial surfaces approximating each other at or near the cutting-edge, with a slight space intervening at the necks. In certain typal forms,—the bilious, for example,—when the front of the arch is flat, the labial surfaces of these teeth form nearly a direct line from side to side ; while in those types in which the arch is well rounded anteriorly, notably in the sanguine temperament, the labial surfaces of these two teeth form a small segment of the arch, so that the mesial extremity of the cutting-edge of each tooth-crown is somewhat in advance of the distal. The lateral incisors are similarly implanted in the alveoli, causing their cutting-edges to project. The roots of these teeth usually have a stronger distal inclination than those of the centrals, and as a result the crowns show a more marked mesial inclination. The mesial surfaces approximate the distal surfaces of the central incisors at or near the cutting-edge. When the front of the arch is flattened, these teeth are but little less prominent than the central incisors, but when the arch is well rounded they continue the segment begun by the centrals, and are necessarily less prominent. While the occlusal surfaces of the teeth are usually considered as forming a perfect plane (see Occlusion of the Teeth), the lateral incisors are generally a trifle shorter than the centrals. The cuspids may be considered as occupying the corners or turning-points of the arch. They are more prominently placed than the adjoining teeth, this feature being increased by the bulging or general convexity of their labial surfaces. The extremity of the occlusal surface of the cuspids —*i. e.*, the point of the cusp—is a trifle below the cutting edge of the laterals and about on a line with that of the centrals. While the apical extremities of the roots of the cuspid teeth are directed away from the median line, the crowns assume almost a perpendicular position, this condition resulting from a bend in the tooth at the neck. Although the perpendicular position is most commonly assumed by the crown, it is not unusual to find either a mesial or distal inclination present. Reference has been made to the cuspid teeth occupying a position which might be termed the turning-points or corners of the arch, and in most instances it may be thus considered ; but in those typal forms— the sanguine, for example—the tooth-line is unbroken and passes over the cutting-edges of the incisors, the summit of the cusps of the cuspids, and is continued backward over the buccal cusps of the posterior teeth. The bicuspids are placed nearly perpendicular in the arch, but occasionally deviate from this by a slight mesial or buccal inclination. The length usually corresponds to that of the central incisors, and their buccal surfaces are slightly less prominent than the corresponding surfaces of the cuspid teeth. The increase in the buccopalatal diameter of the crowns of the bicuspids over that of the incisors and cuspids results in breaking the palatal line of the occlusal surfaces. In the bilious and kindred types, the tooth-line is carried directly backward from the cuspid to the first molar, making the buccal face of the bicuspids equally prominent, but when the arch is well rounded the second bicuspid is slightly more prominent than the first. The first and second molars usually assume a perpendicular position, but are occasionally inclined to

the mesial and buccal. The relative prominence of the buccal
as well as the palatal surfaces of these teeth is also controlled by
the form of the arch. The occlusal surfaces are about on a
level with those of the bicuspids and central incisors, but generally the lack of development in the distal half of the crown
of the second molar results in the production of a slight upward
curve to the tooth-line level at this point (see Occlusion
of the Teeth). On account of the limited accommodations
afforded it, the position of the superior third molar is quite
variable. It may be either to the buccal or to the palatal of the
tooth-line, and is usually strongly inclined to the distal. In

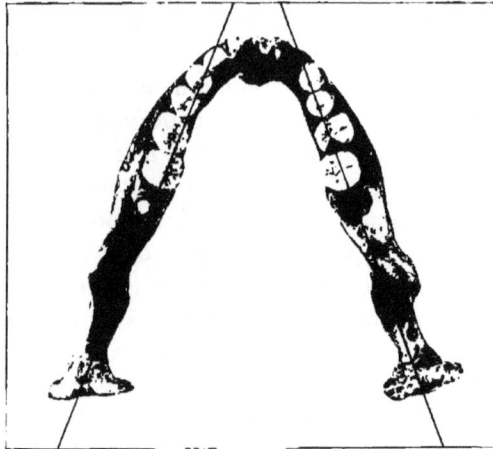

FIG. 33.—THE TOOTH-LINE IN THE LOWER JAW.

those cases in which there is a decided dip to the arch (see
Occlusion of the Teeth), this tooth is relatively shorter than
those anterior to it, but when the tooth-line level is a perfect
plane, the length of this tooth corresponds to the other molars
and bicuspids.

The inferior incisors are placed more nearly in a perpendicular position than the superior, and a reverse condition
exists, in the lateral incisors being a trifle larger than the
centrals. The inferior cuspids are probably more constant in
their position than any class of teeth in the mouth, in nearly
all instances assuming a direct perpendicular. Like the supe-

rior cuspids, they may be said to establish the corners or turning-points of the inferior arch, and are somewhat more prominent in the tooth-line than neighboring teeth. All of the six anterior inferior teeth may be slightly inclined to the mesial. The inferior bicuspids and molars, instead of having the buccal inclination possessed by the corresponding superior teeth, are inclined to the lingual. The first molar seldom deviates either to the mesial or the distal, the second molar is generally inclined to the mesial, while the third molar is strongly inclined to the mesial. In the inferior arch the curve formed by the incisors and cuspids is the segment of a smaller circle than the correspond-

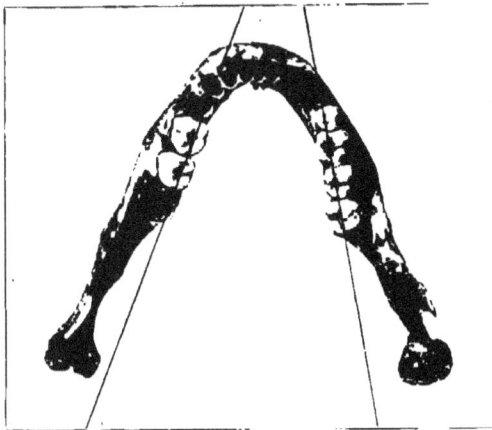

FIG. 34.—THE TOOTH-LINE IN THE LOWER JAW.

ing curve in the superior arch. This curve may be continued over the buccal cusps of the bicuspids and molars, or it may be broken at the cuspid tooth and continued backward in a direct line (Fig. 33). The teeth in the inferior arch are placed directly over the body of the bone as far back as the second bicuspids, while the molars frequently overhang the body of the bone by an extension of the alveoli inward (Fig. 34).

The curve described by the dental arch is quite variable, and this variation is generally referred to in connection with the temperament. Thus, in the sanguine temperament (Fig. 35), the arch is well rounded anteriorly, the circle being continued back-

ward to the region of the molars, where the line is broken by
slightly inclining to the palatal. In this arch the distance in a
straight line from the center of the second molar, on one side,

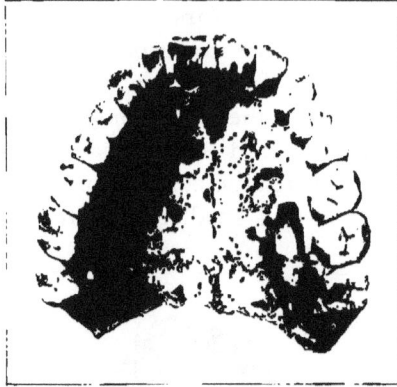

FIG. 35.—SANGUINE.

to the center of the corresponding tooth on the other, is about
equal to the distance from either of these points to the median
line between the central incisors, forming a right-angle triangle.

FIG. 36.—BILIOUS.

In the bilious temperament (Fig. 36) the arch presents a broad
front from cuspid to cuspid, with but little curve ; at these points
it turns abruptly backward, being continued almost in a direct

line to its extremity. In this arch the side of the triangle (represented by the line from molar to molar) is much reduced in length. In the nervous temperament (Fig. 37) the arch is Gothic in form,

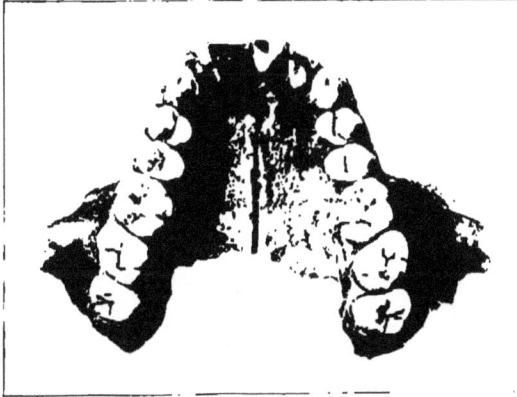

FIG. 37.—NERVOUS.

the segment formed by the anterior teeth being that of a much smaller circle than either of the types previously referred to. The distance from molar to molar is much less than the distance

FIG. 38.—LYMPHATIC.

from molar to median line. In the lymphatic temperament (Fig. 38) the arch is well rounded and broad, the segment being that of a much larger circle than any of the above, the side of the triangle (formed by the line from molar to molar) being of the greatest length.

Interproximate Spaces. — In a mesiodistal direction the crowns of the teeth, as a class, are broader at their occlusal surfaces or cutting-edges than at their necks (Fig. 39). This bell-shaped form of the tooth-crowns causes their proximate surfaces to touch at a point representing their greatest mesio-distal diameter, which is usually near the cutting-edge or occlusal surface. Between this point of contact and the cervical line there exists a V-shaped space, called the interproximate space. These spaces are largest in that class of teeth found in the ner-

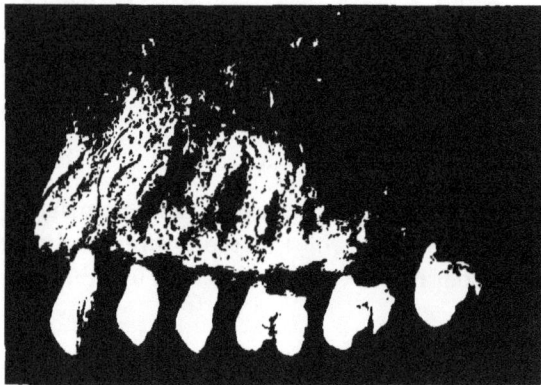

Fig. 39.—Section of Superior Maxilla Showing Interproximate Spaces.

vous and bilious types, where the necks of the teeth are much constricted, and the bell-shaped crown strongly outlined. In teeth of this class the point of contact is slight, and the inter-proximate spaces are only partially occupied by the gum tissue, leaving a free passage between the point of contact and the gin-giva margins. In the sanguine and lymphatic temperaments the proximate surfaces of the teeth are nearer parallel with one another, thus making the point of contact cover a greater extent of surface, and reducing the size of the interproximate spaces.

CHAPTER V.

OCCLUSION OF THE TEETH.

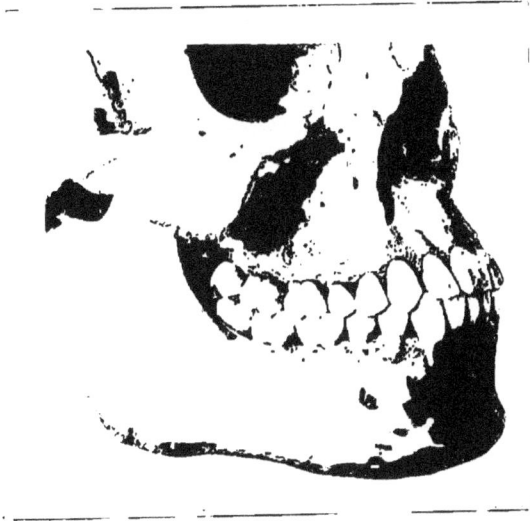

FIG. 40.—THE TEETH IN OCCLUSION.

As stated elsewhere, the teeth are arranged in the mouth in the form of two parabolic curves, one of which occupies the upper half and the other the lower half of the cavity. To properly perform their function it is necessary for the superior and inferior teeth to come in contact, which they are enabled to do by the movement of the lower jaw, and it is the relation existing between the two when thus brought together that constitutes the *occlusion of the teeth*. During mastication the teeth do not only occlude, and remain stationary at a given point until the lower jaw is again depressed, but, through the combined movements of the mandible, the inferior teeth are made to move from side to side, thus grinding or crushing any substance placed between the occlusal surfaces of the bicuspids and molars. This gliding antagonism of the teeth is commonly termed the *articula-tion*, and it is important that a distinction be made between the terms "occlusion" and "articulation," the former referring to the relations existing between the superior and inferior teeth when brought together normally and held firmly in that position, while the latter relates to the various movements of the teeth after being brought together in occlusion. In the majority of instances the segment described by the superior arch is somewhat larger than that formed by the inferior, and the superior teeth project over and are partly outside of those in the inferior arch. Figure 40 presents a labial and buccal view of the teeth in position in the alveoli, and also in occlusion. It will be observed that the superior teeth are not directly antagonistic to those of the same name in the inferior arch. There are two reasons for the presence of this condition: First, the mesiodistal diameter of the superior central incisors is much greater than that of the corresponding inferior teeth; second, the larger circle present in the superior arch. This arrangement provides that each tooth, instead of being antagonized by a single tooth of the opposite jaw, is met in occlusion by a portion of two teeth. The superior central incisor is met in occlusion by the entire cutting-edge of the inferior central incisor and the mesial third of the cutting-edge of the inferior lateral incisor. The superior lateral incisor is met in occlusion by the distal two-thirds of the cutting-edge of the inferior lateral incisor and by the mesial cutting-edge of the inferior cuspid. The superior cuspid is met in occlusion by the distal cutting-edge of the inferior cuspid and by the mesial two-thirds of the buccal cusp of the inferior first bicuspid. The superior first bicuspid is met in occlusion by the remaining or distal third of the inferior first bicuspid and by the mesial two-thirds of the buccal cusp of the inferior second bicuspid. The superior second bicuspid is met in occlusion by the remaining or distal third of the buccal cusp of the inferior second bicuspid, and by the mesial incline of the mesiobuccal cusp of the inferior first molar. The superior first molar is met in occlusion by the distal incline of the mesiobuccal cusp of the inferior first molar, by the entire distal cusp of the same tooth, and by the mesial incline of the mesiobuccal cusp of the inferior second molar. The superior second molar is met in occlusion by the distal incline of the mesiobuccal cusp of the inferior second molar, by the entire distobuccal cusp of the same tooth, and by the mesial incline of the mesiobuccal cusp of the inferior third molar. The superior third molar is met in occlusion by the distal incline of the mesiobuccal cusp of the inferior third molar and by the entire distobuccal cusp of the same tooth, thus being the only tooth in the superior arch with but a single opponent. Likewise each inferior tooth is met in occlusion by two in the superior arch, with the single exception of the central incisor, which occludes with the superior central

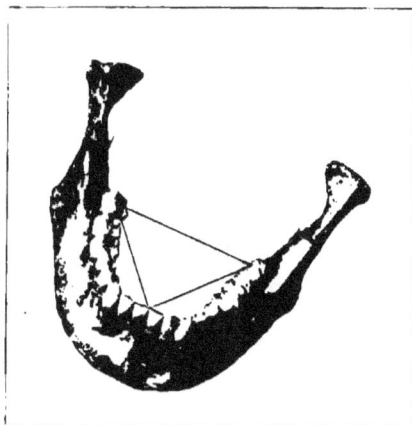

FIG. 41.—THE MANDIBLE AT THE ADULT PERIOD, SHOWING THE EQUILATERAL TRIANGLE DESCRIBED BY THE DENTAL ARCH.

alone. There are many variations from this so-called typical occlusion, as above described, any one of which may be considered normal. In certain types the segmental form of the superior arch is but little greater than that of the inferior, and the cutting-edges of the superior incisors occlude almost directly upon the cutting-edges of the inferior incisors. As a result, all of the superior teeth are forced to the distal, and the relationship between the superior and inferior organs is much altered. When the superior teeth overhang the inferior, the palatal cusps of the superior bicuspids and molars penetrate the fossæ or sulci of the corresponding inferior teeth, when in occlusion, and the

buccal cusps of the inferior bicuspids and molars rest in the
fossæ of their superior opponents.

To assist in the study of the occlusion of the teeth, some reference must be made to the tooth-line level, or plane of occlusion.   For this purpose the lines forming the facial angle
are of value.   These lines are as follows : A fixed line representing the base of the angle may be drawn from the center of
the glenoid fossa, passing forward through the anterior nasal
spine or base of the nose (A, Fig. 42), the angle being completed by a perpendicular line resting upon the labial surface of
the superior incisors, passing upward and touching the most

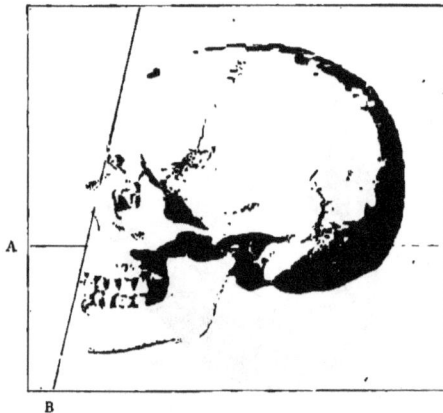

FIG. 42.—LINES SHOWING FACIAL ANGLE, CAUCASIAN OR WHITE RACE.

prominent part of the forehead (B, Fig. 42).   The tooth-line
level is approximately horizontal to this basal line, but instead
of a perfect plane we usually find the superior arch dipping
downward, while the inferior arch will be provided with a corresponding depression.   This dip to the arch is greatest in the
region of the bicuspids, and the extent to which it may exist
varies with the type of tooth and the consequent nature of the
occlusion.   Thus far no reference has been made to what is
commonly termed the *overbite*, and the *cusp forms* in the teeth.
As these two factors exert a dominating influence over the character of the occlusion.   The effects which they produce
will be briefly described.   The overbite is so named from the

superior incisors and cuspids projecting beyond, or overhanging and partly covering, the labial surfaces of the corresponding inferior teeth. This may be a prominent feature in the tooth-occlusion, or it may exist to a very slight degree. Although the overbite is usually referred to as existing in the incisive region alone, it is not confined to these teeth, but is also present in the bicuspids and molars by the buccal cusps of the superior teeth overhanging those of the inferior. The extent of the over-bite is gradually diminished from before backward, the central incisors presenting the greatest amount of overhanging surface,

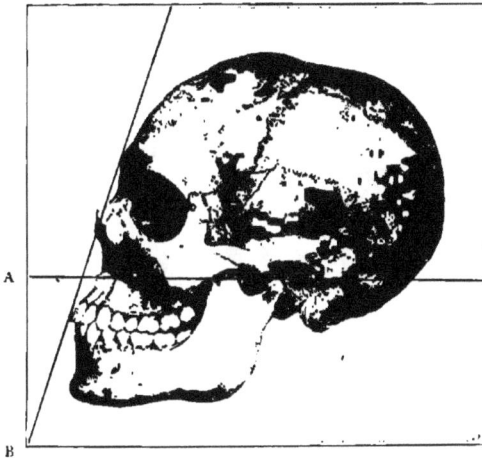

Fig. 43.—Lines showing Facial Angle, Negro or Mixed Races.

which condition is slightly decreased in the laterals, and a cor-responding reduction is continued until the third molars are reached, at which point the overbite is scarcely to be observed. Where the overbite is extensive, as shown in figure 40, the superior incisors overhanging and hiding from view about one-third of the labial surfaces of the inferior incisors, the cusps of the bicuspids and molars will be correspondingly long and pene-trating, the buccal cusps of the superior teeth extending well down over the buccal cusps of the corresponding inferior teeth. In an occlusion of this class, which is usually found in the ner-

vous and bilious types, the dip to the arch will be a prominent
feature, the occlusion will be firm and well locked, and the articular movements will be slight during mastication. In the

FIG. 44.—LYMPHATIC.

FIG. 45.—SANGUINE.

FIG. 46.—BILIOUS.

lymphatic and sanguine temperaments the occlusion is loose and
wandering, greater freedom of movement being permitted by
the short overbite and the corresponding lack of cusp-formation.

Figure 44 represents such an occlusion; the superior arch is but little greater in its segmental outline than the inferior. The cutting-edges of the superior incisors are somewhat more prominent than those of the inferior, but the former do not overlap the labial surfaces of the latter. In an occlusion of this character the dip to the arch is not so pronounced, and the articular movements are much more extensive.

CHAPTER VI.

THE BLOOD- AND NERVE-SUPPLY TO THE TEETH.

THE BLOOD-SUPPLY TO THE TEETH.

Briefly stated, the course of the blood from the heart to the teeth is as follows : From the heart to the aorta, to the common carotid artery, to the external carotid artery, to the internal maxillary artery, from the various branches of which the teeth are supplied.

The Internal Maxillary Artery (Fig. 47).—This artery, otherwise known as the deep facial, is the largest of the two terminal branches of the external carotid. In addition to supplying the teeth, it is distributed to the roof and floor of the mouth, to the maxillary sinus, and to other parts of the face and head. It has its origin from the external carotid artery opposite the condyle of the mandible within the substance of the parotid gland, passes forward between the condyle of the jaw and the sphenomaxillary ligament, from which point it passes obliquely upward and forward between the external and internal pterygoid muscles until it reaches the sphenomaxillary fossa, where its terminal branches are given off. It is divided into three portions—the first or maxillary, the second or pterygoid, and the third or sphenomaxillary. The teeth are supplied from branches of the first and third divisions, the superior teeth receiving their blood-supply from the alveolar or superior maxillary and the infra-orbital branches of the third division, while the inferior teeth are supplied by the inferior dental branch of the first division.

The alveolar or superior maxillary branch arises, in common with the infra-orbital branch, from the internal maxillary as it passes into the sphenomaxillary fossa. It passes downward, in a tortuous manner, in a groove provided for it in the back of the maxilla. In its downward course it gives off the following *branches:* The *antral,* to supply the antrum ; the *dental* (known as the posterior dental arteries), which pass into

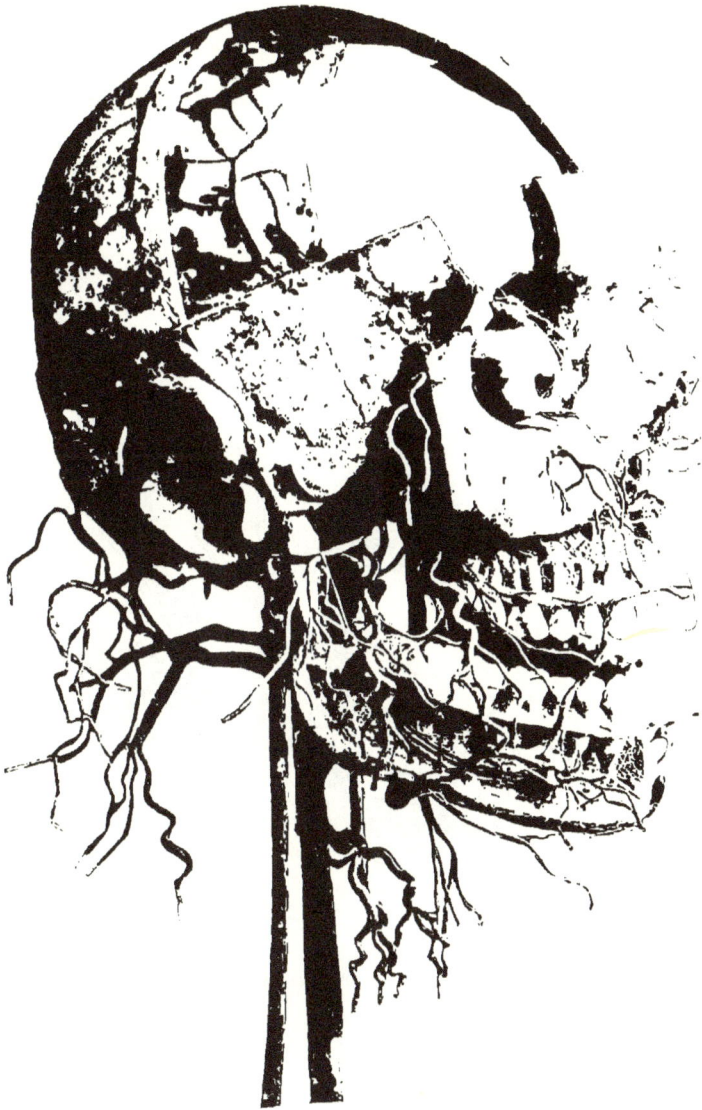

FIG. 46a.—DISSECTION SHOWING BLOOD-SUPPLY TO THE TEETH.

111

the substance of the bone through the posterior dental canals to supply the molar and bicuspid teeth ; the *alveolar* or gingival, to supply the gums ; and the *buccal*, to the lateral walls of the mouth. The anterior superior teeth are supplied through the *infra-orbital branch* of the internal maxillary. This branch arises from the internal maxillary artery, generally, in common with the posterior dental. It passes forward in company with the maxil-

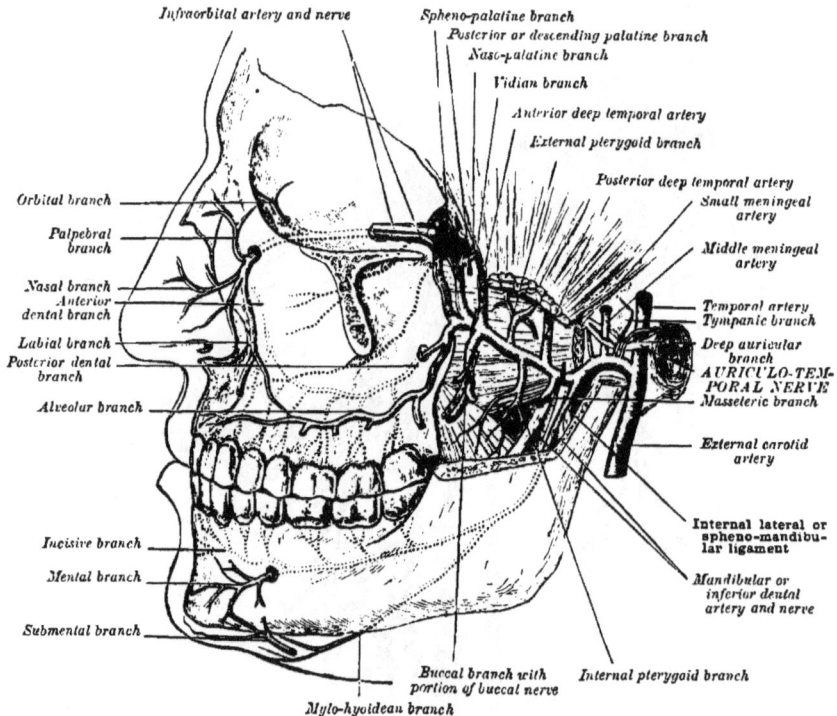

FIG. 47.—SCHEME OF INTERNAL MAXILLARY ARTERY.

lary division of the fifth nerve—first along the groove and then in the canal on the orbital plate of the maxilla, and finally makes its exit upon the face through the infra-orbital foramen. Besides giving off branches to the orbital and nasal cavities, it supplies the incisor and cuspid teeth through its anterior dental branch, which passes downward through a groove in the anterior wall of the maxilla.

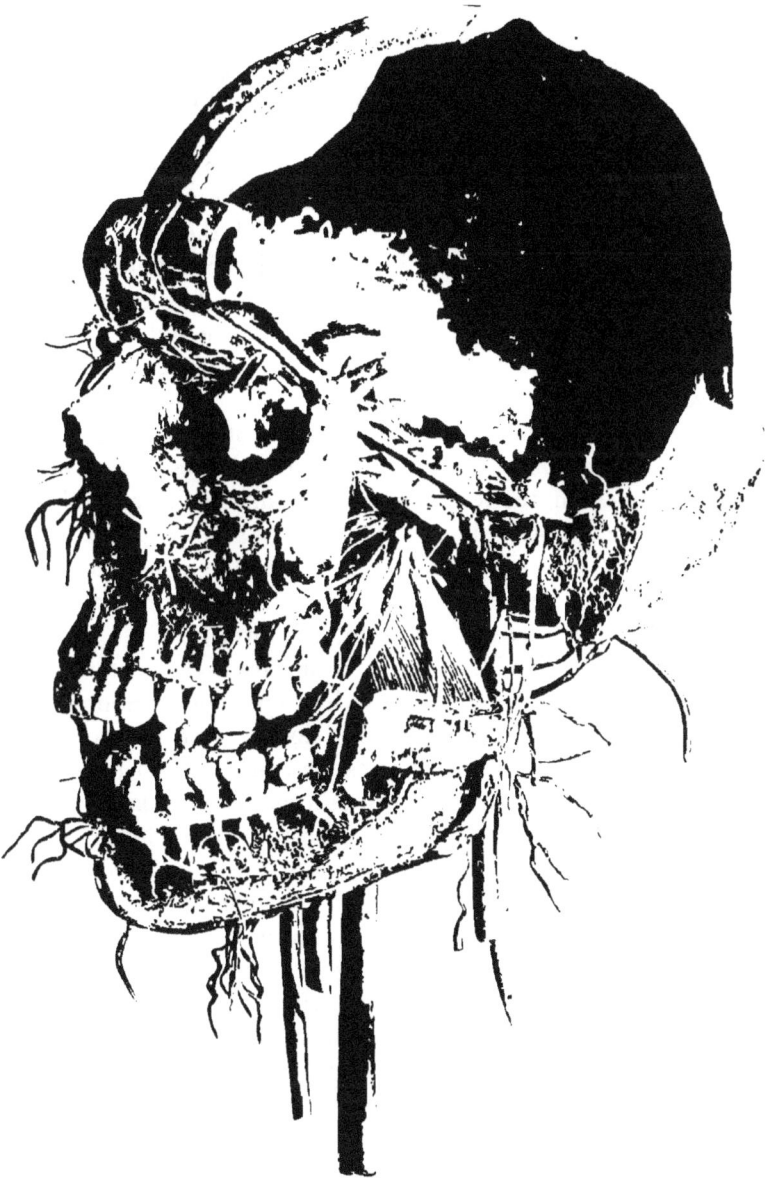

FIG. 47a.—DISSECTION SHOWING NERVE-SUPPLY TO THE TEETH.

The **Inferior Dental Artery** (Fig. 47).—The inferior teeth receive their blood-supply through the inferior dental or mandibular artery. This artery arises from the under part of the internal maxillary as it passes downward and forward between the sphenomaxillary ligament and the neck of the jaw and enters the inferior dental canal through the inferior dental foramen. It passes forward in the canal accompanied by the inferior dental nerve, and in so doing sends off twigs to supply the molar and bicuspid teeth. When the mental foramen is reached, it divides into two *branches*, the *incisive* branch and the *mental* branch.

The incisive branch continues its course within the cancellated structure of the bone, sending off minute branches which supply the anterior teeth, the terminal branches anastomosing with the artery of the opposite side.

The mental branch passes out through the mental foramen accompanied by the mental branches of the inferior dental nerve, and supplies the tissues of the chin and lower lip.

The Veins.—The blood, in returning from the teeth to the heart, is first taken up by the posterior dental and inferior dental veins, which in their course follow closely that of their corresponding arteries. These veins, in conjunction with others which accompany branches of the internal maxillary artery, form the pterygoid plexus. At the posterior confluence of this plexus the returning blood empties into the internal maxillary vein. Accompanied by the internal maxillary artery it passes backward and outward, enters the parotid gland, and finally empties into the temporomaxillary vein midway between the zygoma and the angle of the jaw. After leaving the substance of the parotid gland, the temporomaxillary vein passes downward until near the angle of the jaw, where it divides into two branches, one of which passes downward and slightly forward, uniting with the facial to form the common facial vein, and the other, after passing downward and backward, empties into the external jugular vein. The external jugular vein returns the principal portion of the blood from the teeth, and from its point of beginning it passes almost perpendicularly downward and empties into the subclavian vein, which, by joining with the internal jugular vein, forms the innominate vein, which, in turn, empties into the superior vena cava, thus communicating with the heart.

THE NERVE-SUPPLY TO THE TEETH.

The nerves supplying the teeth are derived from branches of the fifth cranial nerve, otherwise known as the trifacial or trigeminal nerve. The fifth nerve is the largest of the cranial nerves, and consists of two parts, a large root (sensory) and a small root (motor). The larger portion passes into a ganglion (the Gasserian ganglion), frequently compared to the ganglion on the posterior root of the spinal nerve. It arises, or makes its appearance, at the surface of the brain, on the anterior part of the side of the pons Varolii. The sensory root which, through its branches, supplies the teeth, is composed of from 80 to 100 filaments, each inclosed in a neurilemma, the entire bundle being bound together in a single nerve.

The fifth nerve is divided into three divisions : First, or *ophthalmic;* second, or *superior maxillary ;* and third, or *inferior maxillary* (mandibular). The branches which supply the teeth are included in the second and third divisions, the superior teeth being supplied by branches from the superior maxillary nerve, and the inferior teeth by branches from the inferior maxillary nerve.

The Second Division, or Superior Maxillary Nerve (Fig. 48).—This nerve, composed entirely of sensory fibers, is intermediate in size between the inferior maxillary and the ophthalmic divisions. It passes forward from the Gasserian ganglion and leaves the cranium through the foramen rotundum. It traverses the upper part of the sphenomaxillary fossa, and passes into the orbit through the sphenomaxillary fissure ; then passes forward along the infra-orbital groove, and enters the infra-orbital canal, where it receives the name of the infra-orbital nerve. Passing through this canal, it emerges upon the face through the infra-orbital foramen. The superior maxillary nerve, beside supplying the teeth, sends off branches to the dura mater, to the orbit, and terminal branches in three groups— labial, nasal, and palpebral. The branches given off to the teeth are the *posterior superior dental,* the *middle superior dental,* and the *anterior superior dental.*

The posterior superior dental arises from the second division of the fifth nerve, by one or two roots, just before it passes into the infra-orbital canal. It is divided into a superior and an inferior set ; the former passes forward and terminates in the canine fossa, while the latter, usually the largest, enters the posterior dental canals, and, following the line of the alveolar process through minute canals in the bone, sends off twigs to the molar teeth, ending in a plexiform manner by communicating with the

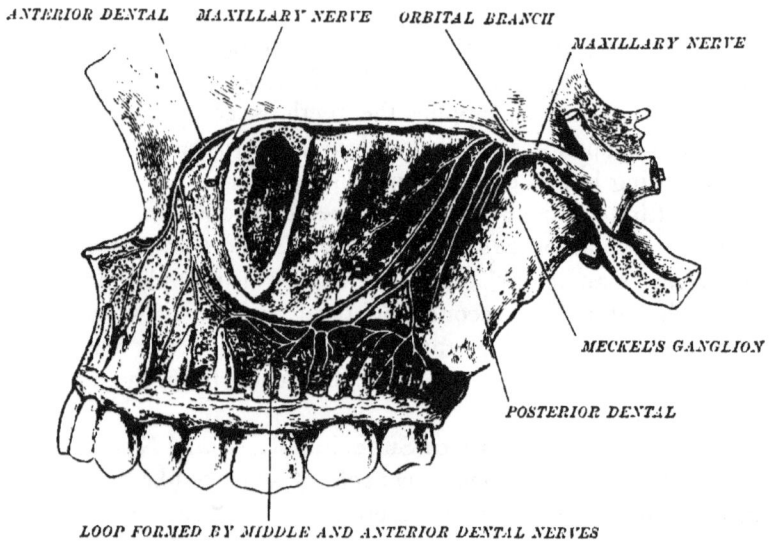

ANTERIOR DENTAL   MAXILLARY NERVE   ORBITAL BRANCH

MAXILLARY NERVE

MECKEL'S GANGLION

POSTERIOR DENTAL

LOOP FORMED BY MIDDLE AND ANTERIOR DENTAL NERVES

FIG. 48.—THE MAXILLARY NERVE, SEEN FROM WITHOUT.—(*Beaunis.*)

middle superior dental nerve. This nerve is also distributed to the gums and adjacent buccal mucous membrane.

Middle Superior Dental Nerve.—The infra-orbital nerve, soon after entering its canal, gives off this branch, which passes outward, downward, and forward over the outer wall of the maxillary sinus, and, after forming plexuses with the posterior dental branches, gives off filaments to supply the bicuspid teeth.

The anterior superior dental nerve, which is the largest of the dental set, is given off from the infra-orbital nerves, enters a

canal close to the infra-orbital foramen, passes over the anterior
wall of the maxillary sinus, and, after communicating with the
middle and posterior dental nerves, divides into ascending and
descending branches, the latter being distributed to the incisor
and cuspid teeth.

The Third Division, or Inferior Maxillary Nerve (Fig.
49).—This is the largest of the three divisions of the fifth nerve,

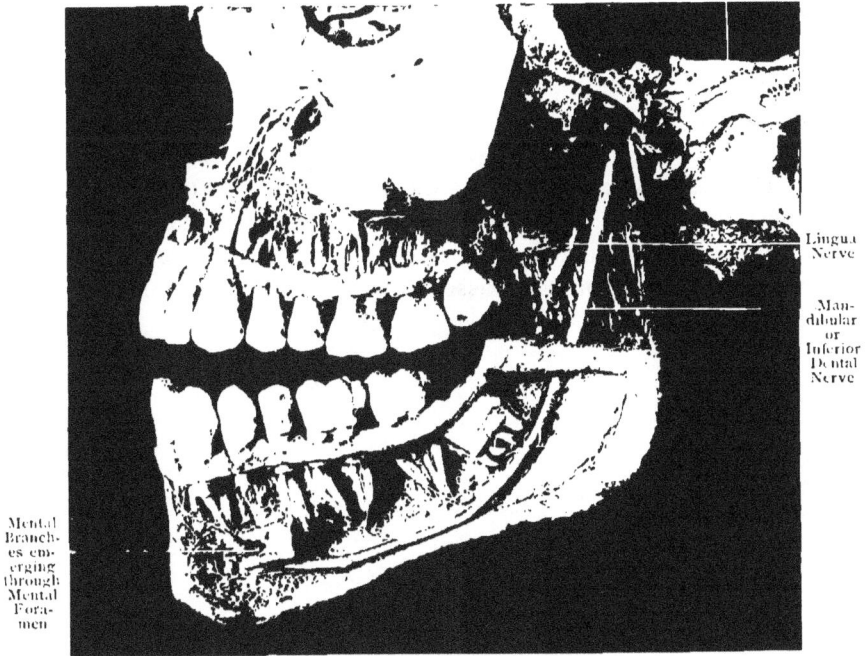

Lingua
Nerve

Man-
dibular
or
Inferior
Dental
Nerve

Mental
Branch-
es em-
erging
through
Mental
Fora-
men

FIG. 49.—DISSECTION SHOWING MANDIBULAR (THIRD) DIVISION OF FIFTH NERVE.

and is both motor and sensory in its function. Besides being
distributed to the inferior teeth, it sends filaments to the lower
portion of the face, the muscles of mastication, the tongue, and
mandible. It arises from the Gasserian ganglion, passes downward, and emerges from the skull through the foramen ovale,
after which it divides into a small anterior (motor) branch and
a large posterior (sensory) branch.

The Inferior Dental Nerve.—This is the largest branch of the

inferior maxillary nerve. From its point of origin it passes
downward internally to the external pterygoid muscle, and, upon
reaching a point between the ramus of the mandible and the
sphenomandibular ligament, it enters the inferior dental canal
through the posterior or inferior dental foramen. Before entering the foramen, two branches are given off, a lingual and a mylohyoid branch. The nerve is accompanied through the inferior
dental canal by the inferior dental artery, and, when the mental
foramen is reached, it terminates by dividing into an *incisive* and a
mental branch. Between the posterior dental foramen and the
mental foramen the nerve gives off a series of twigs to the bicuspid
and molar teeth, and, by communicating with one another within the substance of the bone, form a fine plexus.

The incisive branch follows the incisive arteries through the
substance of that part of the bone between the mental foramen
and the symphysis, and supplies the incisor and bicuspid teeth,
while the mental branch passes forward to supply the chin and
lower lip.

CHAPTER VII

OTHER STRUCTURES WITHIN THE MOUTH.—THE GUMS.—THE MUCOUS MEMBRANE.— THE ALVEOLODENTAL PERIOSTEUM.— GLANDS, DUCTS, ETC.

The Gums (*Gingivæ*).—The gums are formed by a layer of tough vascular tissue, covering the alveoli, closely attached to their periosteum, and provided with a free margin (gingival margin), which is closely molded to the necks of the teeth. They are covered on both aspects by the general mucous membrane of the mouth, that overlying the labial and buccal surfaces being reflected from the lips and cheeks, the palatal surface being continuous with that of the hard palate, and the lingual surface reflected from the under surface of the tongue and floor of the mouth. In the immediate region of the necks of the teeth the gums are especially thin and hard, being closely adherent to the periosteum and peridental membrane in this region. In passing toward the alveolar walls the membrane becomes less firmly attached to the underlying structure, and, when finally passing into the mucous membrane of the cheeks and lips, is quite loose and flabby. This condition also prevails on the lingual aspect, but the palatal surface remains firm throughout, the entire mucous membrane overlying the roof of the mouth being similar in structure and attachment to that portion immediately surrounding the necks of the teeth. In various situations about the labial, buccal, and lingual surfaces of the gums small slender folds of mucous membrane are found extending to the surrounding tissues. These folds, which act as a bridle or curb to the adjacent movable parts, are known as the *frena* of the mouth. The principal frena are found at the median line, and are three in number—the frenum labium superioris,. frenum labium inferioris, and the frenum linguæ. The two former extend from the inner surface of the lips to the gums, to which their extent of attachment is somewhat variable. The frenum labium superioris is usually much larger than the frenum labium inferioris, and its attachment to the gum frequently extends almost to the gingival border. The frenum linguæ extends from the under surface of the tip of the tongue to the lingual surface of the inferior gums. This is a much stronger fold than those connected with the lips. Similar bridles are found in the buccal region, usually near the bicuspid teeth, but they are much smaller than those at the median line. The *gingival margins*, or that portion of the gums embracing the necks of the teeth, present much variety in outline. Instead of encircling the neck of the tooth in a direct line, the margins are made up of a series of semicircles. Using the incisive region for reference, the labial and palatal margins are concave rootward, while the interproximate spaces are partly or completely filled by gum tissue, having the outline reversed or convex in the direction of the crowns of the teeth. The gingival margin is also termed the "free margin of the gum," this name better describing its extent. As previously stated, the gums are attached to the periosteum and peridental membrane; but in most instances, and particularly before the adult period, the margins of the gums extend beyond the peridental membrane, the limit of which is formed by the cervical line. That portion of the gum margin beyond the cervical line is in close contact with the neck of the tooth, but is not adherent to it, the connecting medium, the peridental membrane, not being present to form the attachment. The curvature of the cervical margins, and the nature of the tissues which enter into their construction, are usually considered as strongly indicative of the temperament of the individual. Thus, in the bilious temperament these margins are inclined to angularity and the tissues rather thick and firm. In the sanguine type the outline formed is almost a perfect semicircle, and the tissues are of moderate thickness and firmness. In the nervous type the curvature is strongly parabolic, and the tissues firm and delicate. In the lymphatic the tissues are loose and thick, and the curvature is long and poorly defined. In some instances the interproximate spaces are completely filled by the gingivæ; in others the space is only partly occupied by these tissues. The former condition is present when the proximate surfaces of the teeth are nearly or quite parallel with each other, thus reducing the capacity of the space. The latter condition is present when the crowns of the teeth are bell-shaped and the interproximate spaces extensive. The labial and buccal surfaces of the gums are more or less broken by numerous prominences and depressions, all of which accord with the variations upon the surface of the bone beneath.

Mucous Membrane of the Mouth.—The term "membrane" in a general sense is one applied to thin layers of tissue, somewhat elastic and of a whitish or reddish color. Such tissues are found lining either closed cavities or canals which open externally, absorbing or secreting fluids, and enveloping various organs. The simple membranes are of three varieties, being either mucous, serous, or fibrous. The mucous membranes are so called from the clear viscid fluid (mucus) which they secrete. They line the various cavities or tracts of the body which communicate with the exterior. The three grand divisions of mucous membrane are those lining the digestive, respiratory, and genito-urinary passages. Lining the entire cavity of the mouth we find the beginning of the digestive tract, being continuous with the skin on the exterior and performing many similar functions within. It is soft, smooth, and velvety, of a bright red color, and quite vascular; it is covered on the exterior by a layer of epithelial cells overlying the vascular parts. Immediately beneath this is a network of fibrous connective tissue forming the proper mucous membrane, and still deeper is a third layer, somewhat loose in texture, but composed of fibrous connective tissue, the submucous membrane. The oral mucous membrane, at its point of beginning on the contiguous surfaces of the lips, is endowed with keen sensibility; it is dry, bright red in color, and plentifully supplied with vascular papillæ, in many of which are sensory nerve terminals. Distributed along the line of junction between the integument and the mucous membrane are numerous sebaceous follicles, which, however, are devoid of hair-bulbs. The characteristic dryness of this surface gradually becomes changed to a mucus-secreting one, as that part of the membrane lining the interior of the lips is approached. Distributed over the surface of the labial mucous membrane are a number of minute openings, the mouths of the labial glands, which lie immediately beneath the membrane. The buccal mucous membrane, or that lining the cheeks, is similar to that covering the internal surface of the lips. It is penetrated at various points by the mouths of the buccal glands, which, in general, are smaller and less numerous than the labial glands. In the region of the second molar teeth the membrane is broken by four or five openings of larger size, which communicate with the molar glands. The mucous membrane covering the hard palate is thick and firm, less brilliant in color than that covering the cheeks and lips, and firmly bound down to the periosteum. Running from before backward at the median line, the membrane is formed into a slight fold, the median raphe, while near the anterior portion of the palate are a number of fantastically arranged folds, the rugæ (see General Description of the Mouth). The thin but rather dense fibrous aponeurosis forming the soft palate is covered anteriorly by the oral mucous membrane. Suspended from the center of the free margin of the soft palate is the uvula, which is likewise covered by mucous membrane, and from the base of this, on either side, are two folds of the membrane, which extend outward and downward, forming the anterior and posterior pillars of the fauces. The mucous membrane covering the tongue has already been described in connection with that organ (see p. 40). From the mouth the digestive mucous membrane passes through the fauces, pharynx, and esophagus to the stomach, and is so continued throughout the whole digestive tract. Other prolongations also pass into the ducts of the salivary glands.*

The Alveolodental Periosteum, or Root Membrane.

This membrane invests the roots of the teeth, and at the same time lines the walls of the alveoli. Being reflected from the peri-

---

* For a minute description of the mucous membrane of the mouth see Part ii.

osteum covering the outer alveolar walls, it enters the alveolar sockets as a single membrane, affording nourishment to the bone on one side and to the cementum of the tooth-root on the other. It is a connective tissue of moderate density, and is rich in its nerve- and blood-supply.   The general direction of its fibers is transverse, being attached at one extremity to the alveolar wall and at the other to the cementum of the root.   The connective-tissue fibers are not merely attached to the surface of the calcified structure, but the strength of this attachment is greatly increased by the passage of the fibers into the substance of the bone at one extremity and into the lamellæ of the cementum at the other.   In general, the membrane is more closely adherent to the cementum than to the bone, usually clinging to the latter when removed from its socket.   The nature of the articulation between the tooth-root and the alveolar socket, to the production of which this membrane so largely contributes, is one peculiar to itself.   While there is no marked mobility, there is, nevertheless, sufficient elasticity in the intervening membrane to provide against the severe concussions and lateral strains incident to mastication, the former being provided for by the general elasticity, while the latter is cared for by specially distributed fibers, which serve to return the tooth to its normal position when slightly rotated or laterally displaced.   This elasticity is greatest in youth and up to the meridian of life, after which time it gradually becomes less pronounced.   The membrane is thickest about the apical ends of the roots and in the cervical region, and the distribution of the fibers at these points is somewhat different from those about the body of the root.   In the former location they are spread out fan-like from the apical root surface to meet the surrounding alveolar wall, while in the latter they pass longitudinally over the alveolar margins to unite with the periosteum of the parts.   In conjunction with the functions already mentioned, the peridental membrane is the medium by which all forces applied to the tooth-surface are taken up and conveyed to the brain.   A tooth in a normal condition may be said to possess no sense of touch, unless it be so severely applied as to make its influence felt by the pulp, in which case the sensation becomes one of pain.   The nerves of the membrane act in precisely the same manner as do other sensory-nerve terminals, being influenced by the slightest touch applied to the surface of the tooth-crown.[*]

Blood-supply to the Alveolodental Periosteum.—A very clear idea of the blood-supply to this membrane may be obtained from figure 50.

Entering the alveolar socket as a single arterial branch, the thickest portion of the membrane is gained where a number of smaller twigs are given off, one or more of which enter the pulp-canal of the tooth-root through the apical foramina supplying the pulp, which in turn supplies the tooth-structure within, while the others ramify through the substance of the peridental membrane, through its capillaries, supplying the cementum from without; while passing through the membranous structure, further minute branches are given off which penetrate the walls of the alveolus and anastomose with the arteries which supply the oral mucous membrane, in this manner providing a generous blood-supply to the parts. Further on we shall see that the tooth-pulp and the peridental membrane spring from the same source (see Development of the Teeth), and the blood-supply to the parts during the saccular stage of development is alike distributed to the base of the pulp and to the

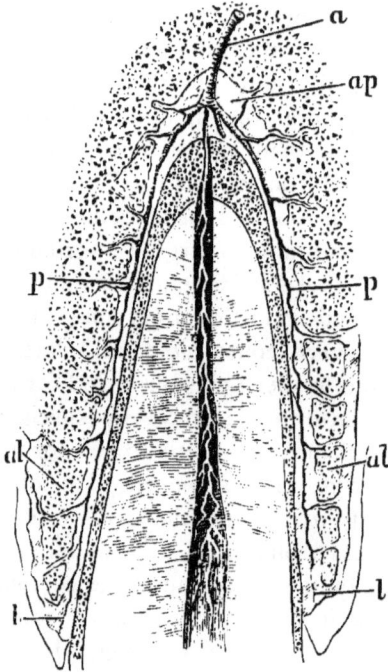

FIG. 50.—ROOT AND MEMBRANE OF TOOTH.
p,p, Peridental membrane; *ap,* apical space; *a,* artery; *al, al,* alveolar process; *l, l,* dental ligament.

---

[*] This tissue is more fully described in Part ii.

follicular walls. After the completion of the developmental process this distribution is but little changed, the blood vessels accommodating themselves to the alterations incident to the generation of the parts.

The Nerve-supply to the Alveolodental Periosteum.— The nerve-supply to this membrane is distributed in a manner similar to the blood-supply, being derived from a single filament given off from the dental nerve and entering the tooth-socket by the side of the blood-vessels, and by numerous filaments which reach the structure by passing through the many minute canals in the substance of the alveolar walls and the intervening septa.

GLANDS OF THE MOUTH.

The glands of the mouth are of two kinds, being either serous or mucous, and, as they differ in the character of their secretions, so they differ in structure. The mucous glands are the most numerous, and are found beneath the mucous membrane of the lips, in the same membrane lining the cheeks, the hard and soft palate, the tonsils, and at the back of the tongue. These glands are quite variable in size, but are all of macroscopical proportions, appearing when examined in this manner as minute whitish bodies. The secretions from the glands are poured into the mouth through small ducts which pass in various directions through the substance of the mucous membrane. Beginning as a single duct for each gland, they pass to the submucous tissue, here branching into two or more smaller ducts terminating in alveoli, the number and size of these depending upon the size of the gland with which they are connected. The glands are variously named according to their location, those occupying the lips being known as *labial glands ;* those of the cheeks, the *buccal glands ;* those of the palate, the *palatal glands*, etc. The mucous glands, although differing in size, are similar when histologically considered.

The Labial Glands.—These are among the largest mucous glands of the mouth, and are more numerous in the upper than in the lower lip. The form, size, and location of these glands may best be studied by dissection, which may readily be accomplished by first removing the integument and muscular tissues from the parts, when they will be brought into view. The glands are irregularly arranged, and are most numerous near the median line. The body of each gland is rounded and held in position by connective tissue, as well as by the duct connecting it with the interior of the mouth.

Besides the mucous glands of the lips, there are present numerous sebaceous glands. These are somewhat smaller, and are situated beneath the mucous membrane covering the contiguous surfaces of the lips, the numerous ducts leading from them opening upon these parts.

The Buccal Glands.—The glands of the cheek, otherwise known as buccal glands, are similar to, but smaller than those of the lips, and are placed between the submucous tissue and the buccinator muscle. These glands also pour their secretions into the mouth through numerous ducts which pass through the buccal mucous membrane.

In the region of the third molar tooth another set of mucous glands open into the mouth, known as the *molar glands*. They are placed between the buccinator and masseter muscles, are similar in construction, and secrete a like fluid to those previously described, being larger than the buccal and smaller than the labial glands.

Palatal Glands.—Situated between the mucous membrane of the hard palate and the periosteum are numerous mucous glands similar to those previously described. Provision is made for the accommodation of these glands by many small depressions in the bony plates. They are irregularly distributed over the surface of each lateral half of the hard palate, but are absent at the median raphe and immediately beneath the rugæ. The mucous membrane of the hard palate is tense and hard, in consequence of which it is not so thick as the buccal and lingual membranes, and the ducts from the numerous glands are therefore not so long. The glands of the soft palate, uvulæ, and fauces are situated beneath the deep layer of mucous membrane covering these parts, in the former structure opening on both the oral and nasal surfaces.

Lingual Glands.—The glands of this organ are of two

kinds—mucous and serous.   The former are chiefly found at the
back part of the tongue, but a few of smaller size are present
near the tip.   The serous variety are to be found only at the
back of the organ, and are closely associated with the taste
organs in this region.   These glands are assisted in performing
their function by being placed between bundles of striped muscular tissue, the activity of which forces the secretions to the
surface by compressing the glands.   While the majority of the

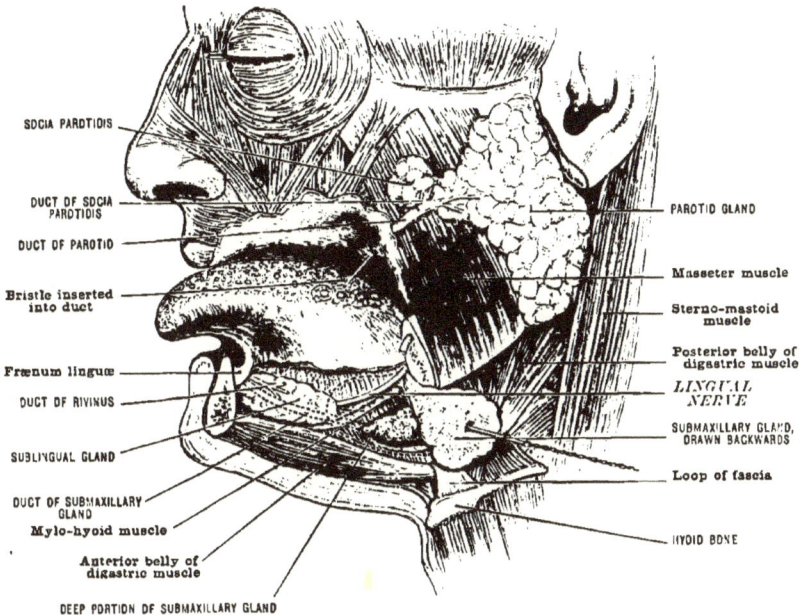

FIG. 51.—THE SALIVARY GLANDS.—(*Morris.*)

lingual glands are present in the circumvallate region, a number
are found distributed beneath the mucous membrane of the
borders and tip of the organ.

The Salivary Glands   (Fig. 51).—These glands, while
outside the mouth, are so closely associated with its functions
that a brief description will be presented.   The chief salivary
glands are six in number, three on each side.   They are named
parotid, submaxillary, and sublingual ; the former, secreting true,

thin, watery saliva, is a true salivary gland, while the latter two
are known as mixed, or mucosalivary glands, secreting both
mucus and saliva.

The Parotid Gland.—This gland is the largest of the three,
and is placed a little below and in front of the ear, having the
following boundaries: Anteriorly, by the ramus of the mandible;
posteriorly, by the styloid and mastoid processes of the temporal
bone; above, by the root of the zygoma, and below, by a line
drawn backward from the angle of the jaw. While the extent
and outline of the gland is somewhat variable, its position is
approximately that outlined in figure 51. Its superficial surface,
somewhat lobulated, is in close relation to the skin and fascia,
while deeply it penetrates well into the neck by two processes,
one of which passes behind the styloid process and beneath the
mastoid process of the temporal bone and the sternomastoid
muscle, while the other passes in front of the styloid process.
Given off from the body of the gland and extending in various
directions are a number of processes or lobes, one extending
forward between the two pterygoid muscles, and known as the
pterygoid lobe; another passing into the glenoid cavity, the
glenoid lobe; while a third passes deeply between the carotid
vessels, and is called the carotid lobe. In many instances there
is an additional lobe, which is detached from the body of the
gland, known as the *socia parotidis*. When present, this lobe is
placed over the parotid duct and empties into it. Passing
through the substance of the gland are a number of arteries and
veins, principal among which are the external carotid, transverse
facial, and internal maxillary, the gland receiving its blood-
supply by branches from these. The internal carotid artery
and internal jugular vein lie close to its internal surface. The
facial nerve and its branches and the great auricular nerve pass
through the gland from before backward, and supply its substance with nerve-force. The weight of this gland is from one-
half to one ounce.

The Parotid Duct.—Leading from the gland to the mouth
is the parotid, or Stenson's duct. After passing through the fat
of the cheek and the fibers of the buccinator muscle, the duct
comes in contact with the deep layer of the oral mucous mem-

brane. After passing between this structure and the cheek tissues for a short distance it enters the mouth opposite the crown of the superior second molar tooth, the orifice of the duct appearing on the surface of the mucous membrane in the form of a small papilla, which may be readily observed with the naked eye. When first given off from the gland a number of small ducts are present, but these soon unite-and form a single canal. The parotid duct is quite dense, of considerable thickness, and is lined by a reflexion of the buccal mucous membrane.

The Submaxillary Gland.—This gland, which receives its name from occupying a position below the maxillary bone, is somewhat smaller than the parotid. It is situated beneath the mylohyoid ridge, and occupies the anterior part of the submaxillary triangle, extending upward to occupy the submaxillary fossa on the lower border of the maxilla. Superficially, it is covered by the skin and a few muscular fibers and the deep fascia. The facial vein and branches of the facial nerve pass over its superficial surface. Deeply, it is in relation with the mylohyoid and hyoglossus muscles, and also with the mylohyoid artery and nerve. The gland receives its blood- and nerve-supply from the arteries and nerves which penetrate it. This gland is separated from the sublingual gland by the mylohyoid muscle, and weighs about two drams.

The Submaxillary Duct.—The submaxillary duct, otherwise known as Wharton's duct, passes forward and inward, opening into the cavity of the mouth on the summit of a small papilla near the frenum linguæ. The duct, which is nearly two inches in length, first passes through the adjacent muscular tissue, and finally beneath the oral mucous membrane to its outlet. Like the parotid duct, it is lined by a reflexion of the oral mucous membrane.

The Sublingual Gland.—This is the smallest of the salivary glands, and is also named from its location beneath the tongue. Its position is beneath the tip of the tongue and the mucous membrane covering this part of the floor of the mouth. It rests in the submaxillary fossa of the maxilla, meeting with its fellow at the median line, and extending as far back as the mylohyoid

9

muscle, which separates it from the submaxillary gland. The gland is supplied with blood from the sublingual and submental arteries, and with nerves from the gustatory. The weight of this gland is about one dram.

The Sublingual Ducts.—This gland communicates with the mouth by one large duct—the duct of Rivini—which springs from the main portion of the gland, and by a number of smaller ducts, eight to twenty in number, which open on the floor of the mouth. The duct of Rivini follows the submaxillary duct, and opens with it at the same papilla. The smaller ducts are given off from a number of little lobes which cluster about the fore part of the gland.

CHAPTER VIII.

A DESCRIPTION OF THE SUPERIOR TEETH IN DETAIL.—CALCIFICA-
TION, ERUPTION, AND AVERAGE MEASUREMENTS.—THEIR SUR-
FACES, RIDGES, FOSSÆ, GROOVES, SULCI, ETC.

SUPERIOR CENTRAL INCISOR.

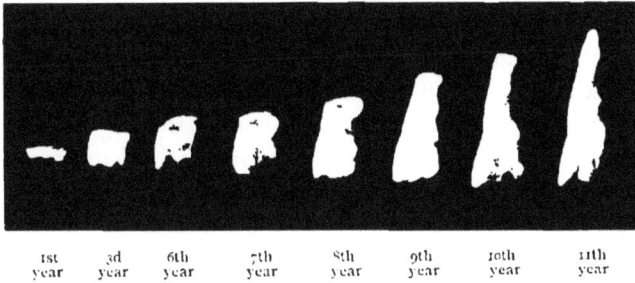

1st year	3d year	6th year	7th year	8th year	9th year	10th year	11th year

FIG. 52.

CALCIFICATION BEGINS, FROM THREE CENTERS, FIRST YEAR AFTER BIRTH.

CALCIFICATION COMPLETED, TENTH TO ELEVENTH YEAR.

ERUPTED, SEVENTH TO EIGHTH YEAR.

AVERAGE LENGTH OF CROWN, .39.

AVERAGE LENGTH OF ROOT, .49.

AVERAGE LENGTH OVER ALL, .88.

During the first year after birth this tooth begins to calcify, this primitive process taking place along the future cutting-edge of the tooth in three distinct lobes or plates, which afterward unite and form three eminences or tubercles, the lines of this union being indicated upon the completed crown by two more or less defined grooves—*developmental grooves*. By the end of the third year the deposit of lime-salts has carried the process of calcification to a point about midway between the cutting-edge and the cervical line. By a continuation of this formative action the calcification of the crown is completed

The measurements given are in fractions of an inch, and are taken from Black's " Dental Anatomy."

between the fifth and sixth year. At the beginning of the
seventh year calcification has progressed to such an extent that
the neck of the tooth and base of the root are fully outlined.
Between the seventh and eighth year the cutting-edge of the
tooth begins to make its appearance through the gum at a point
either to the right or left of the median line, and, by a gradual
separation of the gum tissue, eruption takes place. During
the following year about one-eighth of an inch has been added
to the length of the root. At the end of the eighth year the
root has become calcified to about one-half of its completed
length. During the ninth year, owing to a reduction in the
diameter of the root, the extent of growth in an apical direction
has almost doubled that of the previous year, and a decided
narrowing of the free root margins is to be observed. At the
eleventh year calcification is completed in the outer root walls,
and the apical foramen has been established (Fig. 52.)

The crown of the superior central incisor presents for
examination four surfaces—labial, palatal, mesial, and distal ; two
angles—a mesial and a distal ; and a cutting-edge. The general
form of the crown is that of a double incline plane, or wedge-
shape, the cutting-edge representing the junction of the two
sides of the incline, one of which looks anteriorly (labial) and
the other posteriorly (palatal). The labial side of the incline is
convex, while the palatal is concave from the cutting-edge
toward the root ; but, upon reaching its upper or cervical third, it
presents a slight general convexity. The base of the wedge is
directed upward and partakes of the contour of the neck of the
tooth.

The Labial Surface of the Crown (Fig. 53).—In general
outline this surface resembles an imperfect quadrilateral. The
margins of the surface are the *mesial*, the *distal*, the *cervical*,
and the *incisive*. The mesial margin begins at the lower border
or cutting-edge and passes upward, usually with a slight distal
inclination, gradually uniting with the cervical margin. The
distal margin begins at the cutting-edge and passes upward with
a slight mesial inclination, also joining the cervical margin. Both
of these margins possess more or less general convexity, and,
at their junction with the cutting-edge, form the mesial and

distal angles of the crown. The cervical margin is rounded and gradually passes into the two lateral margins just described. The incisive margin is marked by the cutting-edge, and extends from the mesial angle on one side to the distal angle on the other. These four margins, which assist in giving to the tooth its typal form, are quite variable. This difference is particularly marked on the mesial and distal margins, where, in some cases, there is a decided convergence in the direction of the root, forming what is commonly termed the bell-shaped crown, while in others the same margins will be nearly parallel with each other, making the width of the crown almost as great at the cervical margin as at the cutting-edge. The mesial angle is usually pointed and square, while the distal is much rounded. This surface of the crown is slightly convex from above downward, as well as from side to side, and in the majority of instances is of

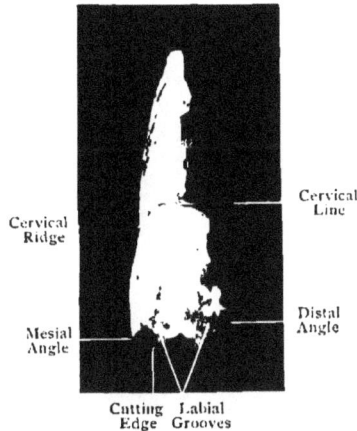

FIG. 53.—SUPERIOR CENTRAL INCISOR, LABIAL SURFACE.

greater vertical than transverse extent. Beginning at the incisive margin are two slight longitudinal depressions or grooves— the *labial grooves*—which are resultant from the developmental lobes previously composing the primitive cutting-edge, and, for this reason, are otherwise known as developmental grooves. In many instances one or more transverse ridges are found upon the cervical portion, but these are supplemental in their nature.

The Palatal Surface of the Crown (Fig. 54).—This surface has its borders formed by three marginal ridges and the cutting-edge. The marginal ridges are pronounced elevations of enamel, and surround the surface upon three sides, the intervening space in many instances being a decided concavity or fossa—the *palatal fossa*. The *mesial marginal ridge* begins at the mesial angle of the crown, passes upward, inward, and backward, following the curvature of the mesial border. The *distal marginal ridge* begins at the distal angle of the crown in a somewhat less pronounced form, passes upward, backward, and inward, following the curvature of the distal surface. Upon reaching the cervical portion of the crown these two margins unite and form the *cervicomarginal ridge*. This ridge may be bold and prominent, or it may be but slightly developed. Near its center it is frequently broken by a depression or pit—the *palatal pit*. In some instances this ridge is deeply penetrating; in others it assumes the form of a fissure, and may completely sever the ridge. This border is sometimes elevated into a slightly developed tubercle or cusp—the *cuspule*. When this is present it has the appearance of being produced by a fold of the enamel, by which it is given more or less prominence near the center of the border, and encircling this enamel fold is a well-marked fissure. The palatal fossa is usually traversed by two longitudinal grooves, which correspond to the developmental grooves of the labial surface.

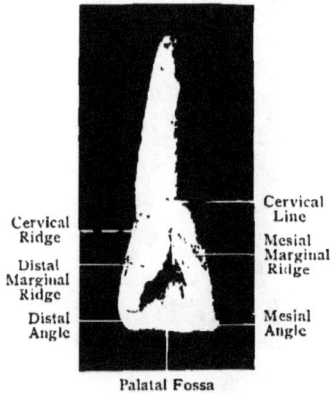

Cervical Ridge
Distal Marginal Ridge
Distal Angle
Cervical Line
Mesial Marginal Ridge
Mesial Angle
Palatal Fossa

FIG. 54.—SUPERIOR CENTRAL INCISOR, PALATAL SURFACE.

When a cuspule is present, the fissure which surrounds it frequently enters into the fossa, or the fossa may be partly covered by the cuspule overhanging its cervical portion. When the cervicopalatal fissure exists, it is not unusual for it to bifurcate and throw a branch along the inner border of each marginal ridge, or it may penetrate the fossa proper and divide it into two parts. The palatal surface is somewhat less in extent than the labial surface, this reduction being principally in a mesiodistal direction, the length of the two surfaces from the cutting-edge to the cervical margin being about equal.

The Mesial Surface of the Crown (Fig. 55).—The outline of this surface resembles an inverted cone or triangle, the lines of which are more or less broken, the apex of the cone terminating at the cutting-edge and the base directed toward the root of the tooth. The base of the cone is made concave by the enamel margin or cervical line. The *margins* of the mesial surface are the *labial*, the *palatal*, and the *cervical*. The labial margin is convex and rounded throughout its entire extent, from the cutting-edge to the cervical line. The contour of this margin varies with the typal form of the crown, in some presenting a decided and well-marked convexity, in others being but slightly curved. The palatal margin is concave and rounded, but the line is much broken. Beginning at the cutting-edge, it is decided and square, this feature usually including the lower third. As it passes upward and the center is approached, the line is more concave and rounded in a mesio-palatal direction, this latter feature increasing upon approaching the cervical line. The cervical margin is that formed by the cervical line. It is usually well defined, being concave or V-shaped, with the point of the V more or less rounded, and with its free ends pointing one in a labial and one in a palatal

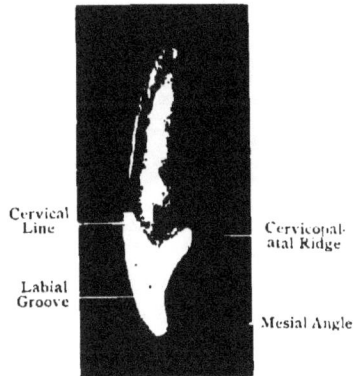

FIG. 55.—SUPERIOR CENTRAL INCISOR, MESIAL SURFACE.

direction, the former being a trifle longer than the latter. The surface between the borders presents a slight general convexity, but with an inclination to flatness near the cervical portion, which is occasionally developed into a slight concavity. Whatever deviations may be present in the borders of this surface, from those assumed in the description just given their union at the cutting-edge will always be in a direct line with the long axis of the tooth.

The Distal Surface of the Crown (Fig. 56).—In a general way, this surface resembles the mesial surface just described. There are, however, one or two minor points of distinction : the borders are all more rounded, the labial border presenting a greater convexity, and the palatal a more perfectly formed concavity. The surface is quite full in the center, from which it slopes away in all directions, thus producing a decided general convexity.

The cervical margin of the surface is almost identical with the cervical margin of the mesial surface. The distance in a direct line from the cervical border to the cutting-edge is a trifle less than the corresponding measurement on the mesial surface. The distal angle is equally constant in its position, and, being connected with the mesial angle in a direct line by the cutting-edge, finds this latter coronal margin always in the labio-palatal center of the crown.

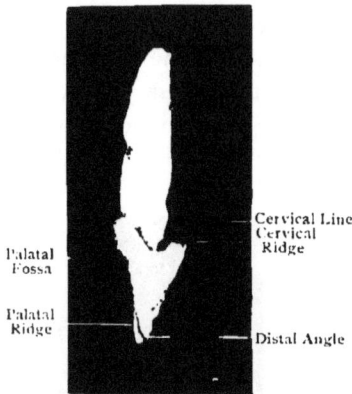

FIG. 56.—SUPERIOR CENTRAL INCISOR, DISTAL SURFACE.

The Cutting-edge of the Central Incisor.—The cutting or incisive edge receives its name from its function, that of cutting or incising the food. It is formed by the junction of the labial and palatal surfaces of the crown, and extends in a direct line from the mesial to the distal surface; at its union with the mesial surface it assists in forming the mesial angle of the crown, and serves the same purpose by its union with the distal surface. In the majority of instances it is an unbroken line. In passing from the mesial to the distal angle it converges slightly in the direction of the root, thus making the crown a trifle shorter on the mesial than on the distal side. In the recently erupted tooth (Fig. 57) the line is broken by the developmental grooves ; these usually disappear by wear, but occasionally traces of their existence remain, and thus permanently break the positive line that would otherwise be present. As the cutting-edge approaches the distal angle of the crown it is inclined to slope away, producing a less positive angle than the corre-

FIG. 57.—A YOUNG SUPERIOR CENTRAL INCISOR, LABIAL SURFACE, SHOWING DEVELOPMENTAL GROOVES—a.

sponding mesial angle. In some instances the cutting-edge is quite thin and inclined to sharpness, in others it is blunt and dull, the former condition being present when there is a decided overbite in occlusion, the latter occurring when this feature is less pronounced. The cutting-edge is frequently referred to as the *occlusal surface* this term being employed to make the description more uniform with the bicuspids and molars, and for this reason is permissible; but it is only in rare instances that the surfaces occlude directly with the opposing teeth, the condition being most frequent in teeth of the lymphatic type, and in cases of malocclusion.

The Cervical Margin.—This margin, which is distinctly outlined by the free extremity of the enamel covering of the crown, also marks the extent of the peridental membrane covering the root of the tooth. The margins formed by this line are those of a double concavity and a double convexity. On the labial and palatal portions it is concave rootward, while on the mesial and distal sides it is convex in this direction. If a line be drawn around the tooth at the extreme upper point of the enamel covering, it will be found to touch only the labial and palatal prolongations, while a space will exist between the line thus drawn and the cervical margins of the proximate surfaces. The character of the cervical curvature varies with the type of the tooth, being more or less pronounced as the case may be. In a typical central incisor the cervical line of the labial surface will usually form the segment of a larger circle than that of the palatal, and, while the mesial convexity may be gracefully curved, the distal may incline to angularity.

The Neck of the Tooth.—The neck of this tooth partakes of a form between that of the crown and root which it joins. It is principally formed by a sudden sloping of the enamel margin to meet the root. It is broader on the labial than on the palatal surface, and is somewhat flattened laterally, with an occasional depression or concavity on its mesial portion. The neck of this tooth is seldom a decided anatomic feature, being less pronounced than upon any other tooth. In the bicuspids and molars both the crown and the roots assist in forming the neck by a constriction of their adjacent parts, while in this tooth the crown alone is instrumental in this direction.

The Root of the Superior Central Incisor.—The root of this tooth is conic in form, its base directed downward, its apex upward. Viewed in transverse sections its outline is that of a rounded triangle, one side of which faces in a labial, one in a mesiopalatal, and one in a distopalatal direction. The labial side is the most flattened, while the two remaining sides are of equal length and oval in form. This triangular outline usually continues throughout the entire length of the root, but in some instances, near the apical end, may have a decided or slight distal curve, included in which will be a more circular form. The taper of the root from the base to the apex is very gradual upon the labial and palatal surfaces, until the apical third is reached, when the two sides converge more rapidly. The mesial and distal surfaces are somewhat flattened and taper very gradually from the base to the apex. In a majority of instances the root is much longer than the crown, but in rare cases its length is barely equal to, or less than, that of the crown.

Bilious Type.—The crown of the superior central incisor in this type is of greater longitudinal than transverse extent; large in size, abounding in angles rather than curves. It possesses neither brilliancy nor transparency of surface, but is slightly inclined to translucency. The labial surface is flat, with more or less decided transverse ridges in the cervical portion. The labial grooves are generally present in the form of well-defined depressions. On account of the angular nature of the tooth, this surface approaches closely to the quadrilateral form. The mesial and distal surfaces are flat, with their margins bold and well defined. The palatal surface also shows the angular nature of the crown in having its marginal ridges squarely set and its developmental grooves definitely outlined. The cutting-edge is rather thin, square, and sharp, the line frequently being imperfectly formed. The mesial and distal angles are both well produced, and the cervical margin, in keeping with the rest of the parts, is inclined to angularity.

Nervous Type.—The central incisor common to this temperament is delicate and graceful in outline. The crown is of medium size, with the length predominating over breadth. The enamel is inclined to transparency, and is of a blue or bluish-gray color, presenting much brilliancy. The labial surface is fairly well rounded, and the labial grooves are present as slightly rounded depressions, which frequently extend well toward the cervical margin, where they gradually disappear. In general outline this surface partakes of the triangular form, the crown of the tooth being broad at the cutting-edge and much constricted at the neck. The mesial and distal surfaces show a convexity in every direction, and the nature of the occlusion is manifest from the decided wedge-shape appearance of the crown providing for a long overbite. Upon the palatal surface but little in the way of detail is to be observed, the entire surface from the cutting-edge to the cervical ridge being smooth and concave. The cuspule previously referred to is occasionally present in this type, breaking the general smoothness of the surface with its prominence. The marginal ridges are poorly defined; the cutting-edge is a sharp, unbroken line; the mesial and distal angles are present in the form of long, graceful curves, rather than definite angles, this being particularly true of the distal angle. The cervical line is decidedly curved, the labial and palatal portions being deeply concave rootward, while the mesial and distal are decidedly convex.

Sanguineous Type.—The crown of the central incisor is usually above the average in size, but is well proportioned, abounding in curves and rounded outlines. The enamel is inclined to translucency, particularly near the cutting-edge. The labial surface is smooth and rounded; the depressions formed by the labial grooves are slightly observable, and extend but a short distance from the cutting-edge. The surface is somewhat greater in longitudinal than in transverse extent, and approaches much nearer to a circular form than the corresponding tooth of other types. The mesial and distal surfaces are well rounded, making the point of contact with approximating teeth near the center of the surface. The palatal surface abounds in heavy rounded lines; the marginal and cervical ridges are particularly prominent, diminishing the extent of the palatal fossa. A cuspule is frequently present in the form of a well-rounded prominence. The cutting-edge is of moderate thickness and slopes away from the center in either direction to assist in forming the rounded mesial and distal angles. The cervical curvature on the labial and palatal surfaces is an unbroken semicircle, while that of the mesial and distal surfaces is less uniform.

Lymphatic Type.—In the central incisor of this typal form the crown is large, but not shapely, and the breadth is equal to, or exceeds, the length. The enamel coloring is muddy or brownish-yellow, and the surface is lacking in brilliancy. The labial surface is flat and smooth, with a faint sign of the labial grooves. The general outline of this surface is that of a circular cone, with the cutting-edge for the base, and the apex formed by the cervical margin. The mesial and distal aspects present a striking contrast to the types previously described, by having a labiopalatal diameter greater than that represented between the cervical line and the cutting-edge. These two surfaces are convex in a labiopalatal direction only, making the point of contact with approximating teeth an extended surface rather than a single point.

The palatal surface is heavy and bulky, frequently to such a degree as to produce a general convexity rather than a concavity, as found in most typal forms. This surface is frequently broken by one or more longitudinal grooves, but is seldom crossed by transverse lines of any kind. The cutting-edge is barely deserving of the name. Although formed by the free borders of the labial and palatal surfaces, these two planes are so far separated at their incisive margins that the space between them, instead of being an edge, becomes a more or less broadened surface, and one upon which the inferior incisors frequently occlude. The line thus formed is straight and direct from the mesial to the distal angle of the crown, both of which are well produced. The cervical curvature is represented by the segment of a much larger circle than that found upon teeth of other types, and the neck of the tooth is heavy and bulky, showing but little constriction at this point.

SUPERIOR LATERAL INCISOR.

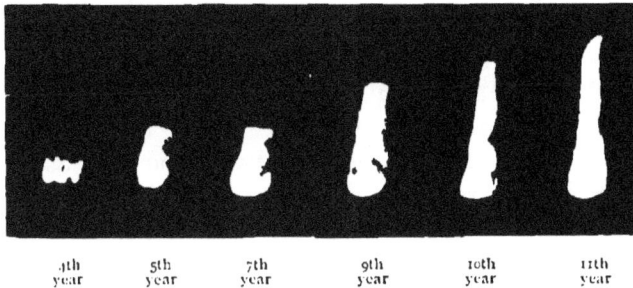

| 4th year | 5th year | 7th year | 9th year | 10th year | 11th year |

FIG. 58.

CALCIFICATION BEGINS, FROM THREE CENTERS, FIRST YEAR AFTER BIRTH.
CALCIFICATION COMPLETED, TENTH TO ELEVENTH YEAR.
AVERAGE LENGTH OF CROWN, .34.
AVERAGE LENGTH OF ROOT, .51.
AVERAGE LENGTH OVER ALL, .85.

Like the central incisor, calcification in this tooth begins during the first year after birth, the process taking place in the same manner, from three centers, along the future cutting-edge, and gradually extending in the direction of the root. By the expiration of the third year the cutting-edge and the angles of the crown are fully formed; the fourth year finds the crown calcified to nearly one-half its completed length; by the fifth year the cervical ridge is reached; while the sixth year usually completes the process of coronal calcification. At the close of the seventh year the base of the root is fully outlined, and during the following year about one-eighth of an inch is added to its length, and still greater progress is made during the ninth year, by which time fully three-fourths of the root length has become calcified. During the tenth year the apical end of the root begins to form by a sudden doubling-over of the free calcifying margins, and by the eleventh year the apical foramen is established (Fig. 58). By the above description it will be observed that at the time of eruption the root of this tooth is only calcified to about one-half of its completed length, and the same may be said of the central incisor; but so much time elapses between the beginning of the eruptive stage and the period at which this phenomenon is completed that the apical foramen is usually established by the time the tooth assumes its permanent position in the jaw.

The crown of the superior lateral incisor, like that of the superior central incisor, presents for examination four surfaces,—labial, palatal, mesial, and distal,—a cervical margin, a cutting-edge, and a mesial and distal angle. The general contour of the crown closely resembles that of the superior central incisor, except that it measures about one-third less from mesial to distal, and is a trifle shorter from the cutting-edge to the cervical line. As in the central incisor, the labial and distal surfaces form a double incline plane, and unite below to form the cutting-edge. The labial side of the incline is convex, while the palatal is concave, but seldom so marked as that found upon the central incisor. The base of the wedge, or double incline, formed by the cervical margin, is correspondingly smaller than that of the crown of the central incisor.

The Labial Surface of the Crown (Fig. 59).—This surface of the crown of the superior lateral incisor is more irregular in outline than the corresponding surface of the central incisor. The margins of the surface are the *mesial, distal, cervical,* and *incisive.* The mesial margin begins at the mesial angle and passes upward with a decided distal inclination to meet the cervical margin. The distal margin is shorter and decidedly more convex than the mesial margin, this variation in outline being still more marked when compared with the corresponding margin of the central incisor. At the cutting-edge these two margins assist in forming the mesial and distal angles of the crown, and by their continuation and union above form the cervical margin.

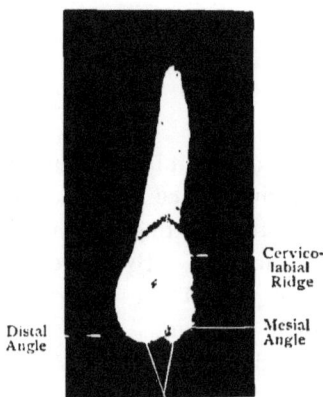

Cervico-labial Ridge

Distal Angle

Mesial Angle

Labial Grooves

FIG. 59.—SUPERIOR LATERAL INCISOR, LABIAL SURFACE.

The incisive margin is formed by the cutting-edge. Like the central incisor, the four margins of this surface vary greatly in the different types; this is particularly true of the two lateral margins, which at times are found to be in the form of a direct line, or even slightly concave, while in others they are both decidedly convex. This surface of the crown shows a greater general convexity than the labial surface of the central incisor, the cervical portion presenting a curve much more decided than that near the cutting-edge. The *labial grooves* are in all respects similar to those described in connection with the central incisor, and extend from the cutting-edge toward the center of the surface, where they grad-ually disappear. Transverse ridges are occasionally found near the cervical portion of the surface.

The Palatal Surface of the Crown (Fig. 60).—This sur-face of the superior lateral incis-or is subject to much variation in form, but presents the same points for examination as the corresponding surface of the central incisor. These consist of the marginal ridges, which are usually more pronounced than those of the central, mak-

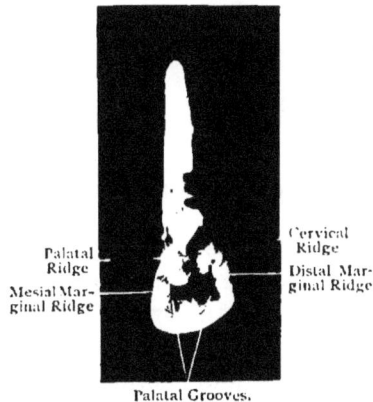

Fig. 60.—SUPERIOR LATERAL INCISOR, PALATAL SURFACE.

ing the concavity or fossa between them small and deep. In some instances the surface will be smooth and flat, with an entire absence of ridges or fossæ. The *distal marginal ridge* is shorter and more bowed than the mesial, and the *cervical ridge* is well marked and proportionately broader and stronger than in the central incisor. In some instances the marginal ridges are but slightly developed, with their cer-vical ends broadened and separated by a deep fissure, giving the appearance of a terminal fold in the enamel. The cervical ridge is frequently broken by a cuspule, which is usually more pronounced than when found upon the central incisor.

The *palatal fossa* may be present as a smooth, unbroken concavity, or it may be subdivided by a longitudinal ridge, which often exists to such an extent as to force the remaining portions of the fossa well against the marginal ridges, where they will be observed as slight depressions rather than marked concavities.

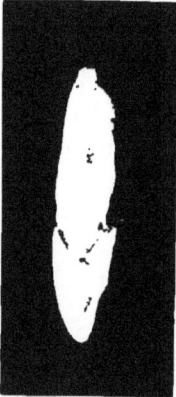

FIG. 61.—SUPERIOR LATERAL INCISOR, MESIAL SURFACE.

The Mesial Surface of the Crown (Fig. 61).—Viewing the crown from this aspect, the outline is that of an inverted cone or triangle. The *palatal margin* of the surface is well defined, and the angle formed by the union of this surface with the palatal surface is moderately acute. The *labial margin* is well rounded, and passes into the labial surface without a decided line of demarcation. The surface on its upper or cervical third is usually flattened and occasionally concave. At the center, and continuing toward the cutting-edge, it is decidedly convex in every direction, thus producing a prominent point of contact with the distal surface of the central incisor.

The Distal Surface of the Crown (Fig. 62).—This surface also shows the characteristic wedge-shape of the crown, and is principally different from the mesial surface in being convex throughout. Near the center it is well rounded and full, providing a point of contact for the mesial surface of the cuspid. The *palatal margin*, while being more decidedly outlined than the labial, is much more rounded than the palatal margin of the mesial surface. From the most prominent point near its center the surface slopes away in every direction, the convexity being most marked near the cutting-edge.

FIG. 62.—SUPERIOR LATERAL INCISOR, DISTAL SURFACE.

The Cutting-edge of the Lateral Incisor.—In the young tooth the cutting-edge presents the three little tubercles common to all incisors, the grooves which divide them passing up over the labial and palatal surfaces and forming the labial and palatal grooves. These soon disappear by wear, usually leaving the cutting-edge in the form of a direct line and connecting the two angles of the crown. Like the central incisor, this margin of the crown may be thin and sharp, or it may be thick and dull.

The Cervical Margin.—This line of demarcation between the crown and the root of the tooth resembles so closely that described in connection with the central incisor that it will only be necessary to mention one or two characteristic differences. The palatal side of the line presents a much smaller curve proportionately, and usually extends a little higher in the direction of the root than that represented upon the labial portion. The mesial and distal portions of the line dip well down, decreasing the length of the crown on these surfaces; the margin on the former surface is usually angular and V-shaped, while on the latter it is circular in form.

The Angles of the Crown.—The angles of the crown are the mesial and the distal, and are formed in the same manner as the same angles of the central incisor. The mesial angle is generally well produced, in most instances being slightly acute; but when the cutting-edge is thin and frail, the angle is frequently much obliterated by wear. That portion of the crown of the superior lateral incisor which is usually referred to as the distal angle is scarcely worthy of the name. It is usually present as a long curve, which begins near the center of the cutting-edge and extends well up on the distal surface. This characteristic outline is sometimes so pronounced as to completely destroy the cutting-edge, the distal surface being carried forward by a long curve ending in the mesial angle.

The Neck of the Superior Lateral Incisor.—In this tooth the neck is usually marked by a constriction much more pronounced than that found in the central incisor. On the labial and palatal surfaces it is principally formed by a sudden sloping of the enamel surface rootward, but on the two lateral surfaces it is formed by a flattening or slight concavity of both the crown and the root.

The Root of the Superior Lateral Incisor.—The root of this tooth is conic in form, and is much more flattened from

10

mesial to distal than the root of the central incisor. At its junction with the crown it is circular in form, the labial portion forming the segment of a larger circle than the palatal, this feature being observed throughout its entire length. The flattening of the mesial and distal sides begins immediately above the neck, and gradually increases as the center of the root-length is approached, where it often develops into a slight longitudinal depression. As the apical end of the root is reached, this longitudinal depression gradually disappears, and the root again becomes circular in form. The thickness of the root is about one-third greater from labial to palatal than from mesial to distal, and, while it is generally classed as a straight root, it is frequently provided with a pronounced distal curve near the apical extremity. In some instances it is found with a double mesiodistal curve.

Bilious Type.—In this type the lateral incisor is frequently poorly developed, the cutting-edge and distal angle are wanting, the crown being in the form of a single conic cusp, the distal surface meeting the mesial at a point near the mesial angle. This form of crown might be classed as one of malformation, but the fact that it most frequently occurs in this temperament would appear to indicate a normal condition. When the crown takes the form common to incisors, it is of greater longitudinal than transverse extent, the angles are well produced, and the mesial and distal surfaces are flat and almost parallel with each other. The labial surface is flat and is frequently broken by transverse ridges near the cervical portion. The palatal surface presents well-marked outlines and margins, a cuspule is seldom present, and the palatal fossa is well marked, but not deep. The cutting-edge is thin and sharp, to provide for the overbite, which is rather long. The cervical border is square and angular.

Nervous Type.—In this typal form the neck of the tooth is a pronounced feature. The crown is long and narrow, the constriction forming the neck beginning well down on the crown and extending over the cervical line to the surface of the root. The labial surface is convex in every direction and the labial groove fairly well defined. The mesial surface is convex near the center and cutting-edge, but often shows a slight concavity on its cervical portion. The distal surface is rounded and smooth. The palatal surface presents a general concavity, a cuspule being more frequently present than in other types. The palatal fossa is deep, and often extends beneath the cervicomarginal ridge in the form of a circular fissure. The cutting-edge is thin and sharp, providing for a long overbite ; the mesial angle is pointed and well formed, while the distal is usually much rounded. The cervical line is well arched, forming the segment of a much smaller circle than that seen on the same tooth of other temperaments.

Sanguineous Type.—The crown is well proportioned, with the length slightly predominating over breadth, all the surfaces being more or less rounded and smooth, showing the crown to be made up of curves rather than angles. The labial surface presents a graceful convexity throughout ; the mesial and distal surfaces are both convex, with their margins poorly defined. The palatal surface shows the rounded nature of the crown in having its fossa and marginal ridges oval and blending one into the other. The cutting-edge is moderately heavy and dull, in keeping with the overbite, which is short. The cervical line is made up of curves rather than angles.

Lymphatic Type.—In this type the crown is generally of greater transverse than longitudinal extent. The neck is poorly produced, the crown and neck uniting without any marked con- striction of the parts. The labial surface is much flattened from mesial to distal, and but slightly convex in the direction of the long axis of the tooth. The mesial and distal surfaces are but little rounded and are nearly parallel with each other, so that the contact with adjoining teeth becomes an extent of surface rather than a single point. The palatal surface is convex above ; but as the cutting-edge is approached it becomes flat, but seldom concave. The marginal ridges are not well shown and the palatal fossa is but a slight depression. The angles of the crown are well produced and the cutting-edge thick and blunt, this marginal surface frequently occluding directly upon the opposing inferior teeth. The curvature of the cervical line is that of a long circle.

SUPERIOR CUSPID.

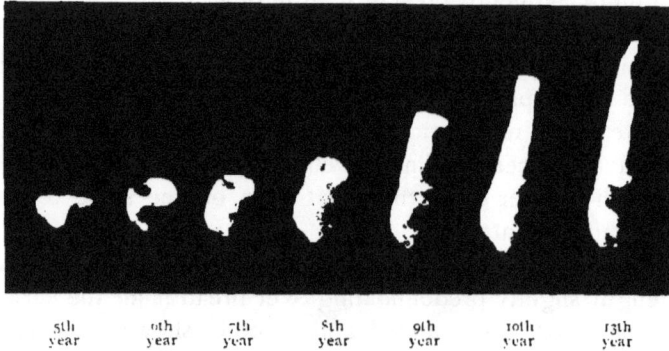

| 5th year | 6th year | 7th year | 8th year | 9th year | 10th year | 13th year |

FIG. 63.

CALCIFICATION BEGINS, FROM THREE CENTERS, THIRD YEAR AFTER BIRTH.
CALCIFICATION COMPLETED, TWELFTH TO THIRTEENTH YEAR.
ERUPTED, TWELFTH TO THIRTEENTH YEAR.
AVERAGE LENGTH OF CROWN, .37.
AVERAGE LENGTH OF ROOT, .68.
AVERAGE LENGTH OVER ALL, 1.05.

About the third year after birth calcification begins in the central lobe, which is gradually extended laterally, until, at the fourth year, it is met by the two lateral lobes, which are somewhat later in beginning, and by the fifth year the three are united, the former eventually establishing the single cusp of the tooth and the latter two the mesial and distal angles. About the sixth year two-thirds of the crown is formed, and by the seventh year the constriction which marks the beginning of the neck of the tooth commences to make its appearance. Between the seventh and eighth year calcification in the crown is completed and the cervical line established, during the following year nearly one-quarter of an inch is added to the length of the root, and by the beginning of the tenth year the root is formed for fully two-thirds of its entire length. Between the twelfth and thirteenth years or at the time of eruption, calcification is completed in the root and the apical foramen established (Fig. 63). In this latter particular the cuspid tooth differs from most of the others in being completely calcified previous to, or about the time of, its eruption, the eruptive process in this tooth, therefore, must, of necessity, differ from that of the other teeth, being fully calcified about the time it makes its appearance through the gum tissue. To reach its final position in the arch the tooth moves bodily downward, the bone filling in behind; while in the incisors, bicuspids, and molars the free calcifying root-extremities remain nearly stationary, the crowns being forced downward as the lime salts are deposited.

The crown of the superior cuspid presents for examination four surfaces,—labial, palatal, mesial, and distal,—two margins,—the cervical margin and the cutting-edge,—and a mesial and a distal angle. In general outline it is of the simplest form, resembling the primitive cone-shaped teeth of many fishes. When viewed by looking directly upon the mesial or distal surface, the wedge-shape common to the incisors is observed. The base of the double incline is, however, much broader proportionately than the corresponding measurement of the incisors. Looking at the crown from a labial or palatal direction, the functional form of the crown, as both a penetrating and incising organ, may be observed in the single cusp from which it derives its name. The cusp, which is formed at the expense of the cutting-edge, divides this latter margin into two distinct portions—the mesial cutting-edge and the distal cutting-edge.

The Labial Surface of the Crown (Fig. 64).—The contour of this surface is that of a broken circle more or less perfectly drawn. It is bounded by five margins—mesial, distal, cervical, mesial-incisive, and distal-incisive. The *mesial margin* is rounded from labial to mesial, and slightly convex from the cutting-edge to the cervical line. The *distal margin* is also rounded from labial to distal, presents a greater convexity, and is somewhat shorter from the cutting-edge to the cervical line than the mesial margin. By a continuation and final union of these two lateral margins the *cervical margin* of the surface is formed, while by their union with the cutting-edges the mesial and distal angles of the crown are established. The *mesial-incisive margin* is usually slightly concave near its center, although in some instances it is convex. The *distal-incisive margin* responds to the same description, although the concavity, when present, is nearest the point of the cusp. From the summit of the cusp these two margins slope away to join the mesial and distal angles, the distal incline being about one-fourth longer than the mesial. This surface is generally of greater longitudinal than transverse extent, its greatest mesiodistal diameter being from angle to angle, or at a point immediately above them. The surface is convex in every direction, and is marked by a central longitudinal ridge, usually well defined—*the labial ridge*. Beginning at the summit of the cusp, this ridge is more or less contracted laterally, but as it passes over the surface in the direction of the root it becomes broadened and flattened, and gradually disappears in the cervical portion. Upon either side of this ridge are the *labial grooves*, well defined at their beginning, but which gradually blend into the surface of the crown as they pass rootward. In some instances these grooves are so strongly defined as to form a decided ridge upon the mesial and distal margins of the surface ;—these are the *labial marginal ridges*.

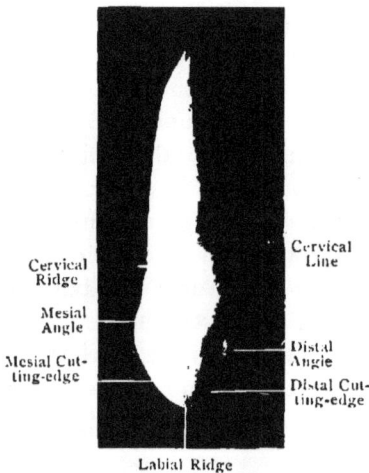

Cervical Ridge
Mesial Angle
Mesial Cutting-edge
Cervical Line
Distal Angle
Distal Cutting-edge
Labial Ridge

FIG. 64.—SUPERIOR CUSPID, LABIAL SURFACE.

The labial ridge and the two labial grooves mark the developmental lines of the crown, the former resulting from the middle lobe, which in this tooth is much the largest of the three, while the latter denotes the line of junction between the middle and the lateral lobes.

The Palatal Surface of the Crown (Fig. 65).—This surface presents nearly the same general outlines as the labial, with the exception of the cervical portion, which is more constricted, tending to produce an oblong or egg-shape. It usually abounds in well-defined ridges and depressions, giving to the tooth a rugged and strong appearance. There is but little general concavity to the surface in passing from the point of the cusp to the root, while it may be flat, concave, or convex. As in the incisors, the margins of this surface are formed by three marginal ridges and by the cutting-edge. The *mesial-marginal ridge* is commonly a well-defined fold of enamel, beginning at the mesial angle and passing upward in the direction of the root, where it unites with the cervical-marginal ridge. It is sometimes quite narrow and rather sharply outlined ; at others, it extends well toward the center of the surface in the form of a well-rounded fold. The *distal-marginal ridge*, which is somewhat shorter than the mesial, begins at the distal angle and passes rootward to meet the cervical ridge. It is well rounded in every direction, but seldom so well produced as the mesial. The *cervical-marginal ridge*, which is formed by a continuation or union of the two former, nearly always partakes of their nature, except when broken by the presence of a cuspule, which

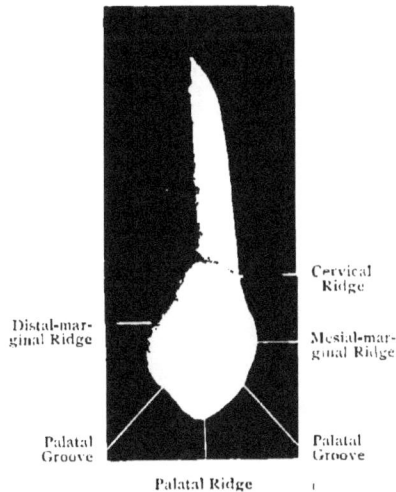

Fig. 65.—Superior Cuspid, Palatal Surface.

Labels: Cervical Ridge; Distal-marginal Ridge; Mesial-marginal Ridge; Palatal Groove; Palatal Groove; Palatal Ridge.

is frequently found upon this tooth (Fig. 68). This small cusp of enamel may be bounded on one or both sides by a fissure, which often extends well under the cervical-marginal ridge, and sometimes completely separates it from the two lateral ridges. Passing through the center of the surface from the summit of the cusp to the base of the cervical-marginal ridge is the *palatal ridge*, which corresponds to the labial ridge of the labial surface. This ridge is usually well produced at or near the point of the cusp, and may continue so throughout, but most frequently becomes reduced in size near the center of the surface. Between this ridge and the mesial- and distal-marginal ridges are two longitudinal depressions—the *palatal grooves*.

Mesial Surface of the Crown (Fig. 66).—In general outline this surface resembles that of the central incisor, excepting that the wedge shape which it describes is more heavily set and blunt, with the surface extending beyond the base of the cone, in the direction of the root, to the extent of about one-third of its entire length. In some cases the base of the cone will be on a line with the cervical margin. The lower two-thirds of the surface, or that nearest the mesial angle, is convex in every direction ; this convexity gradually disappears as the center is approached, beyond which point it is much flattened, usually ending in a slight concavity at the cervical margin. When looking directly upon this surface, its margins will be found within the profile lines, these being represented by the labial ridge anteriorly, and by the palatal and cervical ridges posteriorly. The *margins*, three in number, are the *labial*, which is well rounded and poorly defined ; the *palatal*, more or less distinctly outlined and somewhat irregular ; and the *cervical*, which is represented by the extent of the enamel covering of the crown ; this latter margin being concave in the direction of the root. The most prominent point of this surface serves as a point of contact for the distal surface of the lateral incisor, the extent of contact being much influenced by the type of tooth, but in the cuspid this is usually a single spot rather than an extent of surface.

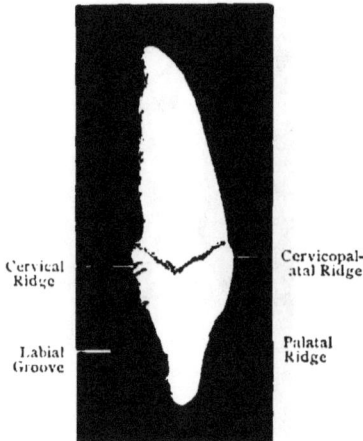

Fig. 66.—Right Superior Cuspid, Mesial Surface.

(labels: Cervical Ridge; Labial Groove; Cervicopalatal Ridge; Palatal Ridge)

The Distal Surface of the Crown (Fig. 67).—This surface in many respects is similar to the mesial, particularly in its general outline. The extent of surface is somewhat less and the

convexity much more marked than that of the mesial surface.
The position of the distal angle, which is the lower boundary
of the surface, being much nearer the cervical line, makes this
surface about one-third shorter than the mesial surface. The
lateral margins of the surface, which are also within the profile
lines, differ from those of the mesial in being more clearly defined. The *cervical margin* differs from that of the mesial surface by having a concavity with much less depth. As stated
above, the surface is decidedly more convex than the mesial,
the point of contact for the mesial surface of the first bicuspid
being almost in the center
of the surface. Near the
cervical margin the surface
is inclined to flatness, and
frequently concave.

The *Cutting-edge, or
Cusp.*—As inferred in the
beginning of this description, the cuspid tooth is
both an incising and a penetrating organ, the latter
function being provided for
by the presence of the single cusp, which divides the
cutting-edge into an anterior or mesial portion and a
posterior or distal portion.
The *mesial cutting-edge*

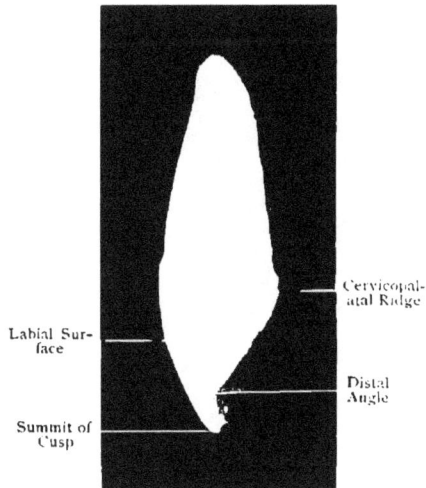

Cervicopal-
atal Ridge

Labial Sur-
face

Distal
Angle

Summit of
Cusp

FIG. 67.—LEFT SUPERIOR CUSPID, DISTAL
SURFACE.

begins at the summit of the cusp and slopes away to meet
the mesial angle, which it assists in forming. The outline of
this edge is usually gracefully curved and unbroken unless
permanently crossed by the labial groove. The *distal cutting-edge* is generally somewhat longer than the mesial.
Immediately after leaving the summit of the cusp it may be
slightly concave; but beyond this point it is well rounded, until
it reaches the distal angle, into which it gradually disappears.
This edge is also frequently broken by the labial groove. In
its entirety the cutting-edge is subject to the same variations as

those of the incisors—*i. e.*, it may be thin and sharp, or it may be thick and blunt.

The Cusp.—The single cusp from which this tooth derives its name is formed by the union of the labial ridge, the palatal ridge, and the mesial and distal cutting-edges. The summit of the cusp is constant in its position, always being in a direct line with the long axis of the tooth, whether it be viewed from a mesial or a palatal direction.

The Cervical Line.—To describe this fully would be to repeat what has already been said in connection with the incisor teeth. This enamel margin differs in one particular only from that of the incisors, and that variation is not a constant one—the palatal portion is frequently extended in the direction of the root, producing a short, positive curve at that point.

The Angles of the Crown.—Owing to the rounded nature of the majority of cuspid crowns, the term angle, as applied to its free extremities, is almost a misnomer, and can only be considered as assisting in description. The mesial angle, which is formed by the union of the marginal ridges of the labial and palatal surfaces with the mesial cutting-edge, is

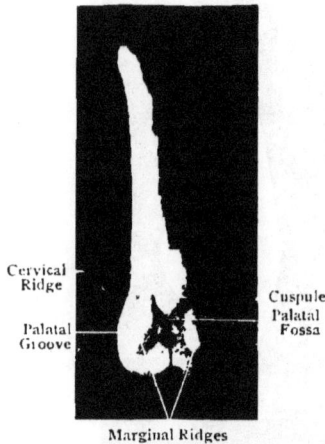

Cervical Ridge

Palatal Groove

Cuspule Palatal Fossa

Marginal Ridges

FIG. 68.—SUPERIOR CUSPID, PALATAL SURFACE, STRONGLY DEVELOPED.

seldom a well-produced angle, usually being rounded in every direction. The distal angle, which is formed in a manner similar to the mesial, is somewhat more deserving of the name, both the labial- and palatal-marginal ridges frequently presenting angularity. The position of the distal angle is usually well toward the center of the crown, and occasionally above this point, and, although it may descend, it is seldom found on a line with the mesial angle.

The Neck of the Superior Cuspid.—This may or may not be a distinctive feature of the cuspid tooth, although when viewed from a labial aspect, the lateral flare or bulging of the crown gives the appearance of a decided constriction between the crown and the root; but, when examined from the mesial surface, this constricted appearance is absent, the contour of the crown passing into that of the root, with the cervical line alone marking the extent of each. The tooth at this point is well rounded anteriorly, flattened laterally, and again rounded posteriorly, the latter forming the segment of a smaller circle than that of the labial surface.

The Root.—This tooth possesses the largest and longest root of any of the teeth, in the latter respect usually exceeding the central incisor by about one-third, and the lateral incisor by one-fourth or more. Like the base of the crown, it is rounded on the labial and palatal surfaces, and is flattened laterally, this form usually being continued throughout its entire length. It gradually diminishes in size from the neck to the apex, and in its entirety forms a perfect cone. C.. the mesial and distal sides it is not only much flattened, but is frequently provided with a longitudinal depression, which is most marked near the center of its length. In some instances this root is possessed of a slight distal curve, which may be gradual from the base to the apex, or it may exist in a more positive way by a sudden distal curve near its apical extremity.

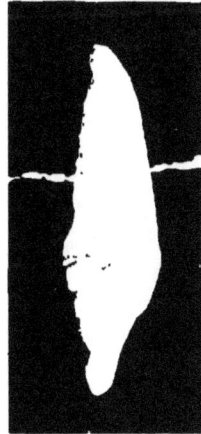

Fig. 69.—Bilious Type, Distal Surface.

Bilious Type (Fig. 69).—The rounded outlines common to the cuspid tooth are less pronounced in this type than in any other, and instead of curves, angles are present. The crown is above the average size, length predominating over breadth, the cusp well formed, and the angles strong. The labial portion is often crossed by a number of transverse ridges near the cervical portion, the labial ridge is bold, as are also the labial-marginal ridges. The mesial and distal surfaces possess no distinguishing features, but the palatal, like the labial, shows the angular nature of the crown in having its margins and ridges squarely set.

A cuspule is more frequently found in this type than any other, and sometimes reaches down to a point corresponding to the transverse center of the crown. The neck is moderately well produced, and the cervical line decidedly V-shaped on its lateral portions, while on the labial and palatal it takes the form of a broken circle. The cutting-edges are rather heavy and square, and are nearly of equal length. In this temperament the cuspid tooth often partakes of the form described in connection with the lateral incisor of the same type—*i. e.*, the absence of the cutting-edge and one or both angles, making the crown a perfect cone.

FIG. 70.—NERVOUS TYPE, LABIAL SUR-FACE.

Nervous Type (Fig. 70).—The crown is of much greater longitudinal than transverse extent, the outlines oval and gracefully formed, and the neck is much constricted from mesial to distal, being made so by the lateral flare of the body of the crown, which is a distinctive feature of this type. The labial ridge is well formed near the summit of the cusp, but usually disappears near the center of the surface. The labial-marginal ridges are seldom present, and the surface in general is convex and smooth. The mesial and distal surfaces show a pronounced convexity near the angles, and often a slight concavity between this point and the cervical line. The palatal surface, while generally showing all the descriptive lines, may be considered smooth ; it is convex from mesial to distal, and slightly concave in the direction of the long axis of the tooth. The cusp is long and penetrating, the distal cutting-edge is much longer than the mesial, and both are inclined to sharpness. The cervical line on the labial and palatal surfaces is deeply arched, frequently giving to the gingival margin a receded appearance.

Sanguineous Type (Fig. 71).—The crown in this type abounds in long curves, the longitudinal and transverse extents are nearly equal, the angles, owing to their circular form, are barely deserving the name, while the cusp and cutting-edges are outlined by one long, oval sweep. The labial surface is prominent and convex, and the developmental grooves are fairly well shown. The mesial and distal surfaces show a moderate general convexity, while the palatal abounds in well-rounded ridges and borders. The constriction forming the neck of the tooth is moderate. The cervical line is in the form of perfectly arched curves, forming on the labial surface the segment of a circle corresponding to the circumference of the crown of the tooth.

Lymphatic Type.—In this the crown is usually greater in its transverse than in its longitudinal measurement; it is lacking in graceful outline, and may best be described as being short, thick, and heavy set. None of the surfaces abound in descriptive lines, although transverse ridges are sometimes present on the labial surface near the cervix. Both the labial and palatal surfaces are convex in every direction, while the mesial and distal are inclined to flatness. The cusp is heavy and blunt, and the cutting-edges, which are nearly of equal length, are thick and dull. The mesial and distal

FIG. 71.—SANGUINEOUS TYPE, DISTAL SURFACE.

angles are well produced. The cervical line approaches near to a direct line encircling the neck of the tooth, the segmental form on the labial surface being that of a much larger circle than any of those previously described. The neck is thick and heavy, and the roots generally short.

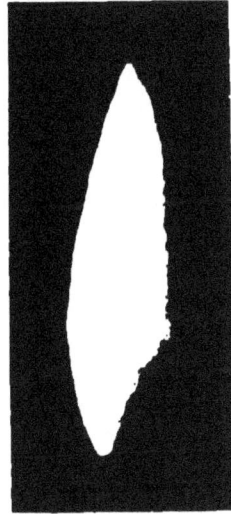

SUPERIOR FIRST BICUSPID.

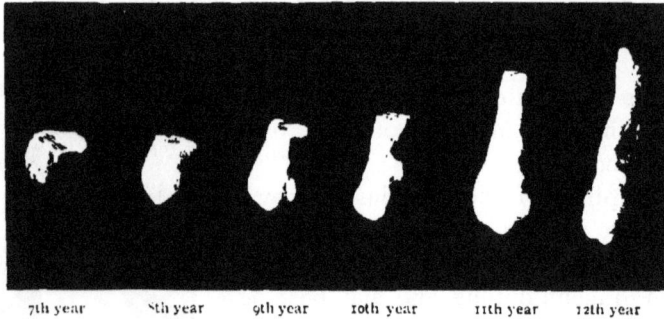

7th year   8th year   9th year   10th year   11th year   12th year

Fig. 72.

CALCIFICATION BEGINS, FROM FOUR CENTERS, ABOUT THE FOURTH YEAR.
CALCIFICATION COMPLETED, ELEVENTH TO TWELFTH YEAR.
ERUPTED, TENTH TO ELEVENTH YEAR.
AVERAGE LENGTH OF CROWN, .32.
AVERAGE LENGTH OF ROOT, .48.
AVERAGE LENGTH OVER ALL, .80.

This tooth, although presenting a crown of vastly different contour, is developed by a process almost identical with that of the incisors and cuspids. As the name implies, it is made up of two cusps, one forming the buccal and the other the palatal half of the crown. Calcification in the buccal cusp begins about the fourth year, the central lobe first receiving the lime salts. During the following year the two lateral lobes begin to calcify, soon followed by a union of the three, thus completing the margins and summit of the cusp. Unlike the incisors, but corresponding to the cuspid, the middle lobe is much the largest of the three, frequently forcing the developmental (buccal) grooves well toward the angles of the crown. The development of the palatal cusp corresponds to the development of the cervical ridge on the incisors and cuspids, calcification taking place in a single lobe and continuing with somewhat greater activity, until a cusp frequently as large as the buccal has been established. Between the fifth and sixth year union between the two cusps takes place, the line of confluence being permanently recorded by a well-defined groove, which traverses the crown from mesial to distal. This groove, although forced to occupy a different position, corresponds with the palatal groove of the incisors and cuspids. After the union of the cusps the process of calcification is continued into the body of the crown, and by the seventh year it is more than half completed. The eighth year usually finds the crown fully formed and the base of the root or roots outlined. As this tooth is generally provided with two roots, the first indication of bifurcation will be observed between the eighth and ninth year by a filling-in near the center of the mesial and distal walls, which finally become united by a thin septum of dentine. After this period the roots calcify separately, and by the middle of the ninth year about one-third of their length is to be observed. During the following year, or at about the time of eruption, the development of the roots has extended to about three-fourths of their completed length, and between the eleventh and twelfth year, or at a time corresponding to that of the crown assuming its final position in the arch, calcification is completed and the apical foramen established (Fig. 72).

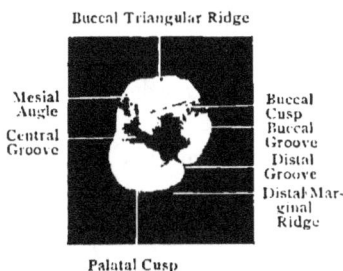

Fig. 73.—SUPERIOR FIRST BICUSPID, OCCLUSAL SURFACE.

The crown of the superior first bicuspid presents for examination five surfaces—buccal, palatal, mesial, distal, and occlusal. In general, the contour of the crown is irregularly quadrilateral, being about one-third greater in its buccopalatal measurement than from mesial to distal. It is somewhat flattened from mesial to distal, but rounded on its buccal and palatal surfaces.

The Occlusal Surface of the Crown (Fig. 73).—The contour of the crown is best observed by a view of this surface, which may be described as trapezoidal or irregularly quadrilateral in form. The four margins of the surface are those which represent the four lateral surfaces,—the buccal, palatal, mesial, and distal,—the latter two being in the form of well-

defined ridges—the *mesial-* and *distal-marginal ridges*. The
buccal margin is formed by the mesial and distal inclines of the
buccal cusp; it has a slight buccal convexity, and at its union
with the proximate surfaces assists in forming the mesial and
distal angles of the crown. The distal half of this margin is
usually somewhat longer than the mesial, and the distal angle
is less pronounced than the mesial. The palatal margin presents
a much greater convexity than the buccal, but the curve formed
is the segment of a much smaller circle. As in the buccal margin, the distal half of this margin is the longest, but unlike the
former, its free extremities pass into the mesial and distal margins without producing angles. The mesial- and distal-marginal
ridges are strong folds of enamel which arise from the mesial
and distal angles and converge slightly as they pass to the palatal, where they are gradually lost in the palatal margin.

The Cusps (Fig. 73).—These are two in number, and are
named, in accordance with their location, buccal and palatal.
The *buccal cusp* is the largest and longest of the two. From
the summit of this cusp four ridges descend—one in a mesial
direction, forming the mesial cutting-edge of the crown ; one in a
distal direction, forming the distal cutting-edge; one to the buccal surface, the *buccal ridge ;* and a fourth, the *buccal triangular
ridge*, descends the central incline. The mesial and distal ridges
enter into the formation of the mesial and distal angles at their
extremities; the latter is slightly longer than the former, and both
are frequently broken near the center by the grooves of development—the *buccal grooves*. The buccal ridge may be well
developed and extend almost to the cervical line, or it may be
slight and disappear near the center of the surface. The buccal
triangular ridge usually ends somewhat abruptly in the central
groove, but in some instances it is continued and joins a similar
ridge from the palatal cusp, this union forming the *transverse
ridge*. The triangular ridge often bifurcates near the center of
its incline, and continues in two distinct but smaller ridges. The
palatal cusp is much less angular than the buccal; the apex is
usually rounded, while the descending ridges are generally three
in number instead of four. The mesial and distal ridges are
nearly of the same length, and pass without interruption into

the mesial- and distal-marginal ridges. The triangular ridge is less clearly defined than its fellow of the buccal cusp, and it is not unusual for it to be entirely wanting. The palatal aspect of the cusp is smooth and rounded, presenting nothing in the form of a ridge in correspondence with the buccal ridge of the buccal cusp. Like the incisors and cuspids, the summits of these cusps are usually in a direct line with the long axis of the tooth.

The *developmental grooves*, all of which are observed upon the occlusal surface, are the central, mesial, distal, two triangular, and two buccal. The central groove is the most marked, is deeply sulcate, and extends through the center of the surface from mesial to distal, ending just within the two marginal ridges in two irregularly formed depressions or pits—the *mesial* and *distal pits*.

This groove marks the line of union between the buccal and palatal lobes. The mesial and distal grooves are not always well defined, but may usually be observed as fine lines passing over the central portion of the mesial- and distal-marginal ridges. The mesial- and distal-triangular grooves begin in the mesial and distal pits, and pass in the direction of the mesial and distal angles, where they are either lost, or may be traced as slight depressions passing over the buccal ridges near the angles, and they may further continue over the buccal surface in the direction of the root. These two grooves, together with the mesial and distal above referred to, form the outlines of the mesiobuccal and distobuccal developmental lobes. The buccal grooves will be described in connection with the buccal surface. Supplemental grooves are seldom found in connection with the buccal half of the occlusal surface, but are occasionally present on the central incline of the palatal cusp.

The Buccal Surface of the Crown (Fig. 74).—In many respects this surface resembles the corresponding or labial surface of the cuspid tooth. It is bounded by four margins—occlusal, mesial, distal, and cervical. The occlusal half of the surface is formed of the buccal cusp, and is cone-shaped, while the cervical half is irregularly quadrilateral in form. The extent of surface from the cervical line to the point of the cusp is usually about one-third greater than the greatest mesiodistal diameter, which is represented by a line drawn from the mesial to the distal angle. The form shown is that of a general convexity, the summit of which is surmounted by a longitudinal ridge—the

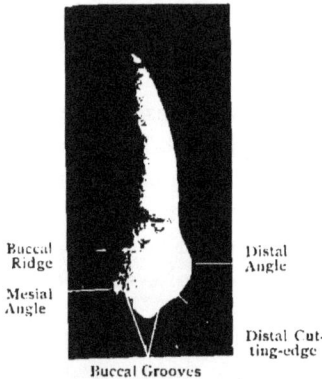

Buccal Ridge

Mesial Angle

Distal Angle

Distal Cutting-edge

Buccal Grooves

FIG. 74.—SUPERIOR FIRST BICUSPID, BUCCAL SURFACE.

buccal ridge. This ridge, which is formed from the central developmental lobe, is most pronounced near the occlusal margin, and gradually disappears near the center of the surface. Upon either side of the buccal ridge are two grooves —the *buccal grooves*—which denote the line of union between the central and the two lateral lobes, and beyond these are the angles of the crown. The buccal ridge springs from the buccal cusp, the summit of which is generally in a direct line with the long axis of the tooth ; when there is a deviation from this the summit is usually thrown a little to the mesial, resulting in a reduction of the length of the mesial cutting-edge.

As previously stated, the greatest mesiodistal diameter of the surface is on a line with the angles of the crown ; this measurement is much reduced at the cervical line, so that the point of contact with adjoining teeth is thrown near the occlusal margins of the crown. This variation in the transverse measurement also results in what is commonly referred to as the " bell shape " of the crown. The cervical margin of the surface is fairly well arched,

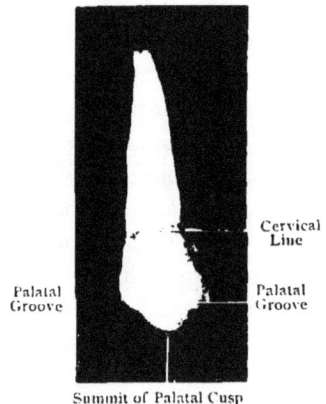

Cervical Line

Palatal Groove

Palatal Groove

Summit of Palatal Cusp

FIG. 75.—SUPERIOR FIRST BICUSPID, PALATAL SURFACE.

but seldom to such a degree as the corresponding margin upon the incisors and cuspids.

The Palatal Surface of the Crown (Fig. 75).—This surface is smooth and decidedly convex, the absence of strongly developed grooves and ridges contributing to the former fact. Like the buccal surface, its greatest transverse measurement is at the base of the cusp in which it terminates. The extent of the surface is about one-third less than that of the buccal, and its occluding and cervical margins are alone well defined, the mesiodistal convexity passing so gradually into these respective surfaces that a positive line of distinction can scarcely be recognized. In passing from the cervical line to the occlusal margin, the surface is rapidly carried toward the center of the crown.

The cervical margin of this surface is usually in the form of a direct line encircling the neck, but occasionally presents a slight concavity in the direction of the root.

The Mesial Surface of the Crown (Fig. 76).—This surface of the crown has three of its borders well defined ; these are the buccal, the occlusal, and the cervical, the remaining or palatal margin passing so gradually into the palatal surface that

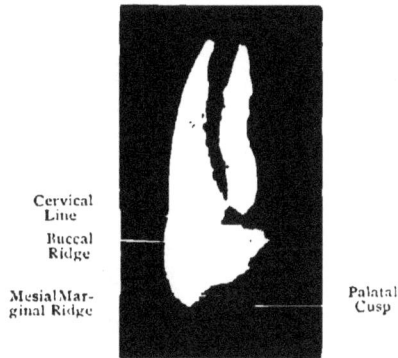

Cervical Line

Buccal Ridge

MesialMarginal Ridge

Palatal Cusp

FIG. 76.—SUPERIOR FIRST BICUSPID, MESIAL SURFACE.

no positive line of demarcation can be given. The buccal margin extends from the mesial angle to the cervical line, and invariably presents a slight buccal inclination, thus increasing the width of the surface on its cervical portion. The occlusal margin is formed by the mesial-marginal ridge, and by a portion of the ridge descending from the palatal cusp. It is irregularly V-shaped, and in many instances is broken in the center by the mesial groove. The cervical margin differs from those of the incisors and cuspids, nearly always being in the form of a straight line from buccal to palatal. The surface in general is flattened, but shows a slight general convexity near the occlusal margin, and frequently a slight concavity immediately below the cervical

line, this form placing the point of contact with the distal surface
of the cuspid near the occlusal margin. The surface is occasionally divided into a buccal and a palatal portion by the mesial
groove, which may extend to the cervical line, but which generally disappears near the center of the surface. The buccal half
of the surface which is formed from the mesial developmental
lobe is inclined to angularity, while the palatal half is decidedly
rounded, particularly in the direction of the occlusal margin.

The Distal Surface of the Crown (Fig. 77).—In general,
this surface resembles the mesial, being flattened and bounded
by three more or less distinct
margins. The slight buccopalatal convexity is not confined to the occlusal portion
of the surface, but is inclined
to extend to the cervical margin, in this particular being at
variance with the mesial surface. This surface passes
into the palatal by a much
longer curve than that shown
on the mesial surface.

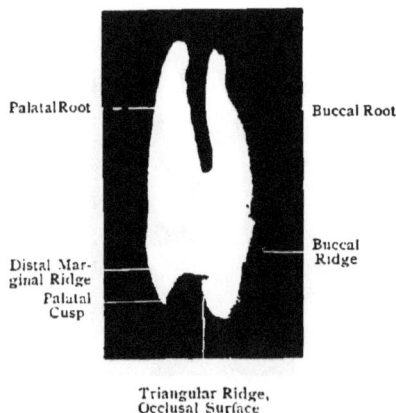

FIG. 77.—SUPERIOR FIRST BICUSPID, DISTAL
SURFACE.

The Angles of the
Crown.—These are two in
number and, as in the teeth
previously described, are
named, according to their
location, *mesial* and *distal*. The mesial angle is formed by
the union of the mesial-marginal ridge and mesial cutting-
edge. It is primarily the product of the mesial developmental
lobe, and is usually well produced. The distal angle, which is
formed in a like manner, is inclined to be more rounded in its
nature.

The Neck of the Tooth.—In most typal forms the neck
of the first superior bicuspid is well defined, particularly upon
the mesial and distal surfaces. Viewing the tooth from a buccal
aspect, the neck is a distinctive feature, but when studied from
the mesial or distal sides, the constriction is scarcely observed,

this being particularly the case if the tooth has but a single root. In general, the neck partakes of the contour of the crown, being convex on the buccal and palatal, and flattened and frequently slightly concave on the mesial and distal.

The Roots of the Superior First Bicuspid.—This tooth is usually developed with two roots, sometimes with only one, and in rare instances it may have three. When two roots are present, one is above the buccal and the other above the palatal half of the crown, and are named, according to their location, *buccal* and *palatal*. In general form the two roots are quite similar, but the buccal is usually a trifle longer than the pal-

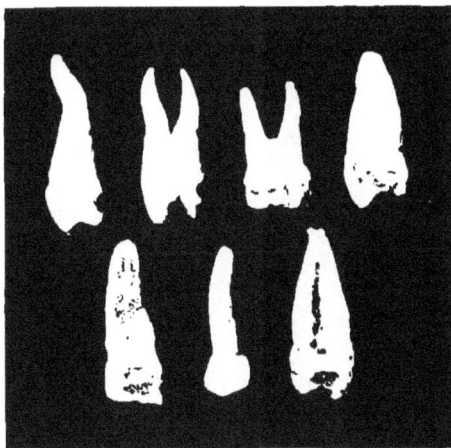

FIG. 78.—TYPES OF BICUSPIDS.

atal. They taper off to a slender apex, and are inclined to curve in various directions near their extremities. The point of bifurcation is generally some distance above the neck, so that the tooth may be said to possess a single root with two branches. Below the bifurcation the root assumes the form of the neck or cervical portion of the crown, but as the bifurcation is approached, the mesial and distal sides present a longitudinal groove, which gradually increases in depth until the single root becomes separated. The curves of the root-branches above referred to are in the buccal branch, first to the buccal and then to the palatal ; while the palatal branch first shows a slight palatal inclination immediately above the point of separation, followed by a gentle buccal curve as the apical end is reached. In some cases the bifurcation begins immediately above the neck of the tooth ; in others it may occur in the apical third ; while a third class is represented by the two roots being united throughout their entire length by a thin septum of dentine and cementum, or, as occasionally happens, by a layer of cementum alone. In this latter instance each root is provided with a distinct canal and foramen. When the tooth has but a single root, it is much flattened from mesial to distal, the flatness being slightly broken by an inclination to convexity. The four surfaces usually converge toward the apical end, which is oblong from buccal to palatal, and generally provided with a slight distal curve. The presence of three roots is so rare that the condition might be classed as a malformation ; but when they do exist, two are usually attached to the buccal and one to the palatal half of the crown, with the point of bifurcation near the neck of the tooth.

Bilious Type (Fig. 79).— The superior first bicuspid of this temperamental type is marked by a crown of moderate length, the neck well pronounced, and the cusps and angles marked by angular outlines.

FIG. 79.—BILIOUS TYPE, DISTAL SURFACE.

The buccal ridge is strongly defined, and the buccal grooves, which extend well up on the buccal surface, cross the mesial and distal cutting-edges, separating them into two distinct parts. The cusps are long and penetrating, and are nearly of equal length, assisting to form the firm and well-locked occlusion common to this type. The mesial and distal surfaces are nearly parallel with each other ; they are seldom convex, so that the approximating teeth are in contact over an extent of surface rather than a single point. The cervical line is but little curved.

Nervous Type (Fig. 80).—In this temperament the bell-shaped crown is strongly observed, the crown being long and the tooth much constricted at its neck. The extreme length of the buccal cusp, and the marked cervical constriction, produces an appearance in the buccal surface resembling the labial surface of the cuspid tooth. The developmental grooves are finely outlined, and the cusps long and penetrating, usually being more pronounced than in any other class. The cutting-edges are sharp and inclined to angularity; the mesial and distal surfaces are convex near their occlusal margins, but near the cervical line a pronounced concavity is observed, which is continued upon the corresponding root-surfaces. This formation forces the point of contact with adjoining teeth well toward the occlusal surface, and results in an extensive V-shaped interproximate space. The cervical line is sharply and gracefully formed, the curvature being well arched.

Sanguineous Type.—The typical superior first bicuspid of this class is provided with a crown well proportioned, its length being somewhat greater than its breadth, but about equal to its buccopalatal measurement. The buccal surface is seldom broken by the buccal grooves, and is strongly convex in every direction. The palatal surface is much more rounded than the same surface of other types. The mesial and distal surfaces are

FIG. 80.—NERV-OUS TYPE, BUCCAL SURFACE.

usually smoothly convex, with an occasional slight concavity immediately below the cervical line. Upon the occlusal surface the grooves are rounded and obscure, rather than sharp and well defined, and the cusps, much less pronounced than in either of the types previously described, are rounded and smooth ; this latter fact is particularly true of the palatal cusp, which is usually much smaller than the buccal. The cutting-edges of the buccal cusp are scarcely deserving of the name, being broad and rounded throughout. The form of the mesial and distal surfaces above described provides for the point of contact near the center of each surface, leaving a slight interproximate space both above and below this point.

Lymphatic Type.—An examination of the superior first bicuspid of the lymphatic type results in finding a tooth vastly different from any of those previously described. The length of the crown from the cervical line to the point of the cusp is less than either the mesiodistal or buccopalatal measurements. In general appearance it is lacking in symmetry, or poorly proportioned. The buccal surface presents a gradual convexity from mesial to distal, and seldom has the buccal ridge well developed. The palatal surface is smoothly convex and passes off into the palatal root without the interposition of a decided neck. The mesial and distal surfaces are flattened and sparingly convex, and are nearly parallel with each other, so that the contact with adjoining teeth is inclined to be distributed over the entire surface, leaving little or no interproximate space. The cusps are short, flat, and rounded, and the occlusal surface much flattened in general, corresponding with the nature of the occlusion, which is loose and wandering. The developmental grooves and ridges are fairly well shown, while the cutting-edges and angles of the crown are smooth and rounded. The neck is less pronounced in this type than in any other, the curvature of the cervical line is very slight, the root is short and heavy set, frequently passing well up toward the apex before bifurcating.

SUPERIOR SECOND BICUSPID.

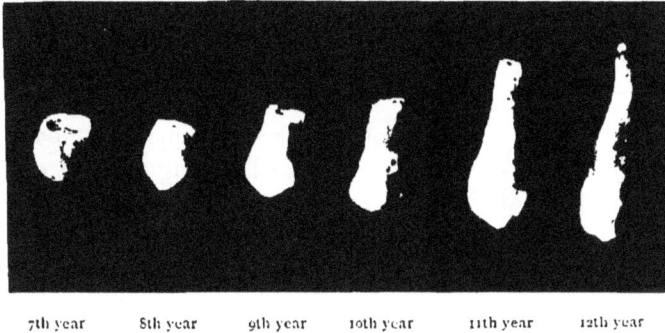

7th year   8th year   9th year   10th year   11th year   12th year

FIG. 81.

CALCIFICATION BEGINS, FROM FOUR CENTERS, ABOUT THE FIFTH YEAR.
CALCIFICATION COMPLETED, ELEVENTH TO TWELFTH YEAR.
ERUPTED, ELEVENTH TO TWELFTH YEAR.
AVERAGE LENGTH OF CROWN, .29.
AVERAGE LENGTH OF ROOT, .55.
AVERAGE LENGTH OVER ALL, .84.

The process of development in this tooth is identical with that of the first bicuspid, calcification in the buccal half of the crown taking place in one central and two lateral lobes, while the palatal half is developed from a single center. The calcifying process is about one year later than that in the first bicuspid, the summit of the buccal cusp receiving its lime salts about the beginning of the fifth year. During the following six months calcification in the lateral lobes, and also in the palatal lobe begins. By the sixth year the occlusal surface and a portion of the crown is completed by a union of the various lobes, and at seven years the crown is calcified for more than two-thirds of its completed length. Between the eighth and ninth year the contour of the crown is established, and the neck of the tooth and outline of the root-base formed. At the tenth year about one-third of the root-length is formed, and during the following year about $\frac{1}{2}$ of an inch is added to it. By the eleventh or twelfth year calcification is completed and the apical foramen established (Fig. 81).

This tooth so closely resembles the first bicuspid that a description in detail will be unnecessary ; there are, however, a few minor points which are at variance and must be described in order to distinguish one from the other. In general, the tooth is a trifle smaller than the first bicuspid, the cusps are somewhat shorter, and the various ridges less distinct. A distinguishing feature of the occlusal surface is found in the diminished length of the central groove (Fig. 82). This groove, as observed in the first bicuspid, extends from mesial to distal for fully three-fourths of the entire distance ; but in the second bicuspid it is diminished by one-third, being thus reduced by the broadened marginal ridges, which force the mesial and distal pits well toward the center of the surface. It is not uncommon to find the triangular grooves joining the central groove directly in the center of the surface, forming a central pit from which may radiate numerous small supplemental grooves and ridges. The summits of both the buccal and palatal cusps are nearer to the mesial than to the distal surface, thus increasing the length of the ridges which descend from them in a distal direction, and decreasing those which pass to the mesial. The buccal surface presents a greater convexity than that of the first bicuspid ; the buccal grooves are usually shallow depressions and are frequently entirely wanting, thus giving the buccal ridge the appearance of extending its margins to the angles of the crown. Unlike the first bicuspid, the mesiodistal diameter of the mesial surface

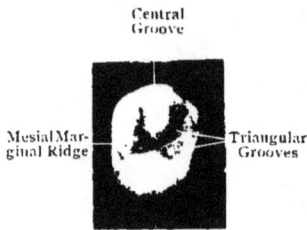

Central Groove — Mesial Marginal Ridge — Triangular Grooves

FIG. 82.—SECOND SUPERIOR BICUSPID, OCCLUSAL SURFACE.

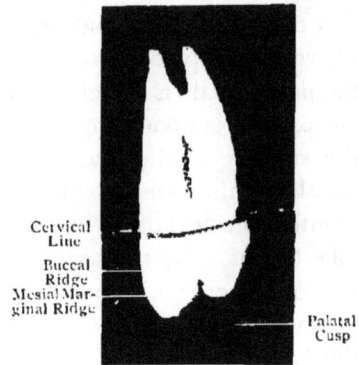

Cervical Line — Buccal Ridge — Mesial Marginal Ridge — Palatal Cusp

FIG. 83.—SUPERIOR SECOND BICUSPID, MESIAL SURFACE. MOST COMMON FORM.

is but little more than the same measurements on the palatal surface. The neck of the tooth is not quite so pronounced as that of the first bicuspid, thus giving less of the bell shape to the crown. One very important difference between this tooth and the first bicuspid is in the root-formation. We have seen that the first bicuspid is generally provided with two roots, while in the second bicuspid a single root is usually present. Like the first bicuspid, there are exceptions to this, the tooth sometimes being provided with two, and in rare in-

stances with three, roots. When the single root is present, it partakes of the form of the crown at its base, being well rounded on the buccal and palatal portions and much flattened on the mesial and distal. The mesiodistal diameter of the root at its base is only about one-third that of the bucco-palatal measurement, this proportionate size continuing throughout its entire length. In passing from the base to the apex the root is gradually diminished in size, finally ending somewhat abruptly in an oblong extremity. In some instances the apical end is round and pointed, resembling the apex of the incisors and cuspids, but when thus formed the root is usually

curved near its apical third and somewhat extended in length. The mesial and distal surfaces are provided with a well-defined longitudinal concavity, extending from the cervical margin to the apex and dividing the root into a buccal and palatal portion. This depression is often so decided that the contour of the single root is almost lost, and in its place the appearance is that of two roots similar to those described in the first bicuspid. The length of the root is usually a little greater than that of the first bicus-pid, but the crown being a trifle shorter, results in producing a tooth the entire length of which is about equal to that of the first bicuspid.

SUPERIOR FIRST MOLAR.

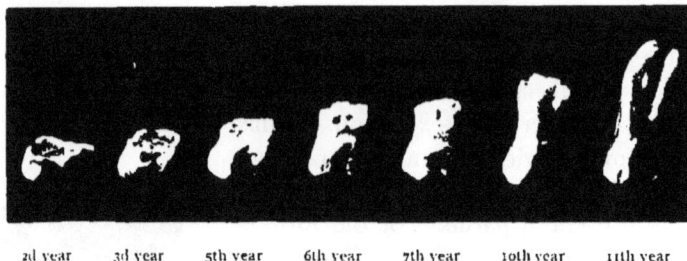

2d year   3d year   5th year   6th year   7th year   10th year   11th year

FIG. 85.

CALCIFICATION BEGINS, FROM FOUR CENTERS, ABOUT ONE MONTH BEFORE BIRTH.
CALCIFICATION COMPLETED, NINTH TO TENTH YEAR.
ERUPTED, SIXTH TO SEVENTH YEAR.
AVERAGE LENGTH OF CROWN, .30.
AVERAGE LENGTH OF ROOT, .51.
AVERAGE LENGTH OVER ALL, .81.

This tooth being the first of the permanent organs to erupt, it precedes all others in the process of calcification, beginning to receive its lime salts as early as the eighth fetal month. The form of the crown being so entirely different from those previously described, embodies a developmental process which is also different, four distinct lobes being present, one for each cusp, these making their appearance during the first year after birth, closely followed by a completion of the occlusal surface by the union of the free calcifying margins, these lines of union being finally represented by the developmental grooves of the occlusal surface. After the completion of this surface, calcification proceeds in the direction of the base of the crown, and at the beginning of the third year about two-thirds of the crown is formed. During the fifth year the contour of the crown is completed, and at the beginning of the eruptive period, or about the sixth year, the developmental process has extended to the base of the roots, and an effort at trifurcation begun. At seven years the three roots with which the tooth is provided are branching out, each into its own socket, subsequent development in each being continued as a separate and distinct process. The tenth year more than half completes the calcifying process in the roots, and at the beginning of the eleventh year the apical foramina are established (Fig. 85). Like the first bicuspid, the crown of this tooth presents for examination five surfaces—occlusal, buccal, palatal, mesial, and distal. In general contour it is irregularly quadrilateral, with the angles of the square more or less rounded, two of its sides convex and two flattened or slightly concave. The length of the crown from the cervical line to the summits of the cusps is about equal to, or slightly less than, its mesiodistal diameter, while the buccopalatal measurement is usually a trifle greater than the mesiodistal.

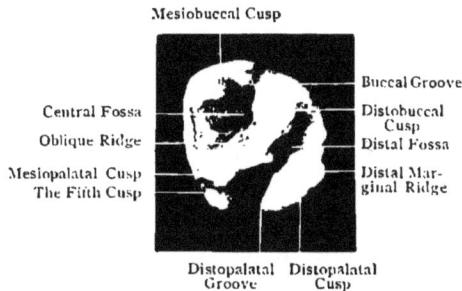

Mesiobuccal Cusp

Central Fossa
Oblique Ridge
Mesiopalatal Cusp
The Fifth Cusp

Buccal Groove
Distobuccal Cusp
Distal Fossa
Distal Marginal Ridge

Distopalatal Groove   Distopalatal Cusp

FIG. 86.—SUPERIOR FIRST MOLAR, OCCLUSAL SURFACE.

The Occlusal Surface of the Crown (Fig. 86). — The coronal outline of this tooth is best studied when looking directly upon this surface, which shows the two convex sides above referred to, represented by the buccal and palatal margins, with the mesial and distal margins more or less flattened. The surface is bounded by these four margins, which are nearly of equal length, the angles formed by their union being more or less rounded, two of which, the mesiobuccal and the distopalatal, are acute angles, while the mesiopalatal and distobuccal are obtuse angles. The surface is divided into four developmental portions—the mesiobuccal, distobuccal, mesiopalatal, and distopalatal. Each one of these parts is surmounted by a well-defined point or cusp, which likewise is named in accordance with its location. These various parts are separated from one another by four developmental grooves—the mesial, the buccal, the distal, and the distopalatal. In the center of the triangle formed by the central incline of the mesiobuccal, distobuccal, and mesiopalatal cusps is a deep depression,—the *central fossa*,—while near the distal margin is a somewhat similar depression—the *distal fossa*. Traversing the surface in various directions are a number of ridges and supplemental grooves, each of which will be described in turn.

The Marginal Ridges of the Occlusal Surface.—These are four in number—the mesial, distal, buccal, and palatal. The *mesial-marginal ridge* is a well-pronounced elevation of enamel which passes from the mesiobuccal angle to the mesiopalatal angle. It is slightly concave in the direction of the root, and is broken near the center of its concavity by the mesial groove, upon either side of which are frequently found one or two small points or tubercles, which are formed either by a division of the mesial developmental groove, or by one or more supplemental grooves. These grooves pass over the ridge and are continued for a short distance on the mesial surface. Descending from the mesiobuccal cusp, the ridge passes in a palatal direction to meet the mesiopalatal cusp, and in so doing has a slight distal inclination until the mesial groove is reached, after passing which it makes a sweeping distal curve and is lost in the palatal margin. This ridge marks the line of junction between the occlusal surface and the mesial surface. The *distal-marginal ridge* in some respects resembles the mesial just described, being concave and ascending in a buccal and palatal direction, with a somewhat rounded outline, to the summits of the distobuccal and distopalatal cusps. The depth of the concavity is usually greater than that of the mesial margin, and is frequently crossed near the center by the distopalatal groove, frequently so marked as to produce a V-shape to the center of the margin. There are occasionally found upon either side of this central groove one or more small tubercles, corresponding to those of the mesial ridge, but they are less frequent and less pronounced. This ridge forms the line of demarcation between the occlusal surface and the distal surface. The *buccal marginal ridge* begins at the mesiobuccal angle, and gradually ascends to the summit of the mesiobuccal cusp, from which it afterward descends in a distal direction to the buccal groove; continuing, it again ascends the distobuccal cusp, after descending from which it ends in the distobuccal angle. The nature of this ridge is a series of cutting-edges, giving to the cusps their angular nature. Besides the buccal groove, which makes a decided break in the center of its course, the ridge is frequently crossed by numerous small supplemental grooves occurring in various locations and forming a series of minute tubercles; this latter condition is most frequently present in young teeth, and is soon obliterated by wear. The course of this ridge is not that of a direct line from mesial to distal, but in its ascent of the mesiobuccal cusp it is inclined to the buccal; in its descent it presents a corresponding return to the palatal, and the same variations are observed in passing over the distal cusp. The *palatal-marginal ridge* begins at the mesiopalatal angle of the crown and passes distally to the distopalatal angle, differing from the three previously described by being heavy and rounded in its nature, more irregular in outline, and divided nearest to its distal extremity instead of in the center of its length. From the point of beginning it makes a curved ascent to the summit of the mesiopalatal cusp; descending from this in a distobuccal direction, it divides, one portion of it passing to join the triangular ridge of the distobuccal cusp, the two uniting to form the *oblique ridge*, the other portion continuing in the direction of the distopalatal cusp, before reaching the base of which it is broken by the distopalatal groove. From this groove the ridge makes a sudden and direct ascent to the summit of the distopalatal cusp, after passing which it gradually descends in a long curve to join the distal-marginal ridge. Like the buccal ridge, it is frequently crossed by numerous supplemental grooves. The ridge forms the palatal margin of the occlusal surface, and gives to the cusps their angularity.

The Cusps (Fig. 86).—These are four in number[*] (see also Fig. 87)—the mesiobuccal, distobuccal, mesiopalatal, and distopalatal.

The Mesiobuccal Cusp (Fig. 86).—In extent of surface this is usually the largest cusp, although it is sometimes exceeded by the mesiopalatal. From the summit of the cusp three ridges descend—the *buccal ridge* to the buccal surface, the *buccal marginal ridge* making a double descent, and the *mesiobuccal triangular ridge*, the latter descending the central incline and ending in the central fossa. The mesial base of the cusp is frequently crossed by one or more supplemental grooves, which begin at the mesial margin and pass in the direction of the central fossa. The central slope of the cusp contributes to the formation of the central fossa, its extent in this direction being controlled by the mesial and buccal grooves, which together form the mesiobuccal triangular groove.

The Distobuccal Cusp (Fig. 86).—This cusp is frequently the smallest in extent of surface, but is usually longer and more pointed than the others. Like the mesiobuccal cusp, three ridges descend from it—the *buccal ridge* to the buccal surface, two which spring from the buccal-marginal ridge, and the *distobuccal triangular ridge*, which descends obliquely toward the distal center of the surface and joins a similar ridge (previously described) from the mesiopalatal cusp, the two forming the oblique ridge. The mesial portion of the base of this cusp assists in forming the central fossa, while a portion of the distal contributes to the formation of the distal fossa. The inner boundary of the cusp is formed by the buccal groove, the distal groove, and by a portion of the distopalatal groove.

The Mesiopalatal Cusp (Fig. 86).—As above stated, this cusp is frequently the largest in extent of surface, and is somewhat rounded, with its summit poorly defined. The ridges which descend from it correspond in name and number to those of the buccal cusps, the palatal-marginal ridge making a double descent, the mesiopalatal ridge descending to the palatal surface, while the central incline is marked by the *mesiopalatal triangular ridge*, which ends in the central fossa. Toward the mesial portion of the cusp one or more small ridges are frequently present, extending from the marginal ridge to the mesial groove. The distal descent of the marginal ridge is bifurcated, one portion making a sweeping curve and joining the transverse ridge from the buccal cusp, forming the oblique ridge previously referred to, the other portion passing in a distal direction and ending at the distopalatal groove. The central incline of this cusp forms the palatal side of the central fossa, and its boundaries are outlined by the mesial, distal, and distopalatal grooves.

The Distopalatal Cusp (Fig. 86).—This cusp is usually the smallest of the four; it is triangular in outline, with the summit nearest to the mesiopalatal portion. The ridges which descend from this cusp are only two in number, one passing in a mesial direction and forming a portion of the palatal-marginal ridge, the other passing to the distal, with a gradual buccal curve, to join the distal-marginal ridge. Of the two remaining inclines, one looks in a distopalatal direction, presenting a surface which is smooth and rounded; the other, sloping by a broad expanse in a mesiobuccal direction, ending in the distopalatal groove, and also assisting to form the distal fossa. This latter incline is often crossed by small supplemental grooves, which take a winding course from the base to the summit of the incline. The inner margin or outline of the cusp is formed by the distopalatal groove.

The Fifth Cusp (Fig. 87).—Although usually referred to as possessing but four cusps, this tooth is frequently developed with five, the additional lobe being situated on the palatal side of the mesiopalatal cusp, about midway between its summit and the neck of the tooth. When present, it is distinctly separate from the main cusp by a well-developed groove—the *mesiopalatal groove*. Both the cusp and the groove may be more or less developed, the former in some instances assuming dimensions corresponding to that of the distopalatal cusp, and the latter sometimes being as well marked as the distopalatal groove. When thus pronounced, the groove begins near the center of the mesial surface, and passes obliquely toward the summit of the mesiopalatal cusp, before reaching which it makes an abrupt turn rootward, and joins the palatal terminal of the distopalatal groove, this union frequently resulting in a well-defined pit—the *palatal pit*. This cusp, as usually found, is small and apparently without function. When occurring on the tooth

12

of one side, it is always present on the corresponding tooth of
the opposite side.   It is never present on any but the superior
first molar.

The Fossæ and Grooves of the Occlusal Surface (Figs.
86 and 87).—The fossæ are two in number—central and distal.
The *central fossa* occupies a position near the center of the surface, and is formed by the central incline of the mesiobuccal,
distobuccal, and mesiopalatal cusp, which usually give to it a
three-sided form.   Connecting the three sides of the fossa, and
in a measure assisting in its construction, is the mesial-marginal
ridge and the oblique ridge.   The depth of this fossa, as well as
that of the distal, is of course regulated by the length of the

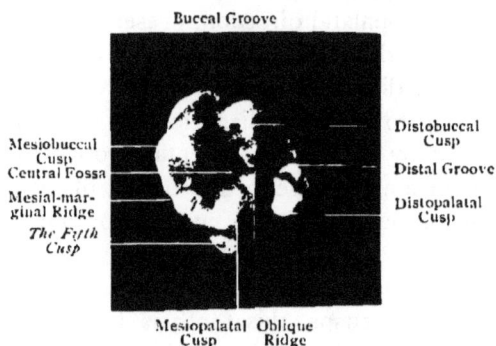

Fig. 87.—Superior First Molar, Occlusal Surface, Strongly Developed, showing
Presence of Fifth Cusp.

cusps, which in turn is much influenced by the temperamental
type of the tooth.   The bottom of the fossa is deeply marked
by two of the grooves of development—the *mesial groove* and
the *buccal groove*.   The former begins on the mesial surface,
passes over the mesial-marginal ridge, and continues in an irregular line to the bottom of the fossa, the latter, beginning near
the center of the buccal surface, enters the fossa by crossing the
buccal-marginal ridge near the center of its length, and also
ends in the *central pit* of the central fossa.   As previously
referred to, the union of these two grooves forms the *mesiobuccal triangular groove*.   From the central pit of this fossa
another groove is given off—the *distal groove*.   It is usually well

defined at its beginning, but as it passes over the oblique ridge it is generally partly obliterated, although occasionally being so marked as to divide this ridge. The *distal fossa* is much smaller than the central, and is of an entirely different form. Its walls are principally formed by the distopalatal incline of the oblique ridge, and the mesiobuccal incline of the distopalatal cusp; a portion of the distal-marginal ridge and the distal incline of the distobuccal cusp also assist in its formation. Like the central fossa, its sides are more or less irregular, from the presence of various grooves and ridges in its vicinity. The greatest length of the fossa is in a disto-palatal direction, and it is traversed by a deep developmental groove—the *distopalatal groove*. When the distal groove crosses the oblique ridge, it usually extends to the bottom of this fossa.

The Buccal Surface of the Crown (Fig. 88).—This surface, which is the result of a union between the mesial and distal developmental lobes, may be divided into a mesial and a distal half. These two portions are

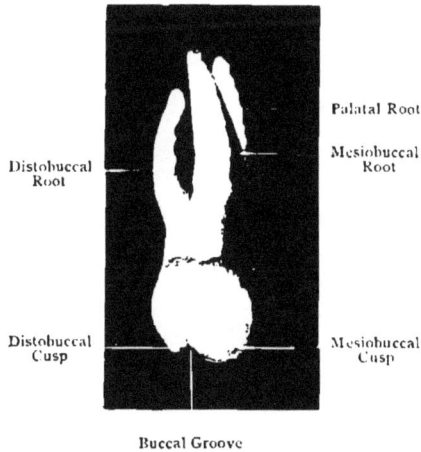

Palatal Root

Mesiobuccal Root

Distobuccal Root

Distobuccal Cusp

Mesiobuccal Cusp

Buccal Groove

FIG. 88.—SUPERIOR FIRST MOLAR, BUCCAL SURFACE.

quite similar in outline, and are separated from each other by the buccal groove, which usually ends near the center or about half way to the cervical line in a decided pit—the *buccal pit*. In some instances this groove is continued to the cervical line, or even beyond this to the bifurcation of the roots. Both the mesial and distal half are provided with a longitudinal ridge (the *buccal ridges*)—one the *mesial buccal ridge* and the other the *distal buccal ridge*. These are similarly formed and descend from the summits of the respective cusps, at which point they are usually well defined, but gradually disappear as they pass toward the cervical line. The location of the buccal groove being a little to the distal of the center of the surface, gives to the mesial portion a somewhat greater extent than the distal. The margins of the surface, which form an irregular quadrilateral, are the mesial, distal, occlusal, and cervical. The mesial and distal margins are rounded, and gradually converge as they pass rootward, making the average diameter of the surface about one-fourth less at the cervical line than at the base of the cusps. In some instances these margins are slightly concave over their cervical portion, and convex on approaching the occlusal margins ; or the mesial may be concave and the distal convex through their entire length ; in some types they appear as straight lines and are parallel with each other. The occlusal margin is formed by the marginal ridges as they pass over the two buccal cusps, being in the form of the letter W. The cervical margin is usually a direct line drawn around the circumference of the tooth, but in some instances deviating slightly · from this. Immediately below the cervical line, and conforming to its general direction, is a rounded fold of enamel—the *cervical ridge*.

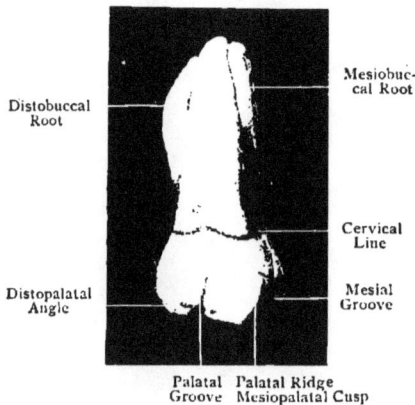

Distobuccal Root

Mesiobuccal Root

Distopalatal Angle

Cervical Line

Mesial Groove

Palatal Groove   Palatal Ridge   Mesiopalatal Cusp

FIG. 89.—SUPERIOR FIRST MOLAR, PALATAL SURFACE.

Palatal Surface of the Crown * (Fig. 89).—In correspondence with the buccal surface, this surface is developed from two lobes,—the mesial- and distal-palatal lobes,—the line of union between the two being recorded by a well-defined groove—the *palatal groove*. This groove, which is a continuation of the distopalatal groove of the occlusal surface, usually ends near the center of the palatal surface in a well-defined pit—the *palatal pit*—or it may

---

* When the fifth cusp is present, the anatomy of the mesial half of this surface becomes somewhat more complex.

continue rootward and gradually disappear. It is located a little
to the distal of the center of the surface, thus making the mesial
a trifle larger than the distal portion. The mesial half of the
surface is smooth and convex ; the palatal incline of the mesiopalatal cusp is seldom provided with a well-defined ridge, although
usually referred to as the *mesiopalatal ridge*. The distal half
of the surface is also smooth and rounded, with the mesiodistal
convexity much more marked than that of the mesial lobe.
The cervical ridge is seldom so pronounced as that of the buccal
surface, but the enamel frequently makes a sudden dip at this
point to meet the cementum of the root. That portion of the
surface immediately below the cervical ridge is smooth and
unbroken, slightly convex
in the direction of the long
axis of the tooth, and
flattened or slightly convex from mesial to distal.
The margins of the sur-
face are the mesial, distal,
occlusal, and cervical. The
surface passes so gradu-
ally into the mesial and
distal that it is somewhat
difficult to define these
margins. In general, the
margins converge slightly
in the direction of the
root. Both the occlusal and cervical margins are similar to the
corresponding margins of the buccal surface.

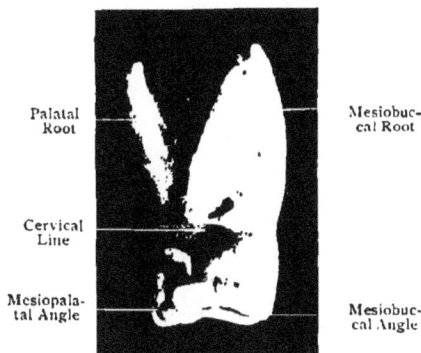

Palatal
Root

Mesiobuc-
cal Root

Cervical
Line

Mesiopala-
tal Angle

Mesiobuc-
cal Angle

FIG. 90.—SUPERIOR FIRST MOLAR, MESIAL
SURFACE.

The Mesial Surface of the Crown (Fig. 90).—This surface is almost an unbroken plane, being smooth and flat. In
some instances it is crossed near the center of its occlusal margin
by a continuation of the mesial groove, but this is seldom so pronounced as to divide the surface. The occlusal third of the
surface is inclined to a slight general convexity, providing a
point of contact for the distal surface of the second bicuspid,
but between this and the cervical line there is often a slight concavity. The margins of the surface are the occlusal, buccal,

palatal, and cervical. The first named is formed by the mesial-marginal ridge of the occlusal surface, and is concave in the direction of the root. The buccal and palatal margins are rounded, and, unlike the lateral margins of the buccal and palatal surfaces, diverge in the direction of the roots. The cervical margin is slightly concave in the direction of the occlusal surface, and its length is much greater than that of any other margin of the crown. This surface is more extensive than either the buccal, palatal, or distal, and is about equal to that of the occlusal. When the fifth cusp is present it alters the form of the palatal margin of this surface by crossing it near the center, and extending for some little distance on the face of the surface.

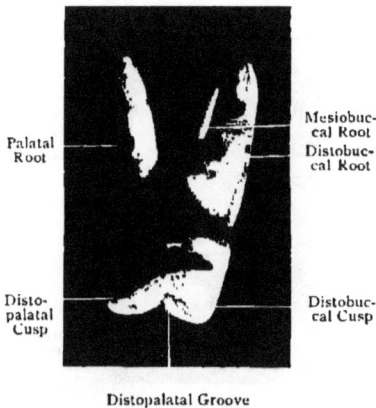

Palatal Root

Disto-palatal Cusp

Mesiobuc-cal Root
Distobuc-cal Root

Distobuc-cal Cusp

Distopalatal Groove

FIG. 91.—SUPERIOR FIRST MOLAR, DISTAL SURFACE.

The Distal Surface of the Crown (Fig. 91). — Taken in its entirety, this surface usually presents a general convexity. The palatal half of the surface is usually somewhat more prominent than the buccal, the latter being flattened and frequently slightly concave, particularly near the cervical portion. In some instances the surface is traversed by a continuation of the distopalatal groove, which, after passing over the distal-marginal ridge, is continued in a longitudinal direction, dividing the surface into two equal parts. Not infrequently this groove, instead of existing as such, is represented as a shallow depression, often extending to the bifurcation of the roots. The margins of the surface are four in number : the occlusal, which closely resembles the corresponding margin of the mesial surface; the buccal, which is not well defined ; the palatal, somewhat angular ; and the cervical, formed by the cervical line.

The Neck of the Tooth.—When looking upon the buccal

surface of this tooth, the constricted portion forming the neck is greatest at a point immediately above the cervical line. Viewed in this direction, the crown is usually bell-shaped, and both the crown and the base of the roots assist in the production of the neck. Viewed from a palatal direction, the neck is a distinctive feature, but is seldom so marked as when examined from the opposite side. When studied from either a mesial or a distal aspect, the neck appears above the cervical line, the prominent fold of enamel immediately adjacent to this line forcing the neck upon the roots.

The Roots of the Superior First Molar.—The roots of this tooth are three in number, two of which are on the buccal side, and are, therefore, called mesial- and distal-buccal roots, and one on the palatal side, known as the palatal root. These three roots are given off from a common base, which is sometimes referred to as the root, while those parts above the point of trifurcation are considered as root-branches. The number, location, and form of the roots of this tooth are, perhaps, more constant than those found in connection with any other cuspidate tooth. The common base from which the roots are given off is similar in contour to the crown of the tooth, excepting in those cases in which the form of the root is carried over this base to meet the neck of the tooth.

The mesiobuccal root (Fig. 88) is flattened from mesial to distal, broad at its base from buccal to palatal, from which point it gradually tapers to the apex. At the base the mesiodistal measurement is less than one-third that of the buccopalatal. In its course it is first inclined to the mesial, but after reaching the center of its length it makes a decided distal curve, which looks almost directly to the distal. The mesial side of this root is decidedly flattened at its base, but as the center of the surface is reached a shallow, longitudinal groove is present, which is continued to the region of the apex. The distal side is also possessed of a similar groove, which extends throughout its entire length. Both the buccal and palatal sides of the root are smoothly convex, the latter being only about half the width of the former.

The distobuccal root (Fig. 88) is much the smallest of the three, and, while inclined to flatness on its mesial and distal sides, it is much more rounded than the mesial root. The mesial side is provided with a slight longitudinal groove, and in rare instances a similar groove exists on the distal side. The buccal and palatal sides are similar to those of the mesial root. The root is generally straight, and tapers gradually from base to apex, ending in a rounded point.

The palatal root (Fig. 89) is usually the largest and longest of the three, and is more rounded in form than either of the buccal roots. The palatal surface is inclined to flatness near its base, and is provided with a well-defined longitudinal groove, which is sometimes independently formed, while at others it is present as a continuation of the palatal groove. This root being the only one given off from the palatal side of the tooth, is constructed with a mesiodistal measurement about equal to that of the base of the crown at this point. From its place of beginning it passes first in a palatal and then in a buccal direction, forming a long curve and ending in a sharp-pointed apex.

Bilious Type.—The superior first molar of this tempera-ment is manifest by a crown with angles well produced, the marginal ridges and cutting-edges of the cusps bold and well marked. The cusps are of medium length, with summits angular and pointed. The developmental grooves are deep and often sulcate, and numerous supplemental grooves are found upon the occlusal surface. The longitudinal and trans-verse measurements of the crown are about equal, and when viewed upon the occlusal surface, the angular nature of its anatomy is noted as a distinctive feature. The neck is fairly well developed, giving a slight bell-shape to the crown. The cervical line is made up of angles rather than curves, and the roots are long and straight.

Nervous Type.—Like the teeth previously described under this class, the crown of this tooth is of greater longitudinal than transverse extent ; the neck is especially well formed, producing a decided bell-shape to the crown. The cusps are long and penetrating, the marginal ridges sharply defined, as are also those ridges upon the central incline of the cusps. The grooves of development are decided and frequently sulcate. The buccal surface is rounded and smooth, with the buccal groove extending well toward the cervical line. The palatal surface also presents a general convexity, and is usually divided by the palatal groove. The mesial surface is convex over its cervical third, and the occlusal margin is a decided convex ridge, serving as a point of contact for the adjoining tooth, and thus forming the characteristic V-shape common to this temperament. It is in this type that the fifth cusp is most frequently present. The cervical line is much curved, and the roots are slim and frail.

Sanguineous Type.—The crown of the superior first molar of this type usually presents a slightly greater longitudinal than transverse extent. The angles of the crown are poorly formed, being rounded and smooth. The cusps are of moderate length, and are rounded in their nature; the marginal ridges, as well as those ridges of the central incline of the cusps, are less distinct than either of the forms previously described. The buccal and palatal surfaces are convex and seldom broken by grooves; the mesial and distal surfaces are convex in every direction, throwing the point of contact with adjoining teeth near the center of the surface. The roots are inclined to be large and oval in form, while the cervical line is a series of long curves.

Lymphatic Type.—In this temperament the crown is much less in its longitudinal than transverse measurement. The neck of the tooth is poorly defined, the crown passing into the root-base without a marked constriction. The mesial and distal surfaces are flattened and nearly parallel with each other, providing a broad contact surface. The buccal and palatal surfaces each present a marked general convexity, the latter being frequently broken by the distopalatal groove. The tooth is provided with cusps which are short and heavy-set; the marginal ridges, as well as all the ridges common to the occlusal surface, are poorly defined. The developmental grooves are shallow and terminate abruptly. There is but little curvature to the cervical line, and the roots are short, heavy-set, and inclined to cluster together.

SUPERIOR SECOND MOLAR.

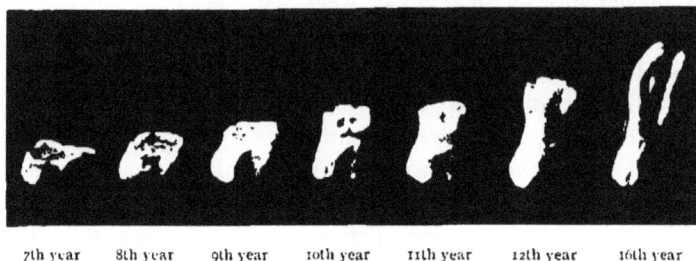

7th year   8th year   9th year   10th year   11th year   12th year   16th year

FIG. 92.

CALCIFICATION BEGINS FROM FOUR CENTERS ABOUT THE FIFTH YEAR.
CALCIFICATION COMPLETED, SIXTEENTH TO EIGHTEENTH YEAR.
ERUPTED, TWELFTH TO FOURTEENTH YEAR.
AVERAGE LENGTH OF CROWN, .28.
AVERAGE LENGTH OF ROOT, .51.
AVERAGE LENGTH OVER ALL, .79.

Calcification in this tooth takes place in precisely the same manner as that of the first molar, but the formative process is much later in beginning, the lime-salts commencing to accumulate in the four separate lobes about the fifth year. At the beginning of the sixth year the formation of the cusps is completed, soon after which they coalesce and the occlusal surface of the crown is established. At the beginning of the eighth year fully two-thirds of the crown is calcified, and the following year the crown and neck are completed and the root-base outlined. By the tenth year the beginning of separate root-development is observed; at the twelfth year, or at the time of eruption, the roots are formed to about one-half of their completed length, the process continuing until the sixteenth or seventeenth year, when calcification is completed and the apical foramina established (Fig. 92). In many respects this tooth closely resembles the first molar previously described, the crown presenting the same number of surfaces similarly named, and also being provided with the same number of roots. Notwithstanding this fact, there are a number of ways in which they are at variance. The crown of the second molar is much smaller than that of

the first, and the quadrilateral outline common to the first molar
is much compressed and broken in the second. The distal
cusps are much smaller proportionately than the mesial cusps,
this being particularly true of the distopalatal cusp. This
reduction in size of the distal cusps gives to that portion of the
occlusal surface a slight distal incline.

Occlusal Surface of the Crown (Fig. 93).—The general
contour of the crown is best studied by viewing it directly upon
the occlusal surface ; this aspect also shows to best advantage
the difference in form between this and the first molar tooth, as
shown in figure 86. The mesial and palatal outlines closely
resemble the corresponding outlines on the first molar, but the

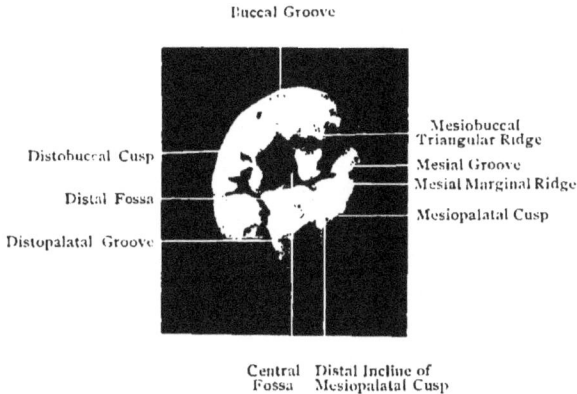

Buccal Groove

Distobuccal Cusp

Distal Fossa

Distopalatal Groove

Mesiobuccal
Triangular Ridge

Mesial Groove
Mesial Marginal Ridge

Mesiopalatal Cusp

Central   Distal Incline of
Fossa   Mesiopalatal Cusp

FIG. 93.—SUPERIOR SECOND MOLAR, OCCLUSAL SURFACE.

buccal and distal are much at variance, the former passing into
the latter without a distinct line of demarcation existing between
the two, this gradual blending of one into the other being at the
expense of the distobuccal angle of the crown, which is poorly
developed. The crown is much compressed in a distobuccal,
mesiopalatal direction, making this measurement of the occlusal
surface about one-third less than the mesiobuccal-distopalatal
measurement. The cusps are much inclined to cluster toward
the center of the surface, this being especially true of those on
the palatal half.

Marginal Ridges of the Occlusal Surface (Fig. 93).—Like the occlusal surface of the first superior molar, this surface of the second molar is bounded by four marginal ridges—the mesial, distal, buccal, and palatal. They are usually less marked than those found on the first molar, and are much more variable in their individual anatomy. *The mesial-marginal ridge* extends from the summit of the mesiobuccal cusp to the summit of the mesiopalatal cusp. It is concave in the direction of the body of the crown, and is broken near its central portion by the mesial groove. In some instances one or more small supplemental grooves are found to cross it. Compared with the mesial-marginal ridge of the first molar, its length is much less and the convexity not so pronounced. The *distal-marginal ridge*, owing to the variation in form and size of the distal cusps, is difficult to describe definitely; suffice it to say that it extends from the summit of the distobuccal to the summit of the distopalatal cusp. The concavity is V-shaped, and is usually crossed near the center by the distopalatal groove. In some instances the distopalatal cusp is almost wanting, in others the distobuccal is but little developed; when either of these conditions are present, the marginal ridge is extended either to the buccal or to the palatal, in a measure taking the place of the missing cusp.

The Buccal-marginal Ridge.—The mesial half of this ridge closely resembles the corresponding margin of the first molar; beginning at the mesiobuccal angle it ascends to the summit of the mesiobuccal cusp, after which it descends by a longer incline to the buccal groove. The distal half of the ridge, unlike that of the first molar, presents much variety, its form being controlled by the character and position of the distobuccal cusp, usually small. As most frequently observed, it ascends to the summit of the cusp, and in so doing it presents a decided palatal inclination. In passing down the distal incline the palatal inclination is increased and gradually passes into the distal marginal ridge.

The Palatal-marginal Ridge.—Branching off from the mesial marginal ridge, this ridge ascends to the summit of the mesio-palatal cusp and descends by a much shorter incline to the distopalatal groove. This portion of the margin is thrown well toward the center of the surface, the location of the cusp carrying it to that point. Like the distal half of the buccal margin, the outline of the distal half of this margin is controlled by the position and form of the distopalatal cusp. In the majority of cases, when the cusp is moderately strong, the ascent from the distobuccal groove to the summit of the cusp is short and abrupt, the descent being somewhat more gradual, and with a decided buccal inclination it passes into the distal-marginal ridge, or ends abruptly at the distal end of the distopalatal groove.

The Cusps and Ridges (Fig. 93).—This tooth is provided with four lobes or cusps, two of which are located on the buccal, and two on the palatal side. They are usually smaller and less angular than the cusps of the first molar. This is particularly true of both the distal cusps, and especially of the distopalatal cusp, which is often quite diminutive and occasionally entirely wanting. When this latter condition exists, the palatal half of the surface is for the most part occupied by what would otherwise be the mesiopalatal cusp, the absence of the distal cusp permitting the distopalatal groove to occupy a position near the extreme distopalatal angle, that portion of the surface which is distal to the groove being a portion of the distal-marginal ridge.

The Mesiobuccal Cusp (Fig. 93).—Like the corresponding cusp of the first molar, this cusp is usually the longest of the four, and in many instances covers a greater extent of surface than any of the others. Its base is outlined by the buccal and mesial grooves, the two together forming the mesiobuccal triangular groove. Descending from its summit to the buccal surface is the mesiobuccal ridge ; the marginal ridge makes a double descent, one in a mesial and one in a distal direction, while sloping toward the central fossa is the mesiobuccal triangular ridge. This cusp is seldom traversed by supplemental grooves such as are found on the corresponding cusp of the first molar.

The Distobuccal Cusp (Fig. 93).—As previously stated, this cusp is not constant in its form; in some instances it is bold and well produced, corresponding closely to the mesiobuccal cusp just described. When thus pronounced it is possessed of ridges, and bounded by grooves which are similar to those described in connection with the first molar. More frequently the cusp is much rounded, its summit being carried well toward the center of the surface. When this formation exists, the buccal ridge is absent, the marginal ridges short and rounded ; the distobuccal triangular ridge which descends from it toward the center of the crown is short and heavy set.

The Mesiopalatal Cusp (Fig. 93).—In the majority of instances this is the largest cusp, particularly when there is a degenerate tendency in the distopalatal cusp. Descending from its summit are a number of ridges, the marginal ridges being given off as already described, the mesiopalatal triangular ridge descending the central incline to the central fossa, and when the cusp has an additional mesiodistal extent by the presence of a diminutive distal cusp, other ridges descend in the same direction. The palatal descent of the cusp is smooth and more rounded than the corresponding surface of the first molar, and is seldom elevated in the form of a definite ridge. The central outline of this cusp is marked by the mesial, the distal, and the distopalatal grooves.

The Distopalatal Cusp (Fig. 93).—In no other cusp do we find such a diversity of form as in the distopalatal cusp of the superior second molar. In some instances it is fully as prominent as its neighbor just described, in others appearing as a mere fold of enamel, and it is not uncommon to find it entirely wanting, the distal-marginal ridge extending to occupy a portion of the space which it should claim. Deductions might be drawn from an average between these two extremes, wherein the existing cusp would be much smaller than any of the others, the summit rounded rather than sharp, but with a decided inclination to occupy the extreme distopalatal angle of the surface, in this latter respect differing from the distobuccal cusp. The mesial and buccal outlines of the cusp are formed by the distopalatal groove, and its mesiopalatal incline contributes to the formation of the distal fossa.

The Fossæ and Grooves of the Occlusal Surface (Fig. 93).—These in name, number, and general form are similar to those of the first molar. The central fossa is never, strictly speaking, in the center of the surface, and is formed by the central incline of the mesiobuccal, distobuccal, and mesiopalatal cusps. It is seldom so deep as the central fossa of the first molar. The distal fossa is more or less pronounced, its size and position being controlled by the extent of development in the distopalatal cusp. The distopalatal groove, which usually crosses the palatal surface of the first molar near its center, is not constant in its location on this tooth, in some cases being near the center, in others near the distopalatal angle of the crown. The buccal groove is never constant in its location, usually crossing the buccal-marginal ridge and passing over the buccal surface near its mesiodistal center, but it is not uncommon to find it forced to the distal by a diminution in the size of the distal cusp.

Buccal Surface of the Crown (Fig. 94).—The most constant difference between this and the corresponding surface of the superior first molar is the wandering location of the buccal groove. While in the majority of instances it may be found near the mesiodistal center of the surface, it is not uncommon to find it passing over the distal third, or even as far posterior as the distobuccal angle. In general, the surface is somewhat more convex and necessarily less extensive than the buccal

FIG. 94.—SUPERIOR SECOND MOLAR, BUCCAL SURFACE.

surface of the first molar. The buccal ridges which descend from the summit of the two buccal cusps are seldom so marked as those on the first molar, and in many instances the distal ridge is wanting. The distal half of the surface frequently passes into the distal surface, by a long gradual sweep, there being no line of demarcation between the two.

The Palatal Surface of the Crown (Fig. 95).—In keeping with the other surfaces just described, the palatal surface differs from the corresponding surface of the first molar in that it presents a greater general convexity. This is particularly true in passing from the cervical line to the occlusal surface. The palatal groove is also less constant in its location. In most instances it is to be found a little to the distal of the center, in others being as far posterior as the extreme distal third of the surface, and in rare instances it is entirely wanting. The general character of this surface, which is smooth and convex, is seldom broken by the presence of well-defined ridges, such as are usually found descending from the palatal cusps of the first molar. As previously referred to, the mesial, distal, and buccal surfaces, as well as the surface under consideration, are proportionately smaller than those of the first molar, and, while this refers to both the transverse and longitudinal measurements, it is particularly applicable to the latter.

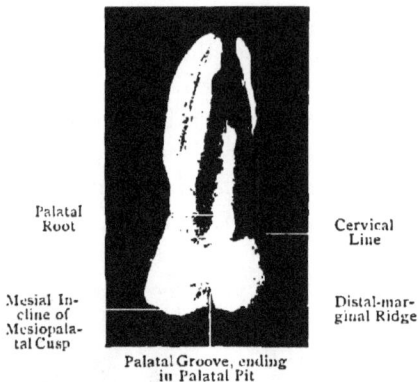

Palatal Root

Cervical Line

Mesial Incline of Mesiopalatal Cusp

Distal-marginal Ridge

Palatal Groove, ending in Palatal Pit

FIG. 95.—SUPERIOR SECOND MOLAR, PALATAL SURFACE.

The Mesial Surface of the Crown (Fig. 96).—Aside from this surface being of less extent than the corresponding surface of the first molar, there are no other differences of importance. In many instances, however, there is a decided tendency for the surface to be concave from buccal to palatal, the convex distal surface of the first molar closely fitting into this concavity. Another variation which is frequently observed is that of the longer and more gradual sweep which it takes in passing into the palatal surface.

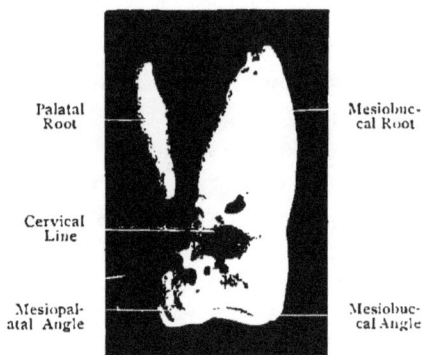

Palatal Root

Mesiobuccal Root

Cervical Line

Mesiopalatal Angle

Mesiobuccal Angle

FIG. 96.—SUPERIOR SECOND MOLAR, MESIAL SURFACE.

The Distal Surface of the Crown (Fig. 97).—This differs from the distal surface of the first molar principally in its more pronounced convexity. Its general form is also much influenced by the nature of the two distal cusps. If one or the other of these is sparingly developed, either the buccal or palatal half of the surface, as the case may be, is quickly rounded off to pass into the deficient lobe.

The Angles of the Crown.—The increased inclination for the crown of this tooth to general convexity dispels, in a measure, the presence of angles, as such, in correspondence with the four corners of the first molar. In some instances the crown is represented as a fairly well-formed quadrilateral, in which case the angles are well defined, but usually this outline is so much broken by a mesiodistal compression that the angular form of the crown is entirely abolished. But, whatever the form of the crown may be, it is well to adhere to the commonly-accepted term, and speak of that point at which the sides of the crown unite as the angles, each being named in accordance with its location.

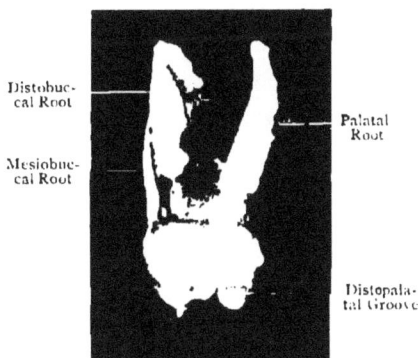

FIG. 97.—SUPERIOR SECOND MOLAR, DISTAL SURFACE.

The Neck of the Tooth.—The principal variation between the neck of this tooth and that of the first molar, is that produced by the greater general convexity of the crown, which contributes to the production of a neck much more constricted. There is also a greater variety in the contour of the neck, incident to the variation in the general outline of the crown.

The Roots of the Superior Second Molar.—These are the same in name and number as those of the first molar—two buccal and one palatal. In many respects they differ from the roots of the first molar. They are much smaller, fre-

13

quently inclined to cluster together, and are often fused, in some instances, all three being united, in others the union existing between but two.   When isolated, each root usually presents a decided distal curve near its apical third.   When the crown is flattened from mesial to distal, as before described, the disto-buccal root is forced to occupy a position much nearer to the palatal than that held by the mesiobuccal root.   The palatal groove seldom passes over the palatal surface of the palatal root, as observed on the first molar.

SUPERIOR THIRD MOLAR.

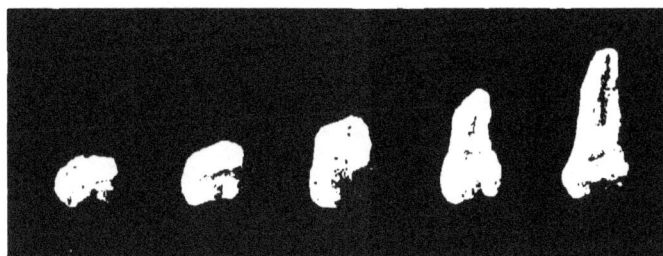

| 10th year | 11th year | 12th year | 14th year | 18th year |

FIG. 98.

CALCIFICATION BEGINS, NINTH YEAR.
CALCIFICATION COMPLETED, EIGHTEENTH TO TWENTIETH YEAR.
ERUPTED, SEVENTEENTH TO TWENTIETH YEAR.
AVERAGE LENGTH OF CROWN, .24.
AVERAGE LENGTH OF ROOT, .44.
AVERAGE LENGTH OVER ALL, .68.

Calcification in this tooth takes place in precisely the same manner as that in the first and second molar, with the exception of the number of lobes, which are sometimes three and sometimes four. The lime-salts begin to accumulate between the eighth and ninth year, and continue with somewhat more activity than that of the first or second molar. Between the ninth and tenth year the three or four cusps, of which the future tooth is to be composed, have coalesced, and by the eleventh year calcification in the crown of the tooth is completed; at the end of the following year the roots, which are variable in number, have made considerable progress; at the fourteenth year they are calcified to about half their length, while at a period between the eighteenth and nineteenth year the formative process is completed (Fig. 98). This tooth, like the cuspid, is usually fully formed before eruption takes place.

This tooth is subject to a greater variety of form than any other tooth in the mouth; in rare instances it is similar in general outline and cusp formation to the first molar, but in a vast majority of cases it is dissimilar, the most constant deviation being its size, which on the average is about one-third less. In the accompanying illustration (Fig. 99) the forms most frequently met with are shown. It will be observed that the contour of the tooth in general is much more rounded than either the first or second molar. The buccal angles of the crown are alone well marked, the mesial and distal surfaces passing into the palatal surface by a long, gradual sweep, and thus obliterating the palatal angles. In the vast majority of instances the tooth is tritubercular, and is usually made so by the absence or diminutive size of the disto-

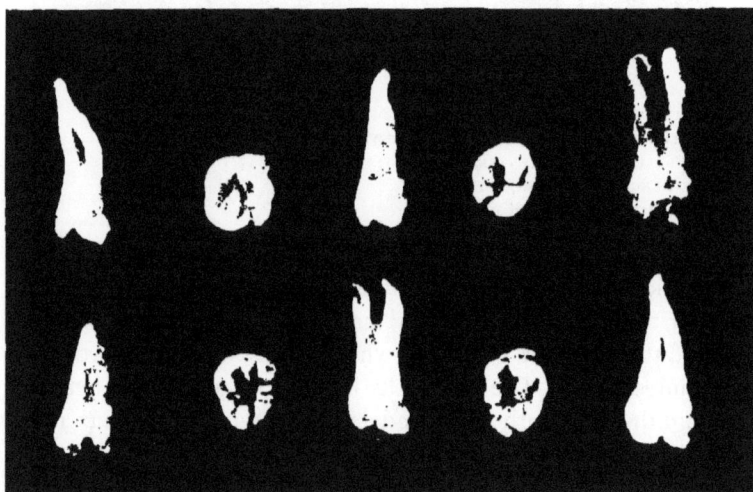

FIG. 99.—VARIOUS TYPES OF SUPERIOR THIRD MOLAR.

palatal cusp. Just as this cusp was inclined to degenerate in the second molar, so we find this retrograde developmental tendency increased in the third molar. With this change in the construction of the occlusal surface, there is a corresponding variation in the grooves, ridges, and fossæ.

Mesial Surface of the Crown (Fig. 100).—In many particulars this surface corresponds in form and outline to the mesial surface of the first molar; it is, however, usually much more convex, seldom presenting a concavity or even a positive flatness. The surface is not only rounded from buccal to palatal, but also from the cervical line to its occlusal margin. Thus formed, a point of contact is provided near the center of the surface. The occlusal margin, the buccal margin, and the cervical margin are almost identical to those of the first molar, but in most instances the palatal margin is wanting, the surface gradually passing into the palatal without a decided line of demarcation.

Distal Surface of the Crown (Fig. 101). — This surface is much less extensive in comparison to the size of the crown than the corresponding surface of either the first or second molars. It is decidedly rounded in every direction and is frequently crossed by the distal developmental groove, and sometimes by one or more supplemental grooves. The general form of the surface is much influenced by the presence or absence of the distopalatal cusp; with the former, the surface is more extensive, presenting less convexity and resembling more closely the distal surface of the first and second molars; with the latter, the extent of the surface is decreased and the convexity increased.

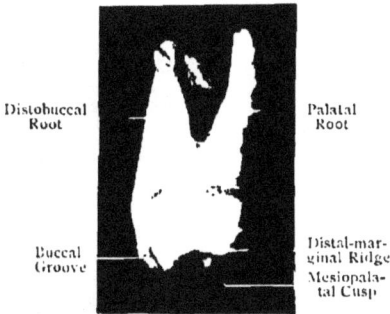

FIG. 100.—SUPERIOR THIRD MOLAR, MESIAL SURFACE.

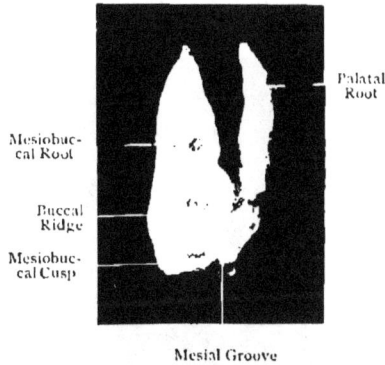

FIG. 101.—SUPERIOR THIRD MOLAR, DISTAL SURFACE.

Buccal Surface of the Crown (Fig. 102).—The mesial portion of this surface is in no way at variance with the mesial portion of the buccal surface of the first or second molar, but much variety of form exists in the distal portion. The buccal groove

which serves to separate these two portions is located well
toward the distal third of the surface, thus reducing the size of
the distal portion to about one-third that of the mesial portion. In general, the surface is but little more convex than
the corresponding sur-
face of the first and second
molar. Its mesial border
is definitely outlined, as
are also the cervical and
occlusal margins, but the
distal margin can not be
definitely located, the surface tending to pass gradually into the distal surface. Like the distal, the
extent of this surface is
much regulated by the
size and shape of the dis-

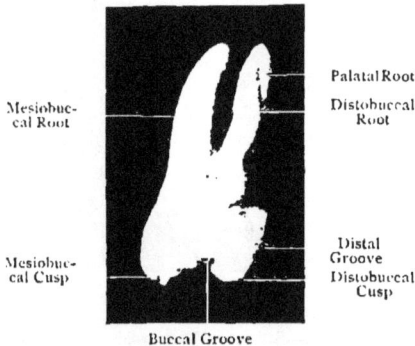

Mesiobuc-
cal Root

Palatal Root

Distobuccal
Root

Mesiobuc-
cal Cusp

Distal
Groove
Distobuccal
Cusp

Buccal Groove

FIG. 102.—SUPERIOR THIRD MOLAR, BUCCAL
SURFACE.

tobuccal cusp, which, like the distopalatal cusp, is inclined to degenerate.

Palatal Surface of the Crown (Fig. 103).—Like the distal
surface previously described, the form of
this surface is much influenced by the presence or absence of the distopalatal cusp.
When this cusp is wanting or but little developed, the surface presented is decidedly
convex and smooth ; in many instances the
mesiodistal circle described is almost a
perfect semicircle, and in passing from the
cervical line to the occlusal margin the surface is carried well toward the center of
the crown by a long gradual sweep toward
the palatal. The palatal groove is usually
absent. The change in form produced by
the presence of the distopalatal cusp is

FIG. 103. — SUPERIOR
THIRD MOLAR, PALATAL
SURFACE.

principally noticeable in a less pronounced convexity and the
presence of the palatal groove, which may be noticed as a slight
depression or as a well-defined groove. This groove, when pres-

ent, is always located near what would represent the distopalatal
angle of the crown: the distopalatal cusp seldom if ever being of
sufficient size to force its location near the center of the surface
as in the first molar.   In some instances this groove is shown
upon the palatal surface when the cusp is not present; in this
case the distal-marginal ridge represents in a measure the cusp
by its bold, heavy development.

Occlusal Surface of the Crown (Fig. 104).—When looking
directly upon this surface, an opportunity is presented to study
the general contour of the crown : the most noticeable difference
in this respect between this tooth and the first molar being observed in its smaller size, and the absence of well-marked
angles.   It will be noted
that the mesial and buc-
cal outlines in a meas-
ure resemble the corres-
ponding outlines of the
first and second molars,
but there is scarcely any
similarity existing when
comparing the distal and
palatal outlines.   In some
instances the crown is
triangular (Fig. 99) ; in
others the mesial- and
buccal-marginal outlines form an obtuse angle, the free ends of
which are joined together by a long semicircle, the latter constituting the distal and palatal outlines (Fig. 99). Again, almost the
reverse of this last-mentioned form is seen, the buccal and distal
outlines constructing the angle, while the semicircular connection
between the two is made up of the mesial and palatal outlines.

The Marginal Ridges.—*The Mesial-marginal Ridge.*—
This ridge is usually well defined, and in most instances is
crossed near its center by the mesial groove, and frequently by
two or more supplemental grooves.   This marginal ridge is
probably the most constant in form, the numerous variations to
which the surface is liable seldom making any material alteration in it.

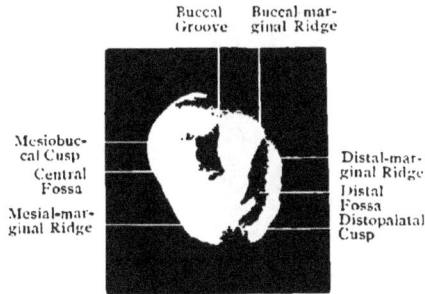

FIG. 104.—SUPERIOR THIRD MOLAR, OCCLUSAL
SURFACE.

The Distal-marginal Ridge.—Unlike the ridge above described, the distal-marginal ridge is most variable in its construction, nearly all of the forms characteristic of the occlusal surface exerting a controlling influence over it. In rare instances the ridge resembles that of the first and second molars, but this form is most frequently interfered with by the absence or diminutive size of the distopalatal cusp, the ridge itself frequently supplying the place of the cusp. In many cases the ridge is elevated near its central part by being reinforced by a portion of the oblique ridge. When the distopalatal cusp is wanting, this ridge not infrequently descends from the summit of the distobuccal cusp to the distal groove and from this point ascends obliquely to the summit of the palatal cusp.

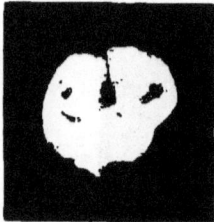

FIG. 105. — SUPERIOR THIRD MOLAR, OCCLUSAL SURFACE, WITH DISTOPALATAL CUSP AND DISTAL FOSSA POORLY DEFINED.

The Buccal-marginal Ridge.—This may be described as similar in most respects to the corresponding margin on the first and second molars, the principal variation being in the distal half, which is much shorter and less pronounced.

The Palatal-marginal Ridge.—Here, again, much variety in outline is noticeable. In nearly all instances the ridge is thrown much nearer the center of the body of the crown, and, when the tooth is bicuspid in form, it simply makes a mesial ascent of the palatal cusp, followed by a gradual incline, and passing into the distal ridge, as above noted. When the distopalatal cusp is present, the ridge is similar to that upon the first and second molars, with the exception of the distal portion, which is less clearly marked.

The Cusps (Fig. 104).—As previously stated, the form most frequently met with is tritubercular, two of the cusps being upon the buccal and one upon the palatal half of the surface.

The Mesiobuccal Cusp (Fig. 104).—This cusp corresponds in nearly every particular to the mesiobuccal cusp of the first and second molars; it is the most constant in size and form of the three. Its summit is usually angular, and the numerous ridges which descend from it are well defined and similar in name and number to those of the first molar.

Distobuccal Cusp (Fig. 104).—The constant inclination to degeneracy in the distal portion of the crown of the tooth is noticeable in this cusp, which is much smaller than the mesio-buccal and scarcely half as large as the corresponding cusp of the first and second molars. In some instances, however, it is inferior only in size, retaining its angularity, being possessed of small but well-defined ridges.

The Palatal Cusp (Fig. 104).—When the three cusps alone are present, this one is much the largest, the extent of the surface covered being all of the palatal half of the crown. The summit of the cusp, which is thrown well toward the center of the body of the crown, is prominent, but seldom angular. Only in rare instances will there be found a palatal ridge descending therefrom, but the central incline is usually marked by a number of wrinkles or folds of enamel resembling minute ridges. The central boundary of this cusp is marked by the mesial and distal developmental grooves.

The Distopalatal Cusp (Fig. 104).—It is to the presence or absence of this cusp that much of the coronal variety may be attributed. When present, it is usually very diminutive in size, and is without definite form. In many instances nature is apparently attempting to cast it off in precisely the same manner in which she is attempting to add to the first molar by a development of the "fifth cusp," the distobuccal cusp appearing to hang to the distopalatal angle of the crown in a manner very similar to the "fifth cusp." When thus situated, it is separated from the body of the crown by a groove, which can not be considered as being upon the occlusal surface. When located in its normal position, it has for its inner boundary the disto-palatal groove.

The Fossæ and Grooves of the Occlusal Surface.—The great variety and *form* common to this surface exerts a controlling influence over the size, number, and position of the grooves and fossæ. In the tritubercular class the central fossa alone is present. The developmental grooves, with the exception of the buccal, are not definitely outlined, but, descending

toward the fossæ from the central incline, are numerous small
ridges divided from each other by a like number of diminutive
supplemental grooves. The distal groove is sometimes well
defined, and crosses over the oblique ridge, which in this type
becomes the distal-marginal ridge. When the distopalatal cusp
is present, all of the ridges and grooves are more pronounced.
In this case the central fossa corresponds more closely to the
central fossa of the other molar teeth, this resemblance increasing just in proportion as the size of the distopalatal cusp increases. The distal fossa, in a vast majority of instances, is
present as a mere pit; the size of this fossa is likewise much
controlled by the extent of development in the distopalatal cusp.
Where the distal-marginal ridge is supplementary to the distopalatal cusp, the distopalatal groove lies between the former and
the oblique ridge. Another peculiarity found only upon the occlusal surface of this tooth is, what appears to be an effort upon the
part of the cusps to cluster toward the center. This is common
only to those teeth possessing three cusps, and accompanying this
form the central fossa shows a number of fantastically arranged
grooves and ridges which ascend the cusps, passing over the
marginal ridges and breaking them into a number of small
tubercles.

Temperamental Types.—The third molar tooth is probably
less influenced by the character and habits of the individual than
any other tooth in the mouth. The inclination to a general degeneracy is no doubt favored by civilization. With a constant
decline in the functional activity, brought about by the present
culinary methods common to civilization, this tooth in a
measure becomes useless, and nature is gradually making an
effort to cast it off. While there are undoubtedly many individuals possessed of the highest mental attainments with the
third molar as fully developed as either the first or the second,
this condition is usually confined to those possessed of little in-
tellectuality. If, in general, the temperament of the subject be
taken into consideration, the cusp-formation on this tooth will
correspond in a relative degree to that on the bicuspids and
molars.

CHAPTER IX

A DESCRIPTION OF THE INFERIOR TEETH IN DETAIL.—CALCIFICA-
TION, ERUPTION, AND AVERAGE MEASUREMENTS.—THEIR SUR·
FACES, RIDGES, FOSSÆ, GROOVES, SULCI, ETC.

THE INFERIOR TEETH.

In most respects the inferior teeth correspond to the superior, but in each class we find a slight variation existing between
the two sets. As compared to the superior incisors, the crowns
of the inferior incisors are more slender and somewhat more
angular in outline. The roots are more slender, proportionately
longer, more flattened laterally, and seldom crooked. The
crowns of the inferior incisors are probably more constant in
form than those of any other teeth in the mouth, seldom varying except in size. The mesiodistal measurement of the crown
of the lateral incisor is a trifle greater than that of the central,
a condition exactly the reverse to that of the superior incisors.
The labial and the lingual surfaces of these teeth are smooth,
and, with the exception of young teeth, show but little trace of
the developmental process by the presence of grooves, fissures,
etc. The outline of the inferior cuspids is almost identical to
that of the corresponding teeth in the superior arch, excepting
that they are in every way more slender. The bicuspids are
proportionately smaller in every direction, have their cusps
much less developed than their superior antagonists, and are
seldom found with more than one root. The crowns of the
inferior molars are somewhat larger than those of the superior,
and are provided with five cusps instead of four,* and they are
attached to the alveolus with two, instead of three, roots. In the
incisors, cuspids, and bicuspids the process of development is
the same as in the corresponding superior teeth, calcification
taking place from the same number of centers along the coronal

---

* The inferior first molar has five cusps in ninety per cent. of cases, while, in the second,
five cusps are present in about fifty per cent.

extremities. In the molars, however, development may proceed from five centers instead of four, as in the superior molars. The manner of development, and the period at which this action takes place, so nearly corresponds with that of the superior teeth that the process will not be repeated.

INFERIOR CENTRAL INCISOR.

CALCIFICATION BEGINS, FIRST YEAR AFTER BIRTH.
CALCIFICATION COMPLETED, ABOUT THE TENTH YEAR.
ERUPTED, SEVENTH TO EIGHTH YEAR.
AVERAGE LENGTH OF CROWN, .34.
AVERAGE LENGTH OF ROOT, .47.
AVERAGE LENGTH OVER ALL, .81.

Like the superior central incisor, this tooth presents for examination four surfaces, a cutting-edge, and various angles, margins, etc. By the union of the labial and lingual surfaces at the cutting-edge the incisive feature is established and the double incline plane common to incisors produced.

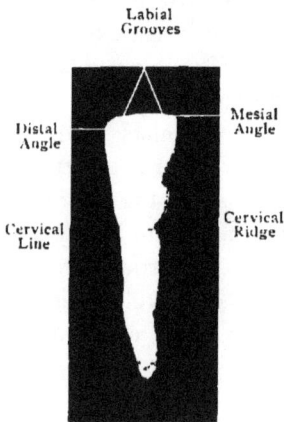

FIG. 106.—INFERIOR INCISOR, RIGHT SIDE, LABIAL SURFACE.

The Labial Surface of the Crown (Fig. 106).—This surface is smooth and convex, its general outline resembling an inverted cone, the base of which is formed by the cutting-edge and the apex by the cervical line. The margins of the surface are, with the exception of the cutting-edge, not so well defined as those of the superior central. Near the cutting-edge the mesial and distal margins pass somewhat abruptly into the respective lateral surfaces, but as the neck of the tooth is approached, they are much rounded. The incisive margin is squarely cut, and is nearly at right angles with the long axis of the tooth. The cervical margin is fairly well defined, and is deeply concave in the direction of the root. Except in very young teeth, this surface is seldom much broken by the labial grooves; but in certain types one or more transverse ridges may be found occupying the cervical third. The mesiodistal diameter at the cutting-edge is about one-third greater than at the cervical line, and, while these measurements are likely to vary in accordance with the temperament of the subject, this variation is not so pronounced as in the superior incisor. The surface is frequently inclined to flatness near the incisive margin, the general convexity becoming more marked as the cervical line is approached.

The Lingual Surface of the Crown (Fig. 107).—In general outline this surface resembles the labial, with the exception of the cervical margin, the lines of which are somewhat more acute. The surface presents a marked concavity from the cutting-edge to the cervical ridge, and also a slight transverse concavity near the incisive margin. All of the margins are more definite than those of the labial surface. The mesial and distal margins are formed by the marginal ridges common to these borders, but these ridges are not so well defined as those of the superior incisors. The cervical-marginal ridge is present as a well-rounded band of enamel, but is never a well-defined cingulum, or cuspule. The depression between these marginal ridges is so slight that it can scarcely be referred to as a fossa, although usually characterized as the lingual fossa. The lingual grooves are generally more pronounced than the corresponding developmental grooves of the labial surface, but end more or less abruptly before reaching the cervical ridge. The mesiodistal measurements of the surface are a trifle less than the corresponding measurements of the labial surface.

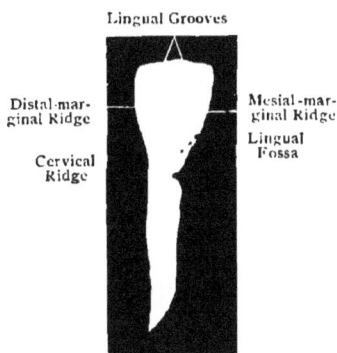

Lingual Grooves

Distal-marginal Ridge

Cervical Ridge

Mesial-marginal Ridge

Lingual Fossa

FIG. 107.—INFERIOR INCISOR, RIGHT SIDE, LINGUAL SURFACE, STRONGLY DEVELOPED.

The Mesial Surface of the Crown (Fig. 108).—The outline of this surface is exactly the reverse of the labial and lingual just described, being a cone, with its base directed downward or in the direction of the root, while its apex is formed by the mesial extremity of the cutting-edge. The cervical margin of the surface, or that represented by the base of the cone, is concave; the labial and lingual margins are rounded over the cervical third and inclined to angularity near the cutting-edge. There is a slight convexity over the entire surface, which is most marked near the center. The lingual half of the cervical portion slopes away to the distal, passing gradually into the lingual surface. By the union of this surface with the cutting-edge and the labial and lingual surfaces, the mesial angle of the crown is formed. This angle is well outlined and reaches out toward the median line, giving to this portion of the crown a prominent appearance. Near the cutting-edge the surface presents a slightly rounded prominence, which provides a point of contact with the corresponding tooth of the opposite side.

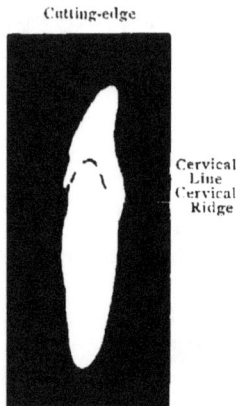

FIG. 108.—INFERIOR INCISOR, MESIAL SURFACE.

The Distal Surface of the Crown (Fig. 109).—In a general way this surface closely resembles that of the opposite or mesial side of the crown. Near the cutting-edge the surface is usually more prominent and presents a more marked convexity, and near the cervical margin it is flattened and sometimes slightly concave. The union of this surface with the cutting-edge and the labial and lingual surfaces forms the distal angle of the crown, which, like the mesial angle, is square and well

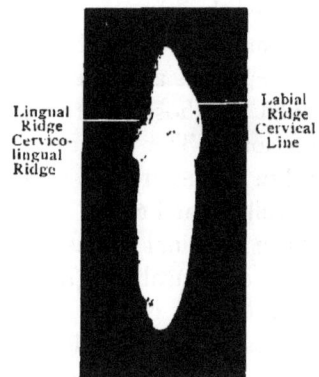

FIG. 109.—INFERIOR INCISOR, DISTAL SURFACE.

defined. The margins of the surface are in no way different from those of the mesial surface.

The Cutting-edge.—In the young tooth this incisive margin is thin and generally divided into three distinct parts (Fig. 110) by the developmental grooves, but these disappear so early that they can scarcely be considered in connection with a description of the fully developed tooth ; in fact, the cutting-edge of this, as well as that of all the inferior incisors, is so susceptible to change by mechanical abrasion that a normal condition is of but short duration. After the disappearance of the primitive cusps, and before further abrasion has taken place, the edge is fairly sharp and placed nearly at right angles with the crown. As in the superior incisors, the cutting-edge is in a line with the long axis of the tooth. The labial margin of the edge is slightly convex, while the lingual is irregularly concave to the same extent.

Developmental Grooves

The Cervical Margin.—This marginal line, which is marked by the extent of the enamel cap, corresponds closely to that of the superior incisors, dipping down with a graceful concavity on the labial and lingual surfaces, with a corresponding convexity on the mesial and distal surfaces. The prominence of this enamel margin, together with the nature of the curvature, is much influenced by the tooth type.

FIG. 110. — YOUNG INFERIOR INCISOR, WITH CUTTING-EDGE, SHOWING THE LINES OF DEVELOPMENT.

The Neck of the Tooth.—A distinctive feature of this tooth is found in the convergence of its mesial and distal surfaces in passing from the cutting-edge rootward, thus being productive of a neck much constricted from mesial to distal. When examined from either the mesial or distal surface, this feature is scarcely noted, the crown passing into the root with little less than the cervical line as a mark of separation. The labial and lingual portions of the neck are rounded and narrow, while the two lateral sides are flat and broad.

The Root.—The root of this tooth is usually smaller than

that of any other tooth in the mouth. It is much flattened from mesial to distal, while the labial and lingual aspects are rounded and narrow. Besides being flattened and broad, the mesial and distal sides are usually found with a longitudinal depression extending from a point near the base of the root almost to its apex. These surfaces gradually taper from the base to the apex, while the labial and lingual first widen from the base and then gradually taper to the apex. The contour of the root-base is generally reduced at the apical extremity, although in some instances the latter is a rounded point. While the root of the tooth is usually straight, there is sometimes a tendency for the apical third to have a slight distal inclination.

INFERIOR LATERAL INCISOR.

CALCIFICATION BEGINS, FIRST YEAR AFTER BIRTH.
CALCIFICATION COMPLETED, TENTH TO ELEVENTH YEAR.
ERUPTS, EIGHTH TO NINTH YEAR.
AVERAGE LENGTH OF CROWN, .35.
AVERAGE LENGTH OF ROOT, .50.
AVERAGE LENGTH OVER ALL, .85.

The crown of this tooth differs from the central incisor in being broader from mesial to distal at the cutting-edge, resulting in a crown more strongly bell-shaped. The cutting-edge, instead of being at right angles to the long axis of the tooth, slopes to the distal at the expense of the distal angle, which is much rounded, while the mesial angle closely resembles the corresponding angle of the central incisors. The labial and mesial surfaces do not differ materially from the corresponding surfaces of the central incisor, excepting that the lingual more frequently shows the lines of development, and the distal is at variance in having that portion which contributes to the formation of the distal angle extended and prominent. The marginal ridges of the lingual surface are probably more definitely outlined than those of the central incisor, and the crown in general presents a stronger appearance. The neck of the tooth is similar to the neck of the central incisor, as is also the root, with the exception of a slight addition to its length.

INFERIOR CUSPID.

CALCIFICATION BEGINS, THIRD YEAR AFTER BIRTH.
  CALCIFICATION COMPLETED, TWELFTH TO THIRTEENTH YEAR.
      ERUPTS, TWELFTH TO THIRTEENTH YEAR.
AVERAGE LENGTH OF CROWN, .40.
    AVERAGE LENGTH OF ROOT, .60.
      AVERAGE LENGTH OVER ALL, 1.00.

There is probably a greater similarity existing between the superior and inferior cuspid teeth than in any other class of teeth in the mouth. Occupying as they do a most prominent position in the dental arch, and being called upon to perform the double function of incis-ing and tearing the food, their crowns are strong and heavy-set, and their roots long and firmly anchored in the alveoli. Like the superior cuspid, the crown of the inferior is sur-mounted by a single cusp, from the summit of which de-scend a mesial and a distal cutting-edge. There is also a labial, lingual, mesial, and dis-tal surface presented for ex-amination.

The Labial Surface of the Crown (Fig. 111).—The crown of the tooth, being a little longer than that of the superior cuspid, gives to this surface the appearance of being more slender, when in reality there is but little difference in the width of the two teeth. This surface is smooth and convex, and, while the labial grooves are usually present, they are not so marked as those found upon the corresponding superior tooth. A pronounced feature of the surface is the labial ridge, which extends from the summit of the cusp to the cervical line, providing additional strength to the crown. Aside from this ridge and the labial

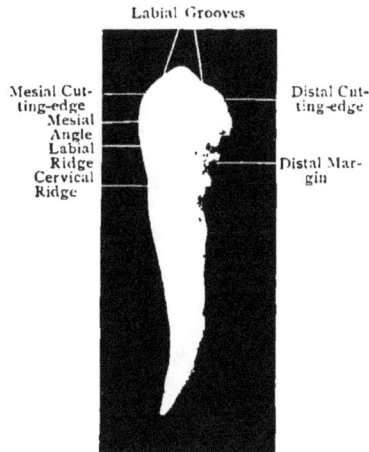

FIG. 111.—INFERIOR CUSPID, LABIAL SUR-FACE.

14

grooves, the surface is occasionally broken by one or more transverse ridges over the cervical portion. The margins of the surface closely resemble those of the superior cuspid, the incisive and mesial being definite in character, while the distal is made equally indefinite by the passing of the labial into the distal surface by a gentle curve.

The Lingual Surface of the Crown (Fig. 112).—The ridges and grooves of this surface are far less bold in their character than those of the superior cuspid. The lingual ridge, which divides the surface into two equal parts, extends from the summit of the cusp to the base of the cervical ridge, while the marginal ridges pass rootward from the angles of the crown, and, uniting, form the cervical ridge. The slight depressions between the lingual ridge and the marginal ridges correspond to the palatal grooves of the superior cuspid, but in this tooth partake more of the nature of fossæ.

Lingual Grooves

Distal Angle
Distal-marginal Ridge

Lingual Ridge
Mesial-marginal Ridge
Cervical Ridge

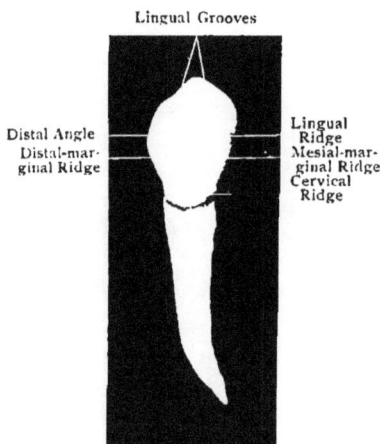

FIG. 112.—INFERIOR CUSPID, LINGUAL SURFACE.

The Mesial Surface of the Crown (Fig. 113).—A peculiarity found in connection with this surface is the general plane existing between the crown and root-surface. In all other teeth the mesial and distal surfaces are found to bulge somewhat beyond the corresponding surface of the root, but this surface of the inferior cuspid is not only usually in a direct line with the mesial surface of the root, but is occasionally inclined to the distal, resulting in a crooked or bent appearance to the tooth. In addition to this individual peculiarity, the surface is flat and passes by a long curve to meet the lingual surface.

The Distal Surface of the Crown (Fig. 114).—This surface is somewhat less in extent than the mesial surface, and, in place of being flat and in line with the root-surface, that portion near the angle of the crown presents a marked convexity, while that near the cervical line is frequently slightly concave. This general form of the surface further assists in producing the distal crook previously referred to. The lingual margin is well defined and somewhat angular, while the surface passes so gradually into the labial that a positive line of demarcation can scarcely be said to exist.

The Cusp and Cutting-edges.—In most respects these are similar to the corresponding parts of the superior cuspid.

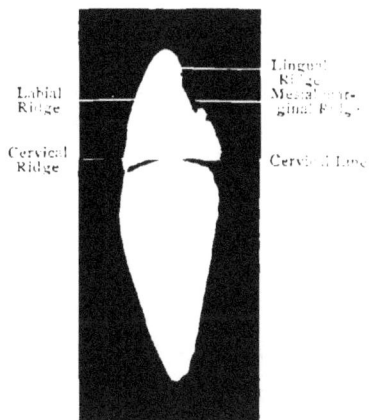

Fig. 113.—INFERIOR CUSPID, MESIAL SURFACE.

Labial Ridge · Cervical Ridge · Lingual Ridge · Mesial marginal Ridge · Cervical Line

The length of the mesial cutting-edge is usually somewhat less than that of the distal, but this difference is seldom so marked as that found in the superior cuspid. The mesial and distal angles of the crown are equally as pronounced as those of the corresponding superior tooth.

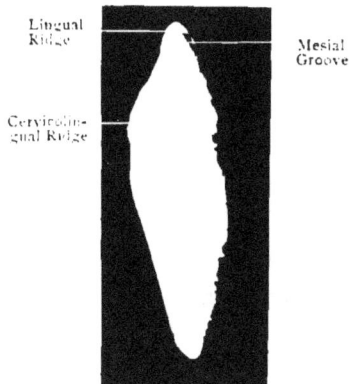

Fig. 114.—INFERIOR CUSPID, DISTAL SURFACE.

Lingual Ridge · Cervicolingual Ridge · Mesial Groove

The Neck of the Tooth. —This is shown by a fairly well-marked constriction, but the passing of the mesial surface of the crown into the mesial surface of the root is not broken by this circular depression. On account of this latter feature the neck of this tooth is somewhat less pronounced than that of the superior.

The Root of the Inferior Cuspid.—The root of this tooth is somewhat shorter and more flattened on its mesial and distal sides than that of the superior cuspid, this lateral flatness frequently amounting to a decided longitudinal depression or groove. As referred to in the description of the crown, the mesial side of the root is continuous in a direct line with this surface of the crown, but as the apical third of the root is approached, there is frequently found a slight distal inclination which affects alike both the mesial and distal sides. The labial and lingual surfaces of the root are abruptly convex and taper very gradually from the cervical line to the apex, while the mesial and distal surfaces taper much more rapidly, the four ending in a slender apex usually flattened from mesial to distal.

THE INFERIOR BICUSPIDS.

In many respects these teeth are similar to the bicuspids of the upper jaw, but there are a few minor points of dissimilarity. They are somewhat shorter and smaller in every respect; their crowns are much more rounded and the cusps are never so strongly developed. Unlike the superior bicuspids, the buccal and lingual cusps are connected by a transverse ridge. The roots are much less flattened from mesial to distal, and are seldom, if ever, bifurcated.

INFERIOR FIRST BICUSPID.

CALCIFICATION BEGINS, ABOUT THE FOURTH YEAR.
CALCIFICATION COMPLETED, ELEVENTH TO TWELFTH YEAR.
ERUPTED, TENTH TO ELEVENTH YEAR.
AVERAGE LENGTH OF CROWN, .30.
AVERAGE LENGTH OF ROOT, .54.
AVERAGE LENGTH OVER ALL, .84.

In general, the crown of this tooth is much more rounded and smaller in all its measurements than that of the superior bicuspid. The buccal surface presents a much greater convexity, which results in forcing the summit of the buccal cusp well toward the center of the long axis of the tooth. The mesio-distal and buccolingual measurements of the crown are nearly equal, and about correspond to the maximum length of the crown. As in the superior bicuspids, the development of this tooth is similar to that of the incisors and cuspids, the buccal cusp being derived from three lobes, while the lingual results from a single center.

The Occlusal Surface of the Crown (Fig. 115).—This surface is so unlike that of the corresponding surface of the superior first bicuspid that a separate description without further comparative reference is required. In general outline the form of a rounded triangle is approached, the buccal margin serving as one side of the triangle, while, by the union of the mesial and distal margins to form the lingual, the remaining sides are established.

The Buccal Cusp.—As previously stated, the summit of this cusp is thrown well toward the center of the surface. Descending from it are four well-defined ridges—the buccal ridge to the buccal surface, the mesial cutting-edge, the distal cutting-edge, and the triangular ridge, the latter descending in a lingual direction to meet the lingual ridge or cusp. This ridge divides the surface into two parts, the center of each being marked by a well-defined pit—the *mesial* and *distal pits.* The mesial and distal cutting-edges are frequently crossed by the buccal grooves, and mark the line of union between the central and two lateral lobes of the buccal cusp. The marginal ridges, one of which begins at the mesial angle and the other at the distal angle, pass to the lingual, where they unite to form the *lingual ridge* or *cusp.*

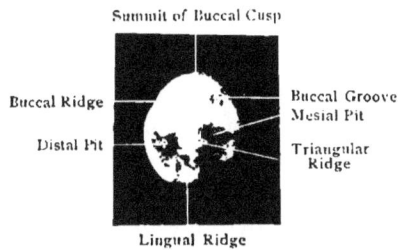

Summit of Buccal Cusp

Buccal Ridge

Distal Pit

Buccal Groove
Mesial Pit

Triangular
Ridge

Lingual Ridge

FIG. 115.—INFERIOR FIRST BICUSPID,
OCCLUSAL SURFACE.

The Lingual Cusp.—This cusp is seldom well developed, and corresponds to the cervical ridge of the incisors and cuspids. The extent of development in the lobe is extremely variable, in some instances amounting to little more than a continuation of the mesial- and distal-marginal ridges, while in others there is a building-up of the enamel in the form of a small tubercle. When this latter condition is present, the triangular ridge of the buccal cusp contributes to its formation. The triangular ridge frequently divides into two or more smaller ridges, which usually end in the mesial pit, but in some instances they continue to the lingual, and divide the lingual ridge into two or more smaller tubercles.

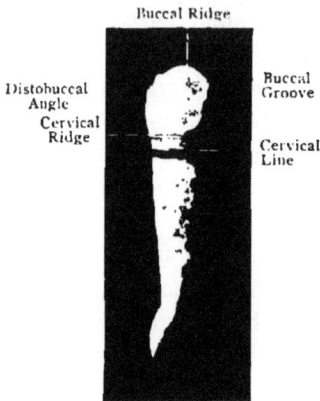

FIG. 116.—RIGHT INFERIOR FIRST BICUSPID, BUCCAL SURFACE.

The Buccal Surface of the Crown (Fig. 116).—This surface is smooth and convex in all directions, and in general outline there is but little variation between it and the corresponding surface of the superior first bicuspid. It is traversed from the point of the cusp to the cervical line with a rounded ridge, the buccal ridge, upon either side of which are the buccal grooves.

Lingual Surface of the Crown (Fig. 117).—This surface is more or less extensive in accordance with the character of the lingual lobe. In most instances the measurement from the summit of the cusp or ridge to the cervical line is about one-half that of the same measurements on the buccal surface. From mesial to distal a well-rounded convexity is present, while from the occlusal margin to the cervical line it is straight or only slightly convex. The surface passes so gradually into the mesial and distal surfaces that no definite lateral margins exist.

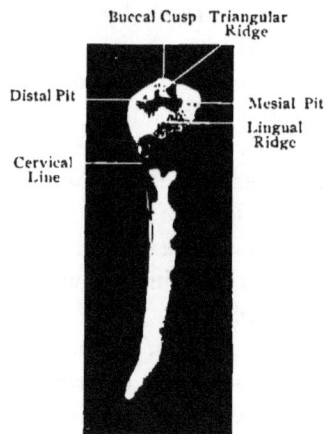

FIG. 117.—LEFT INFERIOR FIRST BICUSPID, LINGUAL SURFACE.

The Mesial Surface of the Crown (Fig. 118).—In the region of the occlusal margin this surface is prominent, with a marked convexity from buccal to lingual; but as the cervical margin is approached, the surface recedes to the distal, and is flattened or is possessed of a slight general convexity. The occlusal and cervical margins alone are well defined, the buccal being gracefully rounded, while the surface passes to the lingual with a long curve.

The Distal Surface of the Crown (Fig. 119).—There is but little difference between this and the mesial surface; the occlusal portion of the surface is somewhat less prominent, resulting in less of the bell-shaped appearance to this side of the crown.

The Neck of the Tooth.— The neck of this tooth is marked by a well-defined constriction, the enamel of the crown suddenly folding in to meet the cementum of the root at the cervical line, forming a band or ridge which completely encircles the tooth. The amount of constriction appears to be evenly distributed between the various parts, so that, viewed in all directions, the neck becomes a distinctive feature of the tooth.

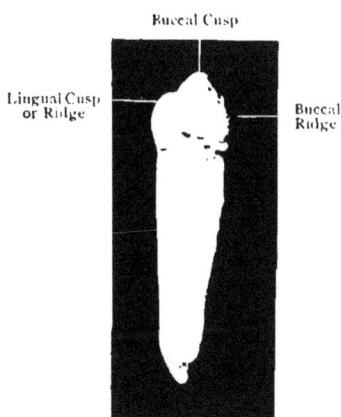

FIG. 118.—LEFT INFERIOR FIRST BICUSPID, MESIAL SURFACE.

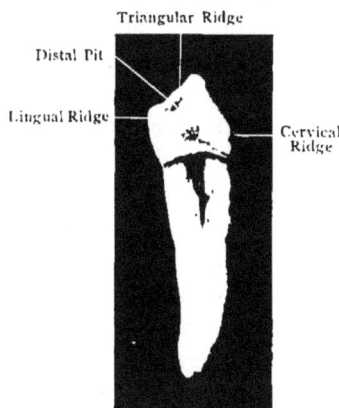

FIG. 119. — RIGHT INFERIOR FIRST BICUSPID, DISTAL SURFACE.

The Root of the Inferior First Bicuspid.—The root of this tooth is usually straight and tapers gradually from base to apex. In rare instances it is bifurcated, and when thus formed, those portions beyond the point of separation are more or less crooked. In the single root the apical third often curves slightly to the distal. The buccal and lingual sides are convex throughout their entire length, while the mesial and distal may be slightly convex, flattened, or provided with a slight longitudinal concavity. In passing from buccal to lingual the mesial and distal sides converge, thus resulting in a narrowing of the lingual side of the root.

INFERIOR SECOND BICUSPID.

CALCIFICATION BEGINS, BETWEEN THE FOURTH AND FIFTH YEAR.
CALCIFICATION COMPLETED, ELEVENTH TO TWELFTH YEAR.
ERUPTS, ELEVENTH TO TWELFTH YEAR.
AVERAGE LENGTH OF CROWN, .31.
AVERAGE LENGTH OF ROOT, .56.
AVERAGE LENGTH OVER ALL, .87.

In general contour this tooth is similar to the inferior first bicuspid, excepting that the crown is somewhat more rounded and the lingual cusp more fully developed, this latter feature causing it to closely resemble the superior bicuspids. The crown of this tooth is frequently a trifle shorter than that of the first inferior bicuspid, but the length of the root generally exceeds that of the latter, making this the longer tooth of the two. **The Occlusal Surface of the Crown** (Fig. 120).— The occlusal surface of this tooth presents a greater variety in form than any other tooth of its class. The general outline of the surface is that of a broken circle, in most instances the mesial and distal margins showing almost as much of a convexity as that of the buccal and lingual. The summit of the buccal cusp usually extends well toward the center of the surface, but it is sometimes forced toward the buccal by an increased development in the lingual cusp. The buccal grooves, which cross the mesial and distal cutting-edges of the buccal cusp, are seldom so well defined as those of the first bicuspid, but they occasionally pass over these marginal ridges and form well-marked grooves, which end in the mesial and distal pits. The triangular ridge of the buccal cusp is usually more prominent than in the

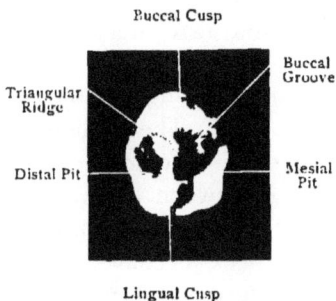

FIG. 120.—INFERIOR SECOND BICUSPID, OCCLUSAL SURFACE.

Buccal Cusp
Buccal Groove
Triangular Ridge
Distal Pit
Mesial Pit
Lingual Cusp

first bicuspid, and divides the surface into two portions, which
are about equal in extent, the center of each portion being provided with a small pit—the mesial and distal pits. As in the first
bicuspid, the mesial- and distal-marginal ridges begin at the
mesial and distal angles of the crown, pass to the lingual, and,
uniting, form the lingual ridge or cusp. The lingual cusp, while
generally well developed, is never so prominent as the buccal.
The lingual lobe is sometimes divided by a groove which passes
from buccal to lingual, thus forming three cusps upon the surface. When this latter condition is present, the mesial and distal
grooves are fully outlined from the
mesial- and distal-marginal ridges
to the center of the surface, where
they unite with the groove previously referred to and form a central pit or fossa. Another form
frequently met with is one in which
the surface closely resembles that
of the superior bicuspids, two well-
defined cusps being present, separated from each other by a central
groove, which passes from mesial
to distal and joins the triangular
grooves at these points. The re
semblance to the superior bicuspids is further increased by the
presence of two small pits, one on the mesial and one on the
distal half of the surface.

Buccal Ridge

Distobuccal
Angle

Cervical
Ridge

Buccal
Groove

Cervical
Line

FIG. 121.— RIGHT INFERIOR SECOND
BICUSPID, BUCCAL SURFACE.

The Buccal Surface of the Crown (Fig. 121).—The
principal variation between this and the buccal surface of the
inferior first bicuspid is that it is less extensive and the buccal
grooves somewhat less defined. It presents a general convexity, which is most pronounced near the center, between
which point and the occlusal margins it is slightly inclined to
flatness. The summit of the buccal cusp is usually to the
mesial of the center of the occlusal margin, so that the mesial
cutting-edge is considerably longer than the distal, this fact also
resulting in forcing the buccal ridge to the mesial of the center

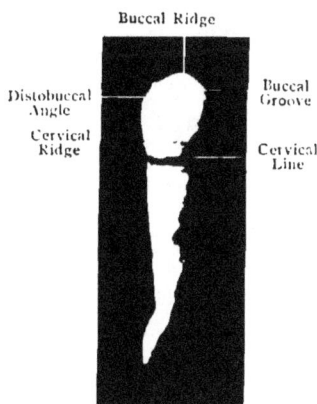

of the surface. The mesial angle of the crown, as observed when looking directly upon the buccal surface, is in a direct line with the mesial side of the root, while the distal angle extends beyond this corresponding line, and gives a prominent or bulging appearance to this section of the crown.

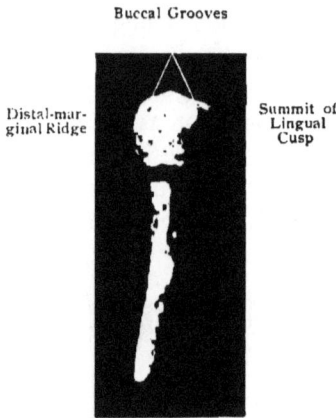

FIG. 122.—INFERIOR SECOND BICUSPID, LINGUAL SURFACE.

The **Lingual Surface of the Crown** (Fig. 122).—Proportionately, this surface is more extensive than the corresponding surface of the first bicuspid, this increase being produced by the additional development of the lingual cusp. It is well rounded from mesial to distal, and passes into these surfaces without the existence of a positive line of separation. From the cervical line to the occlusal margin a slight convexity is present. The general outline of the surface is much influenced by the conditions present upon the occlusal surface.

The **Mesial Surface of the Crown** (Fig. 123).—In the region of the occlusal margins this surface is decidedly convex from buccal to lingual, but in passing toward the cervical line a gradual flatness is apparent, which, however, seldom amounts to a perfect plane. While there is a gradual convergence of this and the distal surface toward the root, it is not so marked as that of the first bicuspid, resulting in less of the bell-shaped appearance to the crown.

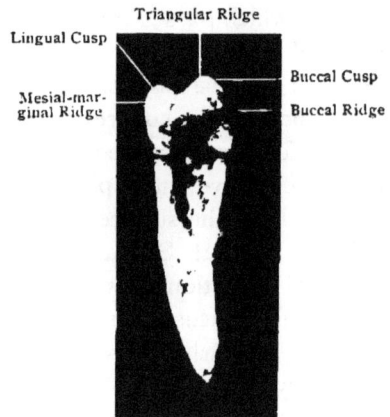

FIG. 123.—INFERIOR SECOND BICUSPID, MESIAL SURFACE.

The Distal Surface of the Crown (Fig. 124).—The description given of the mesial surface applies equally well to this, there being but slight variation existing between the two. Occasionally this surface will present a greater convexity in the region of the occlusal margin, but this is not a constant feature.

The Neck of the Tooth.—The crown of the tooth being somewhat smaller, and the root proportionately larger and longer than that of the first bicuspid, results in diminishing the amount of constriction at the neck, and for that reason this feature is less definite.

The Root of the Tooth.—As previously stated, the root of this tooth is larger and longer than that of the first bicuspid. The mesial and distal sides are flattened and frequently provided with a longitudinal groove. In some instances it is rather blunt, ending in a heavy, rounded apex ; in others it tapers very gradually from the base to the apex, ending in a slim, pointed extremity.

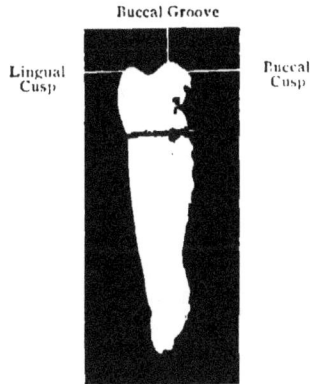

FIG. 124.—INFERIOR SECOND BICUSPID, DISTAL SURFACE.

THE INFERIOR MOLARS.

THE INFERIOR FIRST MOLAR.

CALCIFICATION BEGINS, ABOUT ONE MONTH BEFORE BIRTH.
CALCIFICATION COMPLETED, NINTH TO TENTH YEAR.
ERUPTS, SIXTH TO SEVENTH YEAR.
AVERAGE LENGTH OF CROWN, .30.
AVERAGE LENGTH OF ROOT, .52.
AVERAGE LENGTH OVER ALL, .82.

The process of development in this tooth corresponds to that of the superior first molar, calcification beginning upon the coronal extremities as early as the eighth fetal month, the crown being completely calcified by the fifth year, the roots formed and the apical foramina established by the eleventh year.

There is, however, one important difference between the development of this tooth and the corresponding superior tooth: that of calcification taking place usually from *five centers* instead of four, and, as a result, we find the occlusal surface provided with five well-developed cusps separated from one another by five developmental grooves. When compared with the superior first molar, the crown of this tooth is found to be somewhat less in size; in general outline it is subject to a greater variation and is much more angular in its nature. The mesiodistal measurement of the crown is nearly always greater than the buccolingual, and the length of the crown from the occlusal margins to the cervical line is proportionately less than that of the corresponding superior molar.

The Occlusal Surface of the Crown (Fig. 125).—The general outline of the crown is best studied when looking directly upon this surface. Two principal varieties exist: one in which the sides or margins of the surface appear to be flattened or straightened out, and the other when these same margins are gracefully rounded. In either form the buccal line is the longest, so that the mesial and distal lines converge to meet the lingual.

Buccal Cusp   Buccal Groove

Distobuc- cal Cusp

Mesiobuc- cal Cusp

Distal Groove

Central Fossa

Distal Pit

Mesial Groove

Disto- lingual Cusp   Lingual Groove   Mesio- lingual Cusp

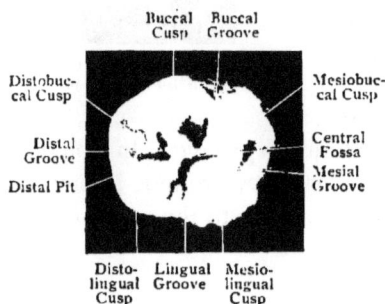

FIG. 125.—INFERIOR FIRST MOLAR, OCCLUSAL SURFACE.

This common form gives to the buccal angles an acute character, while the lingual angles are about equally obtuse. The surface is divided into five distinct or developmental portions, each of which is surmounted by a cusp, named, as their location indicates, mesiobuccal, *buccal*, distobuccal, mesiolingual, and distolingual. Separating these parts are five developmental grooves —the mesial, the distal, the buccal, the lingual, and the distobuccal. The four former cross the marginal ridges from the various surfaces and end in the central˙ fossa, while the latter passes from the distobuccal angle and joins the distal groove, their union being marked by a slight depression or pit—the distal pit. Branching off from the various grooves are a number of supplemental grooves, the presence of which results in the production of a number of smaller ridges.

The Marginal Ridges of the Occlusal Surface.—Properly speaking, these are only two in number, the mesial-marginal ridge and the distal-marginal ridge. Those margins which correspond to the buccal and palatal ridges of the superior molars are so broken by the various cusps and developmental grooves that a definite marginal ridge scarcely exists.

The mesial-marginal ridge is strongly outlined, passing from the mesiobuccal to the mesiolingual angle of the crown in the form of a bold angular ridge. In some instances it is broken near the center by the mesial groove passing over it to reach the mesial surface, in others being further divided by numerous small supplemental grooves.

The distal-marginal ridge is much shorter and less decided than the mesial, and extends from the distobuccal to the distolingual angle. In nearly every instance it is broken by the distal groove, which crosses it to reach the distal surface.

The buccal-marginal ridge is formed by the various ridges which descend in a mesial or distal direction from the three buccal cusps. Near the center the margin is broken by the buccal groove, and it is again broken at its distal third by the distobuccal groove, both of which pass over it to reach the buccal surface of the crown. This is much the longest margin of the surface, and in its entirety presents a gradual buccal convexity.

The lingual-marginal ridge is principally made up of the distal incline from the mesial cusp, and by the mesial incline from the distal cusp. Near the center it is broken by the lingual groove, which passes over it to reach the lingual surface. This margin, unlike the buccal, is not always convex, but in many instances is almost a straight line, extending from the mesial to the distal angle.

The Cusps (Fig. 125).—*The Mesiobuccal Cusp* (Fig. 125).— This is usually the largest, though not always the longest, cusp of the group. It is bounded by the mesial and buccal surfaces and by the mesial and buccal grooves, which together form the mesiobuccal triangular groove. Descending from the summit of this cusp to the distal is a well-defined ridge,—a part of the buccal-marginal ridge,—while in a mesial and lingual direction the descending ridges contribute to both the buccal- and mesial-marginal ridges. Descending toward the center of the surface and ending in the central fossa is the mesiobuccal triangular ridge.

The Buccal Cusp (Fig. 125).—This cusp, which is placed a little to the distal of the center of the buccal surface, is separated from the mesiobuccal cusp by the buccal groove, and from the distobuccal cusp by the distobuccal groove. It is about one-half the size of the mesiobuccal cusp, and a trifle less in length. Descending from it are two ridges, one in a mesial and one in a distal direction, which form a portion of the buccal-marginal ridge; descending to the buccal surface is the buccal ridge, while the central incline gives place to a fourth ridge— the buccal-triangular ridge.

The Distobuccal Cusp (Fig. 125).—This cusp is much the smallest of the five, and is located at the distobuccal portion of the crown, in some instances being nearest the buccal surface, in others forced to the distal by an increase in the size of the buccal cusp. It is separated from the buccal cusp by the disto-buccal groove, and from the distolingual cusp by the distal groove. The ridges which descend from it contribute to both the buccal- and distal-marginal ridges, and descending toward the distal pit is the distobuccal triangular ridge.

The Mesiolingual Cusp (Fig. 125).—This cusp is second in size, and frequently the longest and most pointed. It has for its boundaries the mesial and lingual surfaces, and the mesial and lingual grooves. The ridge which descends from it in a mesiobuccal direction assists in forming the mesial-marginal ridge, while that which passes to the distal forms a part of the lingual-marginal ridge. In the direction of the central fossa a pronounced ridge is present,—the mesiolingual triangular ridge, —which is often supplemented by one or more smaller ridges running in the same direction.

The Distolingual Cusp (Fig. 125).—This cusp usually occupies the distolingual portion of the crown, although sometimes being forced well toward the lingual by the distobuccal cusp.

It is separated from the mesiolingual cusp by the lingual groove, and from the distobuccal by the distal groove. Two of the ridges which descend from it assist in forming the lingual- and distal-marginal ridges, while the one which descends the central incline is the distolingual triangular ridge. The central incline of the mesiobuccal, buccal, mesiolingual, and distolingual cusps contribute to the formation of the central fossa, while the buccal, distobuccal, and distolingual central inclines assist in forming the distal pit or fossa.

The Buccal Surface of the Crown (Fig. 126).—This is the most extensive of the lateral surfaces of the crown. It is convex from mesial to distal, and also from the occlusal margin to the cervical line. The width of the crown from the mesial to the distal angle is always somewhat greater than that at the cervical line, the difference being governed by the typal form of the tooth. A little to the mesial of the center of the surface is the buccal groove, which, after crossing the buccal-marginal ridge, is usually quite deep ; but as it proceeds in the direction of the root it gradually disappears, or it may end abruptly in a well-defined pit —the *buccal pit.* The *distobuccal groove* enters the surface near the distobuccal angle, and gradually becomes less pronounced as it passes rootward. It is seldom so well defined as the buccal groove, and usually ends when about half way to the cervical line. The occlusal margin is made irregular by the presence of the three buccal cusps ; the cervical margin is nearly straight from mesial to distal, and is surmounted throughout by a strong

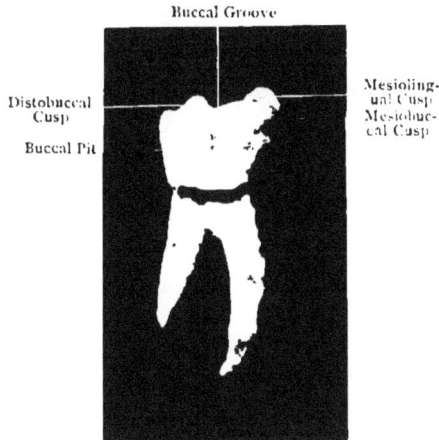

Buccal Groove

Distobuccal Cusp

Buccal Pit

Mesiolingual Cusp
Mesiobuccal Cusp

FIG. 126.—INFERIOR FIRST MOLAR, BUCCAL SURFACE.

enamel fold, the cervicobuccal ridge. The mesial margin s longer than the distal, but neither of them are well defined.

The Lingual Surface of the Crown (Fig. 127).—This sur-

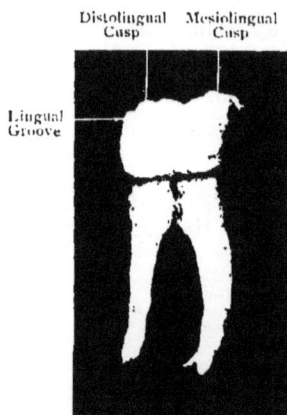

Distolingual Mesiolingual
Cusp   Cusp

Lingual
Groove

FIG. 127.—INFERIOR FIRST MOLAR,
LINGUAL SURFACE.

face is smooth and convex in every direction. It is generally divided into two portions, a mesial and a distal, which are nearly equal in extent. This separation is formed by the *lingual groove*, which is sometimes deep and sulcate, at others shallow, and not infrequently entirely wanting. The surface is nearly one-third less in extent than the buccal, the convergence of the mesial and distal surfaces in passing to the lingual accounting for this difference. The occlusal margin is formed by the double incline of the two lingual cusps; the cervical margin is either straight or slightly concave in the direction of the occlusal surface, while the mesial and distal margins are rounded and poorly defined.

The Mesial Surface of the Crown (Fig. 128).—This surface is inclined to flatness, with a slight bulging near the center, which marks the point of contact with the approximate tooth. It is usually smooth, and unbroken by developmental or other grooves, although the mesial groove occasionally traverses it after crossing the marginal ridge from the occlusal surface.

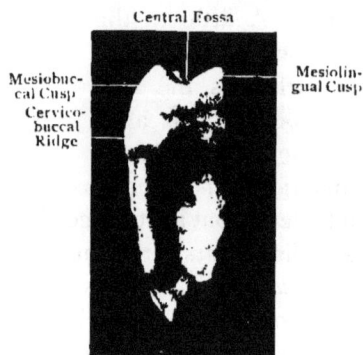

Central Fossa

Mesiobuc-
cal Cusp
Cervico-
buccal
Ridge

Mesiolin-
gual Cusp

FIG. 128.—INFERIOR FIRST MOLAR, MESIAL
SURFACE.

Near the center of the cervical third a slight concavity is often present. The margins of the surface are somewhat irregular, the occlusal margin being made irregularly concave by the ridges which descend from the two mesial cusps ; the cervical margin is slightly concave in the direction of the occlusal surface, and, while the buccal margin inclines to the lingual as the occlusal surface is approached, the lingual is almost perpendicular.

The Distal Surface of the Crown (Fig. 129).—Unlike the mesial, this surface is possessed of a decided convexity in every direction. It is surmounted by a portion of the distobuccal and distolingual cusps, and is frequently broken by the distal groove, which reaches it after crossing the marginal ridge from the occlusal surface. The occlusal margin is irregularly formed of the marginal ridges which descend from the distobuccal and distolingual cusps ; the cervical margin is usually straight, while the buccal and lingual are rounded and indefinite.

The Neck of the Tooth. — One characteristic feature of this tooth is the greater circumference of the crown at the occlusal margin over that at the cervical line, giving a flaring appearance to the crown, and resulting in the production of a neck which is much constricted. This is particularly noticeable when looking upon the

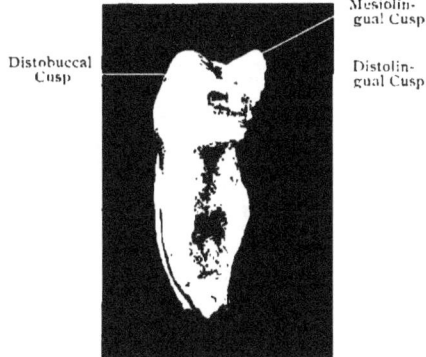

Mesiolingual Cusp

Distobuccal Cusp

Distolingual Cusp

FIG. 129.—INFERIOR FIRST MOLAR, DISTAL SURFACE.

buccal surface of the tooth ; but when looking upon the mesial surface or the distal surface, this feature is not so pronounced, although the rather heavy fold of enamel which surmounts the cervical line contributes much to the formation of the neck from these aspects.

The Roots of the Tooth.—The roots of this tooth are two in number,—one of which is placed beneath the mesial, and the other beneath the distal half of the crown,—and are named the *mesial root* and the *distal root*. The fact that the point of bifurcation is constantly in close proximity to the neck or crown of the tooth is a sufficient reason for the statement that two roots exist,

15

rather than a single root with two branches. The roots are both much flattened from mesial to distal, and broad at the base from buccal to lingual.

The mesial root is usually the largest and longest of the two. After leaving its base it generally inclines to the mesial, but beyond the center of its length it is provided with a distal turn, which in some instances amounts to a decided crook. The center of the mesial side is occupied by a longitudinal depression, as is also the distal side, making this part of the root thin, giving the appearance of an effort to bifurcate, which condition is occasionally present. The buccal and lingual sides of the root are rounded and smooth, and taper gradually to the apex, which is somewhat broadened from buccal to lingual.

The distal root is usually straight, with a more gradual taper throughout, and ending in an apical extremity more pointed than that of the mesial root. A longitudinal depression is also present upon both the mesial and distal sides, but is never so pronounced as that upon the mesial root. The buccal and lingual sides are convex and smooth. The root possesses little or no inclination to bifurcate.

INFERIOR SECOND MOLAR.

CALCIFICATION BEGINS ABOUT THE FIFTH YEAR.
 CALCIFICATION COMPLETED, SIXTEENTH TO SEVENTEENTH YEAR.
  ERUPTS, TWELFTH TO SIXTEENTH YEAR.
AVERAGE LENGTH OF CROWN, .27.
 AVERAGE LENGTH OF ROOT, .50.
  AVERAGE LENGTH OVER ALL, .78.

This molar differs in so many particulars from the inferior first molar that a separate description will be called for. The principal variation is frequently found in the absence of the fifth lobe or cusp,* resulting in the production of an occlusal surface much less complicated.

The Occlusal Surface of the Crown (Fig. 130).—When the crown is studied by looking directly upon this surface, the

---

* When five cusps are present, the anatomy of this surface does not differ from that of the first molar.

variations between this and the first molar are readily noted. Four equally proportioned cusps are observed, separated from each other by four developmental grooves. A single pit or fossa is present, the four grooves arising from this one point. In general outline two principal varieties exist : one in which the opposite sides of the crown are nearly of the same length, and parallel with each other, with the angles rounded ; the other, in which either the buccal or lingual margin is the longest, with the mesial and distal margins converging one way or the other as the case may be. The marginal ridges are formed in a manner similar to those of the first molar, with the exception of the distal portion of the buccal ridge, which is not broken by a developmental groove. Each marginal ridge is divided near its center by one of the grooves of development, the mesial groove crossing the mesial-marginal ridge, the buccal groove crossing the buccal-marginal ridge, the lingual groove crossing the lingual ridge, and the distal groove passing over the distal ridge. In many instances numerous supplemental grooves are present, which in turn form a number of smaller ridges. The four *cusps* are the *mesiobuccal, distobuccal, mesiolingual,* and *distolingual.* In a general way they are similar to the cusps of the first inferior molar, excepting that they are somewhat larger and probably less pointed and less angular. Each cusp is provided with a number of ridges, which descend from the summit to the base, two of these contributing to the formation of the marginal ridges, one passing to the buccal or lingual, and one, the triangular ridge, descending the central incline of each cusp. The names given to these various ridges are identical with those of the first molar.

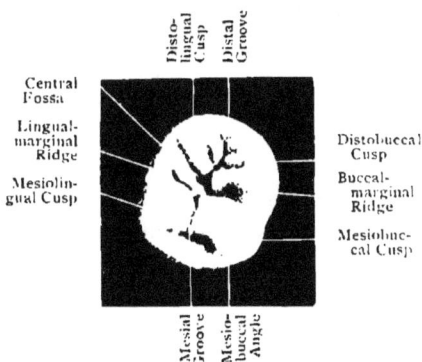

FIG. 130.—INFERIOR SECOND MOLAR, OCCLUSAL SURFACE.

The Buccal Surface of the Crown (Fig. 131).—The principal difference between this and the corresponding surface of the first molar is that produced by the absence of the fifth cusp, the surface being divided into two parts instead of three. The single division is caused by the buccal groove, which reaches the surface after crossing the buccal-marginal ridge from the occlusal surface. The position of this groove is usually a little to the mesial of the center of the surface. Like the buccal groove of the first molar, it may disappear gradually as it passes toward the cervical line, or it may end in a well-marked pit—the buccal pit.

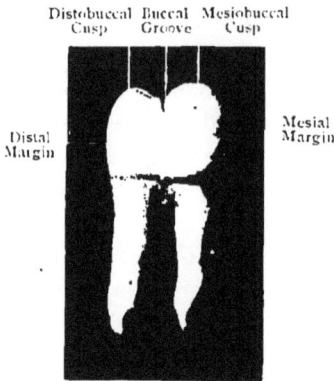

Fig. 131.—INFERIOR SECOND MOLAR, BUCCAL SURFACE.

(labels) Distobuccal Cusp — Buccal Groove — Mesiobuccal Cusp — Mesial Margin — Distal Margin

The Lingual Surface of the Crown (Fig. 132).—This surface so closely resembles the corresponding surface of the first molar that it is somewhat difficult to distinguish one from the other. The occlusal margin may be a trifle less irregular, and in some instances more extensive, than the buccal surface, this latter feature seldom occurring in the first molar.

The Mesial Surface of the Crown (Fig. 133).— This surface corresponds to the mesial surface of the first molar, being flattened or slightly convex from buccal to lingual, with an inclination to a slight depression or concavity near the cervical margin.

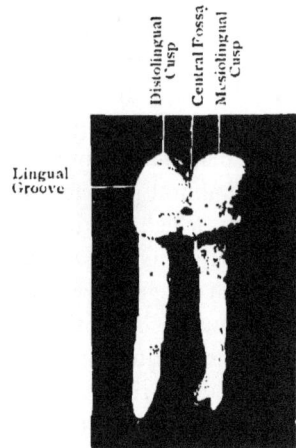

FIG. 132.— INFERIOR SECOND MOLAR, LINGUAL SURFACE.

(labels) Distolingual Cusp — Central Fossa — Mesiolingual Cusp — Lingual Groove

The Distal Surface of the Crown (Fig. 134).—On account of the absence of the fifth cusp, this surface is less complex than that of the first molar. It is convex in all directions; in most instances smooth, in others broken by the distal groove, which

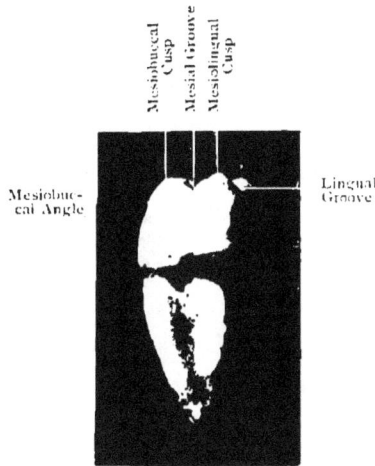

FIG. 133.—INFERIOR SECOND MOLAR, MESIAL SURFACE.

reaches it after crossing the distal-marginal ridge from the occlusal surface.

The Roots of the Tooth.—Like the first molar, these are

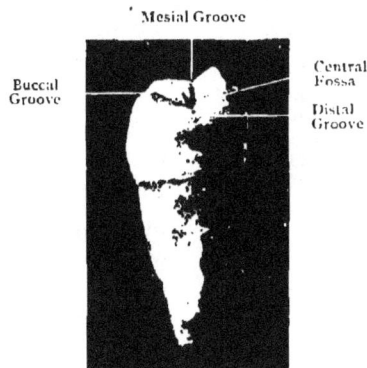

FIG. 134.—INFERIOR SECOND MOLAR, DISTAL SURFACE.

two in number, a mesial and a distal. They are much less constant in form, are often closer together, and in some instances

united. When the two roots exist,—which may be considered
the normal condition,—they are less flattened upon their mesial
and distal sides, with a longitudinal depression wanting or but
slightly apparent. These roots, therefore, are more rounded in
general, taper more gradually from neck to apex, and end in a
rounded apex, this latter extremity often being provided with a
slight distal curve.

INFERIOR THIRD MOLAR.

CALCIFICATION BEGINS, EIGHTH TO NINTH YEAR.
CALCIFICATION COMPLETED, EIGHTEENTH TO TWENTIETH YEAR.
ERUPTS, SIXTEENTH TO TWENTIETH YEAR.
AVERAGE LENGTH OF CROWN, .26.
AVERAGE LENGTH OF ROOTS, .36.
AVERAGE LENGTH OVER ALL, .62.

This tooth is probably subject to a greater variety in form
than any other tooth in the mouth. There are, however, two
varieties which are most frequently met with. In one the crown
of the tooth is similar to the inferior second molar, being provided with four cusps, which
are separated from one another
by four developmental grooves
(Fig. 135). The other is similar to the inferior first molar,
having five cusps and five developmental grooves.

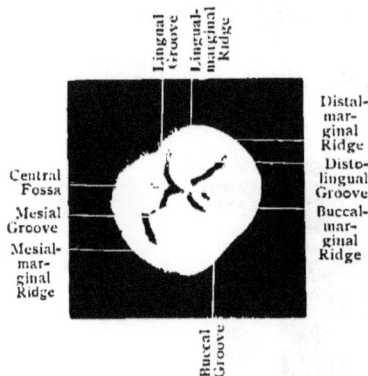

FIG. 135.—INFERIOR THIRD MOLAR,
OCCLUSAL SURFACE.

While these two forms are
those most commonly met with,
the occlusal surface may be
so broken by numerous supplemental and developmental
grooves that even six or eight
well-defined cusps may be present. Whatever complications
may exist upon the occlusal surface, a central fossa is usually present, from which radiate the various developmental grooves.
When the central fossa is absent, the space which it should occupy
is usually taken up by a rounded cusp, by the interference of which
the grooves are prevented from uniting, and their course is much

distorted. Along with these variations, the tooth is subject to much variety in size. In some instances the crown is one-third less in circumference than that of either the first or second inferior molar, while in others it is a trifle greater. The increase in the size of the crown is generally accompanied by an increase in the number of cusps. One feature very common to the crown is its inclination to the circular form, almost resulting in the absence of the angles common to molars in general. The marginal ridges are, of course, subject to the ever-varying conditions to be found upon the occlusal surface ; in general,

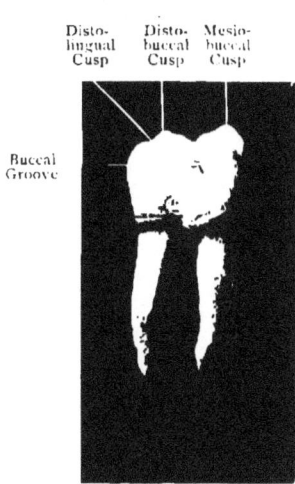

Disto-
lingual
Cusp

Disto-
buccal
Cusp

Mesio-
buccal
Cusp

Distolin-
gual Cusp

Lingual
Groove

Mesiolin-
gual Cusp

Buccal
Groove

Cervical
Line

FIG. 136.—INFERIOR THIRD
MOLAR, BUCCAL SURFACE.

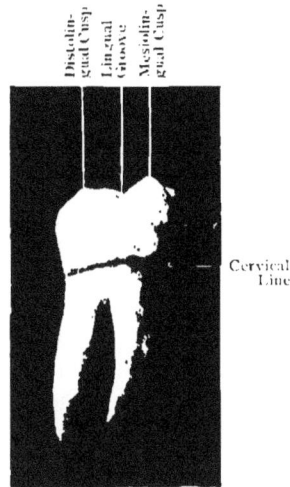

FIG. 137.—INFERIOR THIRD
MOLAR, LINGUAL SURFACE.

they are poorly defined, and are frequently crossed by numerous small supplemental grooves, dividing them into many minute tubercles. The latter are smooth and strongly convex, with their general outlines much influenced by the number of cusps.

The Roots of the Tooth.—While this tooth is strongly inclined to be two-rooted, in imitation of other inferior molars, this condition is by no means the common one. Like the crown, the roots are probably more variable than those of any other tooth. A single conic root may be present, or a mesial and a distal root may exist ; again, the mesial root may bifurcate, thus resulting in three. In some instances, four, or even five, branches may be given off from a common base. When more than two roots are present, they are usually much twisted

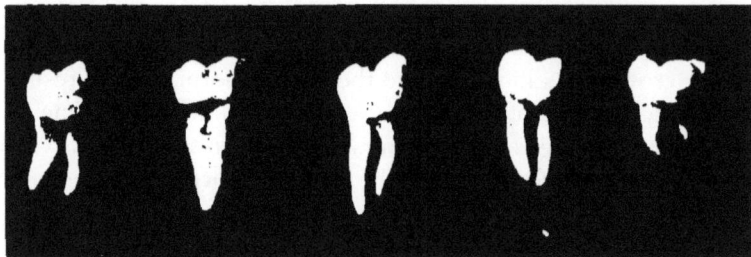

FIG. 138.—TYPES OF INFERIOR THIRD MOLARS.

or crooked, and, while generally inclined to the distal, are liable to branch in various directions.

A better idea of the variations in this tooth may be had from the accompanying illustration (Fig. 138).

CHAPTER X.

THE PULP-CAVITIES OF THE TEETH.

In the preceding chapters the study of the teeth has been confined to their external forms; it will now be necessary to learn something of their internal anatomy, and for this purpose various dissections of each individual tooth must be made.

Dissections.—First, a longitudinal dissection of each tooth should be made by sawing or filing from labial to palatal in the anterior teeth, or from buccal to palatal in the posterior teeth. Second, a longitudinal dissection by sawing from mesial to distal. Third, numerous transverse dissections by sawing through the crown or root at various points.

These dissections will expose to view a central cavity with outlines closely corresponding to those of the tooth itself. This is called the *pulp-cavity*, and in the vital tooth contains the formative and life-sustaining substance of the dentin, the *dental pulp*. The pulp-cavity is divided into two principal parts, that portion within the crown of the tooth being the *pulp-chamber*, while that traversing the root is the *pulp-canal*. At the apex of the root the canal ends in a small foramen, the *apical foramen*, which transmits the blood-vessels and nerves to the pulp. The pulp-chamber occupies the center of the crown and is always a single cavity; the pulp-canals are prolongations from this central cavity, and are usually one for each root, although in some instances two or more canals are present in a single root. The form of the pulp-chamber varies with the shape of the crown, the outline of the cutting-edge in the incisor teeth being reproduced in that part of the chamber nearest to the cutting-edge, while in the bicuspids and molars the occlusal surface is reproduced on the wall of the pulp-chamber, immediately beneath it, the lateral walls corresponding to the various sides of the tooth. In the incisors and cuspids the pulp-chamber passes so gradually into the pulp-canal that a positive line of demarcation between the two is not observed. In the bicuspids and molars the canals may be readily distinguished by a sudden constriction and branching out of the cavity into the various roots, which prolongations gradually decrease in size until the apical foramen is reached. The size of the pulp-cavity is much influenced by the age of the tooth, its functional activity, character of the occlusion, etc. The tooth-pulp, as the formative organ of the dentin, gradually decreases in size as the tooth develops (see Development of the Teeth), and as a result of this action the youngest teeth are provided with the largest pulp-cavities. At the time of eruption of the tooth, the diameter of the pulp-cavity is about equal to one-half the diameter of the crown, while the length of the canal must, of necessity, accord with the extent of root-calcification. As the growth of the tooth proceeds, the diameter of both the chamber and canal is gradually diminished until the formative process is completed; this gradual reduction in size is continued during the life of the tooth, and if permitted to proceed until old age, the chamber and canal may become almost obliterated. It must be remembered that while the diameter of the root-canal is diminished with the growth of the tooth, its length increases, continuing to do so until the time of complete root-calcification. During the period of root-development the diameter of the root-canal is greatest at the free or apical end of the root, at which point it presents a funnel-shaped opening (Fig. 139). As the root continues to calcify, this funnel-shaped extremity of the canal advances in the direction of calcification, and finally, as the formative process nears completion, the mouth of the funnel gradually disappears and the apical foramen is established. The various lobes of the teeth are penetrated by a prolongation of the pulp-cavity, these being called the *horns of the pulp-chamber*. The depth to which the horn penetrates the lobe varies in accordance with the form of the latter. If the tooth is one provided with long, penetrating cusps, the horns of the pulp-chamber will also be long, but if the cusps be poorly formed, the horns of the chamber will be short. In the anterior teeth, when the lobal construction is outlined by well-marked developmental grooves, the horns of the pulp-chamber will be three in number and directed toward the cutting-edge. These are most marked in young teeth, and gradually disappear as age advances. The functional activity of the teeth also serves to materially reduce the size of the pulp-chamber. Thus, when opposing teeth occlude squarely and firmly against each other, with more or less rubbing or sliding during mastication, the external surface is prone to rapid abrasion, and, as a direct result of this external change, the pulp-chamber undergoes a corresponding alteration by a growth of secondary dentin about its walls.

THE PULP-CAVITIES OF THE SUPERIOR TEETH.

Superior Central Incisor.—Figure 139 represents a number of labiopalatal sections presenting the relative size and shape of the pulp-cavity in the superior central incisor at various

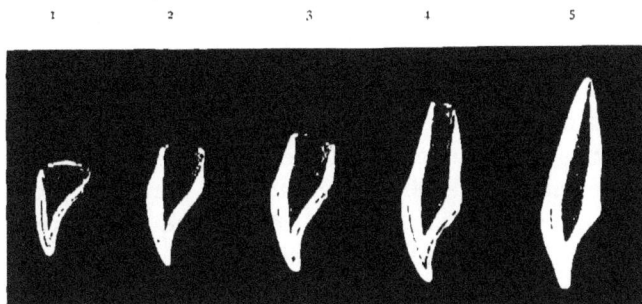

FIG. 139.—THE PULP-CAVITY IN THE SUPERIOR CENTRAL INCISOR, FROM THE SIXTH TO THE TENTH YEAR.

ages. In No. 1 the condition existing at about the sixth year, or at a time immediately prior to the eruption of the tooth, is shown. The tooth-crown is fully formed and calcified; the cervical line may be observed, as well as a small portion of the root-wall. The pulp-chamber, which is represented by the dark portion of the cut, occupies about one-third of the diameter of the crown at its greatest width. The pulp-chamber at this age, when viewed in this direction, forms almost a perfect cone, the base of which is directed upward or toward the future extremity of the root, and its apex downward in the direction of the cutting-

edge of the tooth. The apex of the cone may end somewhat
abruptly, or it may be lengthened into a slender, horn-like projection, extending well toward the cutting-edge. No. 2 represents the same tooth about the seventh year, or at a time
shortly after its eruption. The pulp-chamber has become
slightly reduced in its basal diameter, while but little change
has taken place in the apex. That portion of the pulp-cavity
above the cervical line represents a part of the future pulp-
canal. At this age the canal is a direct continuation of the
conic pulp-chamber, ending above in a broad, funnel-shaped
extremity. No. 3 shows the condition of the cavity about
the eighth year. The diameter of the pulp-chamber is considerably diminished, the apex has slightly receded, and the
horn-like projection has partly disappeared. The increase in
the length of the canal is about $\frac{3}{16}$ of an inch over its length
at seven years. The two parallel sides of the canal have
lengthened proportionately, and the funnel-shaped extremity
is reduced in diameter owing to the gradual narrowing of the
root-walls. No. 4 gives the relative size of the pulp-chamber
and canal at the ninth year, or at a time when root-calcification is nearing completion. The decrease in the capacity
of the chamber is readily apparent; the horn-like projection
has disappeared and the parallel sides of the canal are partly
extended into the chamber, thus shortening the cone. In the
canal a greater reduction has taken place in its diameter,
while its length has increased about $\frac{1}{4}$ of an inch over that
at eight years, and the diameter of the funnel-shaped opening is but little greater than that of the body of the canal.
No. 5, which represents the tooth about the tenth year, shows
the calcific action in the root completed, and the apical foramen
established. A glance at the illustration will show the gradual
decrease in the capacity of the pulp-cavity and the completion
of its growth in an apical direction. At this stage of development the fan-shaped extremity of the canal gradually disappears,
and for the first time in the life of the tooth the canal partakes
of the external root form throughout its entire extent.

Figure 140, A, represents the size and form of the average
pulp-cavity in the adult superior central incisor. In its entirety

it represents a double cone, with a common base near the cervical line, the pulp-cavity forming one cone and the pulp-canal the other. At this common base the cavity assumes its largest diameter, which measurement is approximately equal to one-fourth the labiopalatal diameter of the tooth. The extent and form of the lower cone, or that represented by the pulp-chamber, varies in the adult tooth with the tooth type. Thus, in the nervous type the cone is long and narrow, with the apex ending in a hair-like projection. In the tooth of the lymphatic temperament the cone is prone to be wide, with its apex ending abruptly. In the sanguine and bilious types the form and extent of the cone does not partake of either of the foregoing extremes, but, in keeping with the outline of the crowns, is inter-

C   B   A

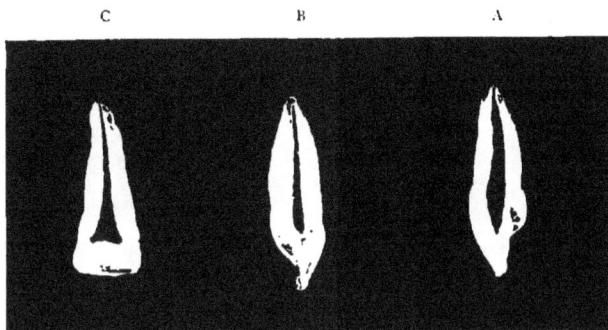

FIG. 140.—PULP CAVITY IN THE SUPERIOR CENTRAL INCISOR.

mediate between them. Figure 140, B, represents the average condition of the pulp-cavity in the central incisor in advanced age, and shows a general reduction in the size of both chamber and canal. A further study of the pulp-chamber and canal may be made by a mesiodistal section made through the long axis of the tooth (Fig. 140, C). The outline of the cavity, viewed in this way, closely follows the outline of the crown and root of the tooth. There is no distinct division between the chamber and the canal, the former gradually blending into the latter. The outline of the entire cavity is that of a single cone, with its base directed toward the cutting-edge and its apex in the direction of the apical extremity of the root. The lower margin of the

pulp-chamber, or that nearest the cutting-edge of the crown, is
broad from mesial to distal and thin from labial to palatal. This
margin in the average adult tooth is about on a line with the
center of the labial surface of the crown, and the lateral walls of
the cavity as they pass upward converge slightly, and finally
blend into the walls of the canal at a point somewhat beyond
the cervical line. During the early life of the tooth the margin
of the chamber nearest the cutting-edge presents three well-
defined horns, corresponding to the three rudimentary lobes
found upon the cutting-edge at this period. These horns
rapidly disappear, and are seldom found after the fifteenth
year. In certain tooth types, however, the mesial and distal
horns may continue present until adult age, and even into
middle life, but when this occurs it is not the result of the

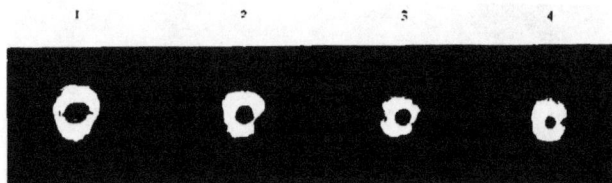

FIG. 141.—TRANSVERSE SECTIONS, ROOT OF SUPERIOR CENTRAL INCISORS, SLIGHTLY
ENLARGED.

temporary tooth form, but is occasioned by the permanent
angular outline of the crown.

Figure 141 represents a number of transverse sections of a
superior central incisor, showing the outline and relative size of
the pulp-cavity in passing from the base of the crown toward
the apex of the root. No. 1 shows the outline of the cavity at
the cervical line; No. 2 represents the condition $\frac{1}{3}$ of an inch
nearer the apex of the root; No. 3 is from the center of the
root length, while No. 4 is from the region of the apex.

Superior Lateral Incisor.—The pulp-cavity in the superior
lateral incisor is so nearly identical with that of the central that it
will only be necessary to call attention to one or two points
which are at variance. Figure 142 shows the five stages as represented by the growth of the tooth. In general it will be
observed that the cavity is much smaller than that of the

central incisor, but this difference is to be accounted for in the smaller proportions of the tooth. No. 1 shows the condition of the crown and pulp-cavity about the sixth year, the pulp-

FIG. 142.—PULP-CAVITY IN THE SUPERIOR LATERAL INCISOR, FROM THE SIXTH TO THE TENTH YEAR.

cavity occupying a large portion of the partly calcified tooth-crown. No. 2 represents the conditions present at the seventh

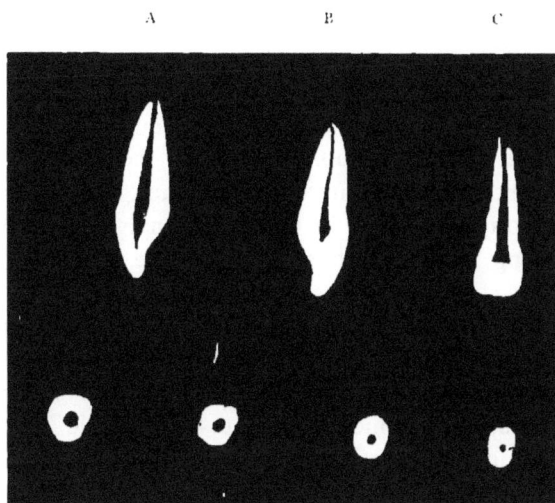

FIG. 143.

year, or about the time of the eruption of the tooth. The pulp-chamber at this age resembles a perfect cone, the base of which reaches to the root-walls, and faintly outlines the beginning of the future pulp-canal. In No. 3, at eight years, the length of the root has increased about $\frac{3}{10}$ of an inch, and the parallel sides of the walls of the pulp-canal have made their appearance. In No. 4, at nine years, by the growth of the root the canal has considerably increased in length and at the same time much decreased in diameter, while in No. 5, at ten years, the root is completely formed, the apical foramen established, and the maximum size of the entire pulp-cavity in the fully formed tooth shown.

Figure 143, A, shows the average condition of the pulp-cavity in the superior lateral incisor at adult age, while figure 143, B, represents the same tooth in old age. In a mesiodistal section—figure 143, C—a very close resemblance to the pulp-cavity in the central incisor will be noticed. While the pulp-cavity is smaller than that of the central incisor, it is usually a trifle larger in proportion to the size of the tooth. Owing to the marked constriction at the neck of this tooth, there is occasionally found a slight line of distinction between the pulp-chamber and canal, but in the majority of instances this is not to be observed. The horns of the pulp-chamber are in every respect similar to those of the central incisor, excepting when they exist permanently, in which case the mesial horn is usually the longest. By the transverse sections shown in figure 143, the gradual decrease in size and change in form in the root-canal is presented, the sections being similar to those made in the root of the central incisor.

Superior Cuspid.—The pulp-cavity of this tooth is in general similar to that of the incisors, excepting that the coronal extremity of the chamber is conic and inclined to a horn-like projection which penetrates the single cusp of the tooth-crown in the direction of its summit. Figure 144 represents a number of labiopalatal sections. No. 1 shows the condition of the pulp-cavity about the seventh year, or fully five years before the eruption of the tooth. The pulp-chamber partakes of the cone shape previously referred to, but the margins, instead of being straight lines, are somewhat bowed or concave, thus conforming more closely to the outline of the crown. The central horn of the chamber is proportionately longer than that of the incisors, in correspondence with the cusp of the tooth. At this age the formative process has barely extended to the root-walls; therefore, the width of the cavity is about equal to its length. In No. 2, at eight years, an increase in the capacity of the chamber over that of the incisors is shown, this being the result of the greater bulk in the tooth-crown. The cone-like outline of the chamber is somewhat broken by an effort of its margins to follow the outline of the crown. In No. 3, at nine years, the principal change has taken place in the canal, which has lengthened fully $\frac{3}{16}$ of an inch, and the funnel-shaped extremity, instead of joining with the

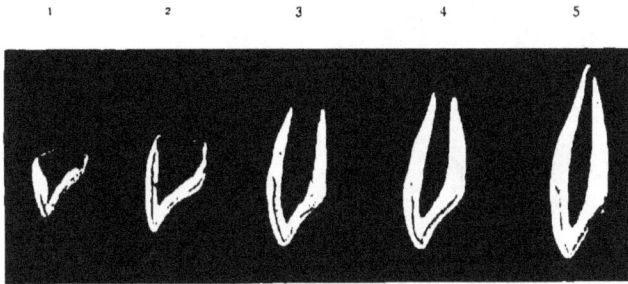

FIG. 144.—PULP-CAVITY IN THE SUPERIOR CUSPID, FROM THE SEVENTH TO THE TWELFTH YEAR.

pulp-chamber direct, is continued below by two parallel walls to the true beginning of this cavity. At ten years, No. 4, a more marked transformation has taken place in both portions of the cavity. The diameter of the chamber at the cervical line has diminished, as has also the length of the cone. The increase in the length of the root, which has been proportionately greater than that of the preceding year, has extended the length of the canal about $\frac{3}{16}$ of an inch. The walls of the canal are no longer parallel with each other, but are inclined to follow the root-outlines. The funnel-shaped opening is much reduced both in length and breadth. No. 5 represents the condition at the time of the eruption of the tooth, or about

16

the twelfth year. The general outline of the pulp-cavity is
that of a double cone, with a common base at a point nearly
corresponding to the cervical line. The diameter in both the
chamber and canal has considerably decreased, while the central
horn in the former has further receded. The calcification of
the root externally is about complete and the foramen formed.
In this particular the cuspid tooth differs from the incisors, and
in fact from all other teeth, in having its root-calcification about
completed and the apical foramen established at or soon after the

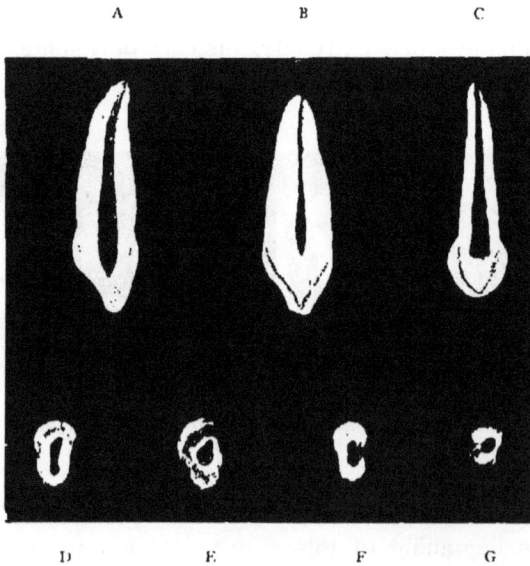

Fig. 145.

time of its eruption. Figure 145, A, gives an idea of the capacity
of the pulp-cavity in the superior cuspid at maturity, while figure
145, B, shows the condition in advanced age. Figure 145, C, is a
mesiodistal section of a matured superior cuspid. The coronal
extremity of the pulp-chamber is square, and but little inclined
to follow the outline of the mesial and distal cutting-edges.
The chamber passes into the canal without a mark of separation, and the latter gradually diminishes in diameter as the apex
of the root is approached. At its point of beginning the canal

is sometimes inclined to flatness from mesial to distal, but in passing toward the apex this tendency disappears, and it becomes more circular in outline. In figure 145, D, a transverse section through the tooth at the cervical line is shown, which gives an idea of the proportionate size and form of the canal in the adult tooth, while E, F, and G represent transverse sections through the root of the same tooth at various points between the cervical line and the apex of the root.

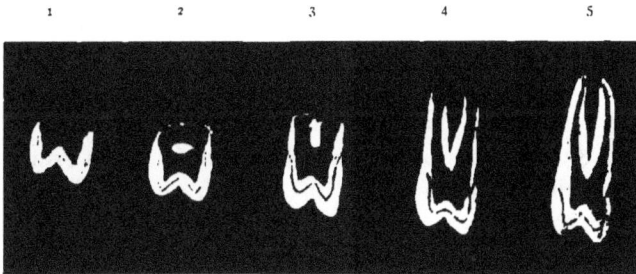

FIG. 146.—PULP-CAVITIES OF THE SUPERIOR FIRST BICUSPID, FROM THE SEVENTH TO THE TWELFTH YEAR.

Superior First Bicuspid.—The study of the pulp-cavity in this tooth differs in many particulars from that of the incisors and cuspids. First, the line of distinction between the pulp-chamber and the root-canal or canals is, in most instances, definitely marked by the bifurcation of the roots and a corresponding branching of the pulp-cavity into two fine canals, one of which occupies the center of each root. This division of the cavity brings the center of the pulp-chamber almost on a level with the cervical line. In figure 146, No. 1 shows the partly calcified crown of the superior first bicuspid at the seventh year. A portion of the pulp-chamber alone may be studied at this period, and this is found to be somewhat irregular in outline, with a broadened, funnel-shaped opening above, and two small, cone-like projections below, pointing into either cusp of the crown. These latter projections are the horns of the pulp-chamber, and are named in accordance with the cusp which they occupy. In very young teeth it is not unusual to find these horns penetrating the dentin almost to the enamel-wall. No. 2, at eight years, shows the crown fully calcified and the outline of the base of the roots established.   The horns of the pulp-chamber have slightly receded, and the branching of the canals is made manifest by the central deposit of dentin.   In No. 3, at nine years, the capacity of the pulp-chamber is much decreased, and appears to have receded bodily rootward.   The roots are calcified to about one-third their full length, and the canals which traverse them are each provided with the funnel-shaped opening at their free cal-cifying extremities.   In No. 4, at ten years, the decrease in the size of the pulp-chamber is not only caused by the deposit of dentin upon the occlusal and lateral walls, but from the direction of the roots as well.   The diameter of the root-canals is much less than at nine years, but the walls are as yet parallel.   No. 5 shows the roots fully formed and the apical foramina established, which condition occurs about the twelfth year.   The horns of the pulp-chamber have receded somewhat, and the center of this cavity is now almost on a level with the cervical line.   The canals have assumed the form of the roots themselves, and their diam-eter is much diminished.   The illustration shows the propor-tionate maximum size of the chamber and canals in this tooth after completion of surface calcification.   It will be observed that the foramina are proportionately smaller than those of the incisors and cuspids at a corresponding period, this condition resulting from the lesser diameter of the roots.

Figure 147, A, illustrates the approximate size and form of the pulp-chamber and canals at adult age, and attention is called to the appearance of the horns of the pulp-chamber.   It will be observed that the horn which penetrates the buccal cusp is larger and more pointed than that directed toward the palatal cusp; this condition is fully explained by the buccal cusp being proportionately larger and longer than the palatal.   In the same figure, B represents the pulp-cavity in the first superior bicuspid at advanced age.   The foregoing description applies only to the two-rooted bicuspids, but as many of these teeth have but one root, an additional description will be necessary.   When a single root is present, many varieties in the outline of the pulp-cavity will be presented; this variation, however, seldom affects the capacity or form of the pulp-chamber.   Two distinct canals may exist in the single root (Fig. 147, C), branching off from the chamber, one from the buccal and one from the palatal portion. These canals gradually taper in the direction of the apex of the root, and may end in a single foramen, or in distinct foramina. Occasionally the canals will unite before reaching the root-apex, and continue as a single canal ending in a single foramen, or they may communicate at one point and again diverge and finally end in separate foramina. In some instances the pulp-canal appears to be a direct continuation of the pulp-chamber, extending throughout the length

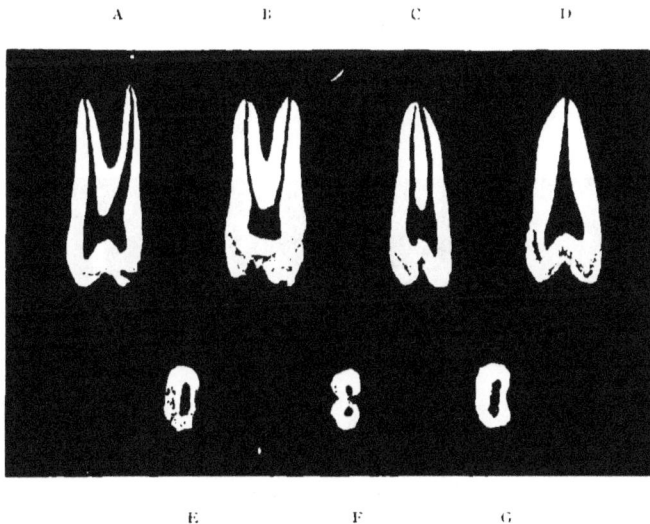

FIG. 147.—PULP-CAVITIES OF THE SUPERIOR FIRST BICUSPID, ENLARGED ABOUT ONE-THIRD.

of the root in the form of a flattened canal, with its greatest diameter from buccal to palatal (Fig. 147, D). When two separate canals exist in the single root, the outward appearance of the root indicates a near approach to two roots; when the single flattened canal is present, the root is also flattened and shows no sign of bifurcation. Reference has been made to the horns of the pulp-chamber, and in this connection it will be well to speak of the extent to which they may exist. In that type of tooth provided with long penetrating cusps the

horns will dip well down into the cusp occasionally to the full
depth of the dentin, and in rare instances may penetrate the
enamel. In those teeth lacking in cusp-formation the length
of the horns will be correspondingly reduced, and may be
entirely wanting. In the two-rooted superior first bicuspid the
floor of the pulp-chamber, or that part of the cavity directed
rootward, is prominent and rounded in the center, from
which point it gradually slopes toward the entrances to the
canals, one of which arises from the extreme buccal margin, and
the other from the extreme palatal margin. Figure 147, E,
represents a transverse section of the two roots immediately
below the point of bifurcation. F represents a section of the two
roots midway between the cervical line and the apical extremity,
while G is a transverse section of D at the cervical line.

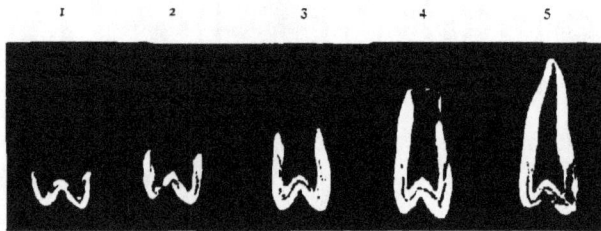

FIG. 148.— PULP-CAVITIES IN THE SUPERIOR SECOND BICUSPID, FROM THE SEVENTH TO
THE TWELFTH YEAR.

Superior Second Bicuspid.—The pulp-cavity of this tooth
is in many respects similar to that of the first bicuspid, the principal variations being in the horns of the chamber, which are
proportionately smaller in correspondence with the diminution
in cusp-formation. There is usually no positive line of demarcation between the chamber and canal, the latter being quite
large, and broad from buccal to palatal. The extent of the pulp-
chamber is sometimes well defined by the presence of two root-
canals, similar to those described in connection with the first
bicuspid. In rare instances the tooth may possess two roots,
each of which would be traversed by a canal. Figure 148 represents the various stages of the development of the pulp-cavity,
as shown by a longitudinal section from buccal to palatal. No. 1

shows the condition at seven years, or at a time when a portion
of the crown only is calcified, in consequence of which the pulp-
chamber alone can be studied at this period. The buccal and
palatal horns of the chamber may be observed penetrating the
dentin in the direction of their respective cusps. No. 2 shows the
advance made in the formative process by the eighth year, or at a
time immediately prior to the eruption of the tooth ; the outline
of the chamber is completed and walls of the future pulp-canal
faintly outlined.   At this period there has been but little change
in the horns of the pulp-cavity. No. 3 shows the condition of the

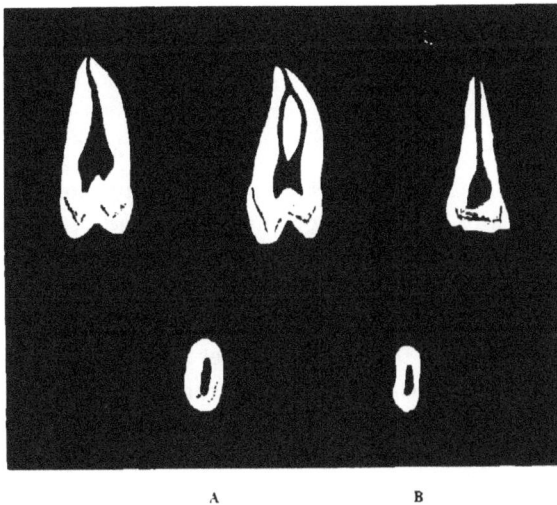

FIG. 149.—SECTIONS OF SUPERIOR SECOND BICUSPID, SLIGHTLY ENLARGED.

tooth at the ninth year, or at the beginning of its eruptive period.
The diameter of the chamber has somewhat decreased, the horns
have slightly receded, the funnel-shaped extremity of the cavity
has advanced beyond the cervical line, and is now confined to
the canal alone.   No. 4 represents the condition of the pulp-
cavity about the tenth year. A gradual decrease in the diameter
of both the chamber and canal is observed, and the horns
of the pulp-cavity are growing less prominent.   The length of
the root having increased nearly one-quarter of an inch, we find
a corresponding addition to the length of the canal.   No. 5

shows the maximum size of the pulp-cavity in the superior
second bicuspid, which condition accompanies the completion of
the external calcific action at the twelfth year. As previously
stated, the cavity, in its entirety, presents no line of separation
between the chamber and canal, but gradually tapers from its
broadened base in the crown to its ending at the apex of the
root. In this tooth, as well as in all those previously described,
the apical foramen at the time of completion of root-calcification
is comparatively large, and readily penetrated during operations
upon it. Figure 149 illustrates a number of sections of a
superior second bicuspid at maturity. The same figure also
represents two transverse sections, A being at the cervical
line, B midway between the cervical line and the apex of the
root.

PULP-CAVITIES OF THE SUPERIOR MOLARS.

The inner anatomy of the molar teeth being much more complicated than any of those previously described, it will be found
necessary to make a number of dissections in various directions
in order to obtain a comprehensive idea of the location and
form of the different parts of the pulp-cavity. The line of demarcation between the pulp-chamber and canals is always
definite, the former occupying a central position in the crown
and seldom extending beyond the cervical line, while the latter
are given off from the floor of the chamber and penetrate the
various roots, their entrances being marked by small funnel-
shaped openings in the floor of the chamber. In the matured
tooth the form of the chamber usually corresponds to that of the
crown of the tooth. The lateral walls of the chamber are four in
number, and are named according to their location—mesial, distal, buccal, and palatal. The average thickness of these walls
at maturity is about equal to the diameter of the pulp-chamber.
In that type of tooth common to the lymphatic temperament
where there is but little constriction at the neck, resulting in the
various sides of the tooth-crown being nearly parallel with each
other, the pulp-chamber is nearly quadrilateral in form ; but in
those teeth marked by a decided constriction at the neck, most
marked in the nervous temperament, the extent of surface cov-

ered by the floor of the chamber is much less than that occupied by the occlusal portion. In the former class, the entrances to the various canals are much farther apart than in the latter. The occluding wall is usually much thicker than the lateral walls, and is penetrated by the horns of the pulp-chamber, one of which extends into each cusp. As in the bicuspids, the extent to which the horns penetrate the cusps is controlled by the prominence of the latter. The floor of the pulp-chamber is irregularly rounded, being high in the center and gradually falling away in the direction of the canals. The entrances to the root-canals, three in number, are placed in the form of an irregular triangle, called the molar triangle. The mesial side of the triangle is usually the longest, the distal next in length, and the buccal the shortest.

In young teeth the entrances to the canals are usually in the form of funnel-shaped openings, are comparatively easy of access, but after maturity may disappear and be but little larger than the canals themselves. To properly study the position occupied by the entrance to the canals on the floor of the pulp-chamber, a transverse section of the tooth should be made at a point somewhat above the cer-

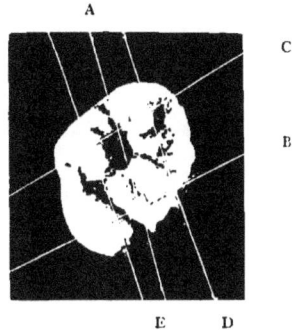

Fig. 150.

vical line, at the same time preserving both the crown and the roots of the tooth for comparison. The entrance to the palatal canal, which is usually the largest and most readily accessible, may be located by a line drawn through the center of the occlusal surface of the crown (Fig. 150) from buccal to palatal, A, and by another line drawn from mesial to distal almost parallel with the palatal-marginal ridge, passing through the summits of the mesiopalatal and distopalatal cusps, B; the point at which these two lines intersect will mark the approximate location of the palatal canal. The entrance to the mesiobuccal canal may be located by a line drawn from the inner side of the mesiobuccal angle to a corresponding position near the distobuccal angle, C. This

should be intersected by a line drawn from the summit of the
mesiobuccal cusp to the summit of the mesiopalatal cusp, D,
the point at which these two lines cross marking the entrance
to the mesiobuccal canal. The location of the entrance to the
distobuccal canal is found by the line C, which is intersected
by another line, E, drawn from the summit of the distobuccal
cusp to a corresponding point on the distopalatal cusp. The
nearer the tooth-crown approaches to the quadrilateral, the
nearer will the molar triangle approach the equilateral.

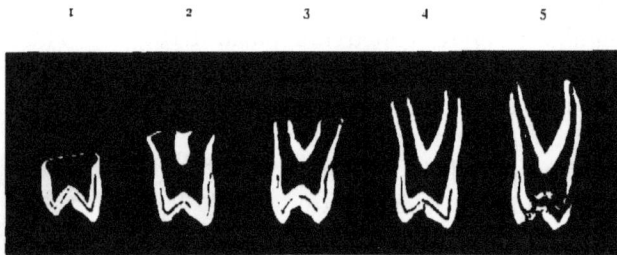

Fig. 151.— Pulp-cavity of Superior First Molar, from the Fifth to the Ninth
Year. Palatal and Mesiobuccal Canals.

Superior First Molar.—In the dissection of this tooth, the
pulp-chamber and two of the root-canals only can be shown,
but these will be sufficient to pursue the study with intelligence.
Figure 151 shows a number of longitudinal sections, made in
such a manner as to expose the palatal canal, usually the
largest, and the mesiobuccal canal. No. 1 illustrates the
approximate size and form of the pulp-chamber at the fifth year.
At this period the chamber occupies a large proportion of the
center of the tooth-crown. Two of the four horns are seen, one
of which penetrates the mesiobuccal cusp, and one the mesiopalatal cusp. In many instances the horns of the molar teeth
are quite slender, penetrating the dentin to a greater depth
than that shown in the illustration, in the form of minute hair-
like projections, which in some instances reach almost to the
enamel walls.

No. 2 illustrates the condition of the pulp-cavity at the sixth
year, or at the time of eruption. The outline of the pulp-

chamber is completed, and the floor has begun to make its appearance by a central deposit of dentin. It will be observed that the lateral walls of the chamber are somewhat less in thickness than the occluding wall, a condition which will become more pronounced as the tooth develops. With the beginning of the formative process in the floor of the chamber we find the trifurcation of the roots established, and the beginning of the canals outlined. The canals at this period are quite similar to those of the bicuspid, being provided with a funnel shaped extremity, which extends from the free calcifying margins of the roots to the floor of the chamber. No. 3 shows the change which has taken place at the seventh year. While the pulp-chamber is somewhat reduced in size, but little change is noticeable in its outline. By this time the floor of the chamber has become an important factor in the tooth development. By the constant lateral extension of this central deposit of dentin the floor of the chamber is gradually spread out, this alteration being at the expense of the entrances to the root-canals, which become reduced in diameter as the floor is extended. The horns of the chamber are slightly less prominent, but this part of the cavity has the appearance of having receded bodily rootward. The roots have advanced somewhat beyond the point of trifurcation, and a definite outline has been given to the canals. At this period the diameter of the root-wall is about equal to the diameter of the pulp-canal. Along with the gradual decrease in the diameter of the roots, there is observed a corresponding decrease in the width of the funnel-shaped extremities of the canals.

At the eighth year, No. 4, a gradual reduction in the capacity of both the chamber and canal is noted. Accompanying the above condition there is found a corresponding increase in the thickness of the surrounding walls. The horns of the pulp-chamber are much reduced in size, and the form of the chamber more closely resembles that of the general contour of the tooth-crown. The increase in the length of the roots is proportionately greater than that of previous years, in consequence of which the length of the canals is increased to a greater degree. In No. 5 the maximum size of the chamber and canals

is apparent, which condition takes place about the ninth year,
or at a time when calcification of the tooth is completed externally. In some instances, owing to the additional length
of the palatal root, the apical foramen may not be established
before the tenth year. At this latter period it is safe to assume
that all three canals have completed their longitudinal extent,
and the foramina, although proportionately large, have been
established, so that a more definite description of each canal
may be given. The *palatal canal* (Fig. 152, A) is usually the

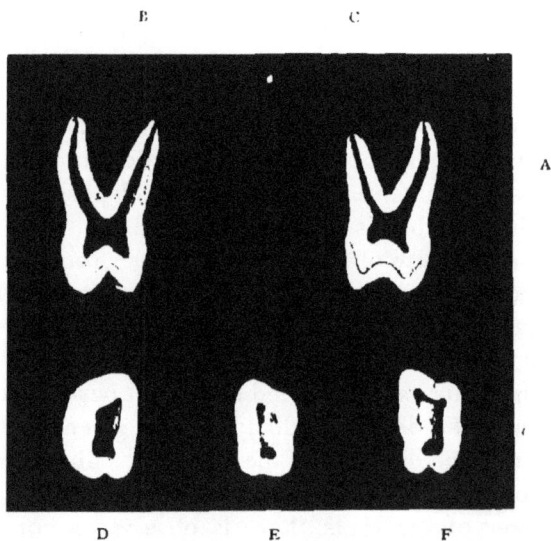

FIG. 152.—PULP-CAVITIES OF SUPERIOR FIRST MOLAR, SLIGHTLY ENLARGED.

largest, and branches off from the floor of the chamber, near the
mesiodistal center of the extreme palatal margin, the entrance
in the average tooth being well defined by a circular, funnel-
shaped opening. The direction of this canal is usually upward
and slightly inward, until the apical extremity is approached, at
which point it is inclined to the buccal. The circular form presented at the beginning of the canal is generally continued
throughout its entire length, in this respect differing from the
two buccal canals. The average length of the palatal canal is
about ½ of an inch. The *mesiobuccal* canal (Fig. 152, B)

branches off from the floor of the chamber, at its extreme mesio-buccal angle, and the entrance, instead of being funnel-shaped and easy of access, is flattened from mesial to distal, and frequently difficult to enter. This flattened form continues throughout its course, which for the distance of $\frac{1}{8}$ of an inch is in a buccal and mesial direction; beyond this point it is usually inclined to the buccal, until the upper third of the root is reached, where it turns rather abruptly to the distal. This canal is generally a trifle shorter than the palatal, averaging about $\frac{3}{8}$ to $\frac{7}{16}$ of an inch. The *distobuccal* canal (Fig. 152, C) branches off from the floor of the chamber at the extreme distobuccal angle. In those teeth which nearest approach the quadrilateral form, the entrance to this canal will be farther from the center of the tooth, the molar triangle in this instance being almost an equilateral. It sometimes happens that the entrance to this canal is directly in the floor of the pulp-chamber, near to, but not directly against, its buccodistal angle. The entrance is usually abrupt, seldom being funnel-shaped, making it by far the most difficult of access. It is inclined to be circular in form, and more or less tortuous in its course. Immediately above the point of beginning it is inclined toward the buccal and distal; near its center it may incline slightly to the mesial; and finally, at its upper third, turns somewhat abruptly in a distobuccal direction. This canal is usually the shortest of the three, its average length being about $\frac{3}{8}$ of an inch.

Figure 152 also illustrates a number of transverse sections of this tooth, D being made at the cervical line, looking toward the crown, E looking toward the roots, while F represents a transverse section at a point immediately above the floor of the pulp-chamber.

Superior Second Molar.—In many respects the pulp-chamber of this tooth is similar to that of the first molar, but there are a few variations which must be briefly described. First, the outline of the tooth-crown being much more flattened from mesial to distal, a corresponding variation is noted in the form of the pulp-chamber, increasing the length of the mesial side of the molar triangle, and decreasing the length of the buccal and distal sides. The chamber is more or less flattened

from mesial to distal, making it somewhat oblong from buccal
to palatal. Second, on account of a reduction in the prominence of the cusps, the horns of the cavity are usually
somewhat less pronounced than those of the first molar.
Third, the floor of the cavity is less convex, and slopes more
gradually toward the entrances of the various canals. In a
general way, the rules given for ascertaining the approximate
location of the entrances to the canals in the first molar apply
to this tooth. The comparative size and form of the pulp-
chamber and canals during the development of the tooth is
shown in figure 153, extending from the ninth to the sixteenth or
seventeenth year, at which latter period the crown and roots of
the tooth are fully calcified externally.

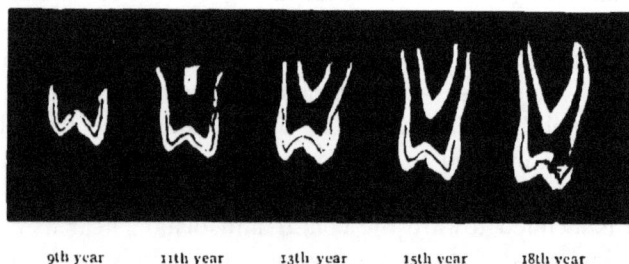

9th year   11th year   13th year   15th year   18th year

Fig. 153.—Pulp-cavities in the Superior Third Molar, from the Ninth to the
Eighteenth Year.

Superior Third Molar.—In this tooth the conditions are so
variable that a description of the pulp-cavity taken from a single
tooth would be insufficient. In the majority of instances the
outline of the tooth-crown approaches the triangular form, and
in consequence the pulp-chamber is triangular rather than
quadrilateral or oblong. The mesial border of the chamber is
the longest, the distal next in length, and the buccal the shortest
of the three. The horns are generally less in number and much
less pronounced than those of either the first or second molars.
The floor of the chamber may be broken by irregularities similar to those previously described, or it may be entirely absent,
this latter condition occurring when the tooth has but a single
root accompanied by a single canal. The various stages of development having been given in connection with the general

description of the tooth, no attempt will be made to describe this
by longitudinal sections, the complications in root-forms making
such a proceeding impracticable.   Instead of so doing, the space
will be devoted to a brief description of the variety of pulp-
canals found in this tooth.   Probably the most frequent condition
is that which resembles the first and second molars—*i. e.*, three
canals branching off from the chamber in as many different
roots, two to the buccal and one to the palatal.   When the three
canals exist, the entrances to them will be well beyond the
cervical line, where they will be found clustered much closer
together than those of the first and second molars, this difference
in their location being so marked that the diagram previously
given can not be depended upon in an attempt to locate them.

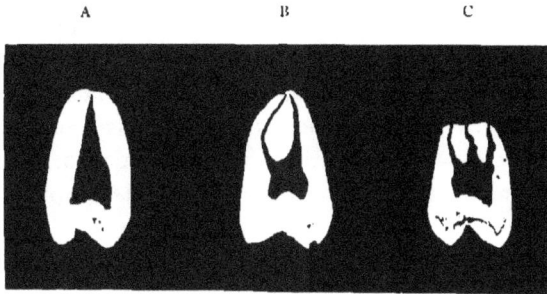

FIG. 154.—LONGITUDINAL SECTIONS, SUPERIOR THIRD MOLAR, SLIGHTLY ENLARGED.

The usual course of these canals is first slightly mesial, then
distal, and finally in a distopalatal direction.   On account of
the pulp-chamber extending well beyond the cervical line, the
canals are much shorter than those of the first or second
molars, their average length being less than ½ of an inch.
Another form frequently met with is that of the flattened
single canal, occurring when the tooth has but a single
root, which shows no signs of trifurcating (Fig. 154, A).   In
this instance the pulp-chamber gradually passes into the canal,
and the chamber is without a floor.   Such a canal is shaped
like the chamber at its point of beginning; but as it passes
toward the apex it becomes flattened in the direction of the
smallest diameter of the root.   But little difficulty is experienced

in entering such a canal, and usually it is readily followed to its apex. Another condition frequently met with in the single-rooted third molar is that of one or more canals branching off from the floor of the chamber, their course through the root-substance being without regard to the external contour of the root (Fig. 154, B). These canals, which may exist to the number of five or six, are usually very minute, and in some instances may pass from the floor of the chamber to the apex of the root almost in a direct line, and end in distinct foramina, or they may take a tortuous course, and when near the apex unite, ending in a single foramen. When the tooth is provided with four, five, or even six small roots, as sometimes occurs, each root will be traversed by a minute canal, the entrances to these being variously placed about the floor and lateral margins of the pulp-chamber (Fig. 154, C). In all operations upon this tooth it must be recalled that it is the last to be calcified, and conse-quently the canals and foramina are proportionately larger than in the other teeth; at the same time, it possesses one advantage over the others—*i. e.* (with the single exception of the cuspid), being fully calcified at or about the time of its eruption.

PULP-CAVITIES OF THE INFERIOR TEETH.

A   B   C

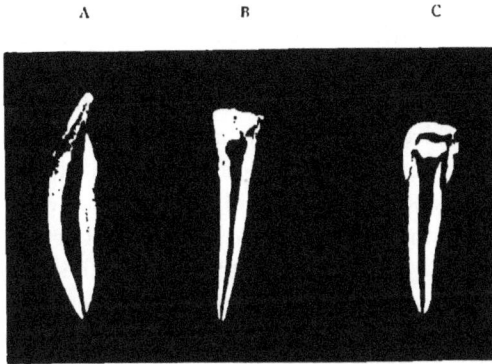

FIG. 155.—PULP-CAVITIES OF THE INFERIOR INCISOR.

The outline of the pulp-cavities of the inferior teeth, like those of the superior, corresponds to the general tooth contour. The comparative size of the cavity at various stages of tooth development will not be repeated in this description, the conditions being similar to those in the superior teeth (see also Development of the Teeth).

Inferior Incisors.—The pulp-cavities of the inferior central and lateral incisors are so nearly alike that a single description will answer for both. Figure 155, A, represents a labiolingual section of an inferior incisor, showing the most frequent form of the pulp-cavity. The tooth from which the section was prepared was one about middle life, the cavity in younger teeth being proportionately larger, while a gradual decrease in diameter would be noted with advancing age. There is no mark of distinction between the pulp-chamber and canal, so that an imaginary separation would have to be made at the cervical line, or slightly below that point. Taken in its entirety, the cavity presents the form of a double cone, the common base of which is slightly to rootward of the cervical line. The chamber penetrates the crown fully half way to the cutting-edge, at which point it ends in a thin, fan-like margin (best observed in

17

mesiodistal section), while the canal gradually decreases in
size until the apical foramen is reached. Although this is the
most common form of the pulp-cavity in the inferior incisors, it
is by no means the constant condition. The tooth is not infrequently provided with a medium-sized pulp-chamber, which
extends somewhat below the cervical line, beyond which point it
branches into two fine canals, which are continued separately
until the apical third of the root is approached, when they again
unite, and finally end in a single foramen. Figure 155, C,
represents a mesiodistal section of a young inferior incisor, in
which the three small horns of the pulp-chamber are apparent.
At this period the fan-shaped extremity of the pulp-chamber
occupies about one-half of the mesiodistal diameter of the crown,
and the horn-like projections extend well toward the enamel cap.
Figure 155, B, shows the average size and form of the pulp-
cavity at maturity, by a mesiodistal section through the long axis
of the tooth. In this it will be observed that the horns of the
pulp-chamber have disappeared, and that the capacity of the
cavity in general is much reduced.

Inferior Cuspids.—The pulp-chamber and canal in this
tooth, while usually conforming to the general contour of the
tooth, are frequently found to vary greatly, both in outline and in
size. The most common form, however, is that shown in figure
156, A, a labiolingual section of an adult tooth. The chamber
and canal have no line of demarcation, and unite at a common
base considerably below the cervical line, the former penetrating
the crown of the tooth to a point about midway between the
cervical line and the summit of the cusp, at which point it ends
in a sharp, hair-like projection. Accompanying this common
form there is much variation in size, even in teeth of the same
age. Figure 156, B, shows another labiolingual section of an
adult inferior cuspid, in which the pulp-cavity fails to accurately
follow the outline of the tooth, and its capacity is much less
than that shown at A. The root of this tooth is in most
instances circular, in which case the canal will be similarly
formed ; but occasionally the root will be much flattened from
mesial to distal, and as a result of this the canal will also be
much flattened. The canal of this tooth is seldom divided. In

figure 156, C, a mesiodistal section of an inferior cuspid is
shown, and it will be observed that the fan-shaped extremity
of the chamber common to the incisors is absent, the cavity

A   C   B

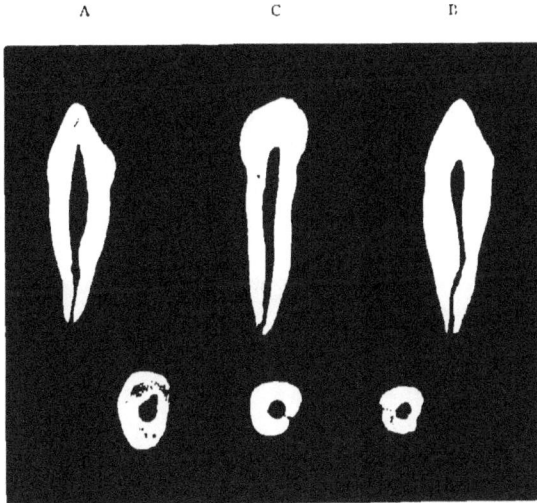

FIG. 156.—PULP-CAVITIES OF THE INFERIOR CUSPIDS.

ending rather abruptly, or by a fine line near the center of the
crown.   The same illustration also shows a number of transverse sections, which will give an idea of the form of the cavity
at various parts of the tooth.

Inferior Bicuspids.—The pulp-cavities of these teeth may
be best described collectively, thus affording an opportunity for
comparison. Unlike the superior bicuspids, it is seldom that
the canals are definitely separated from the chambers. That
part of the cavity within the crown, however, is usually quite
wide from buccal to lingual, and unites with the canal by a long,
funnel-shaped constriction. The center of the pulp-chamber
may be considered as being about on a level with the cervical
line. In the first bicuspid the pulp-cavity is provided with a
single horn, which extends with more or less prominence in the
direction of the buccal cusp. That part of the chamber facing
the lingual cusp is usually rounded off. In the second bicuspid

the occlusal wall of the pulp-chamber generally presents a different form; two well-defined horns are usually present, of which the buccal is the longest; or the chamber may be prominently rounded at these points. The pulp-chamber of the second bicuspid is generally larger than that of the first. The canals of these teeth are usually circular throughout, and are readily penetrated until the apical third is reached, beyond which point they are extremely small. In some instances the canal divides near the center of the root, and is continued as two canals, ending in distinct foramina, or, after separating, they may again unite, and end in a single foramen. In Figure 157 the aver-

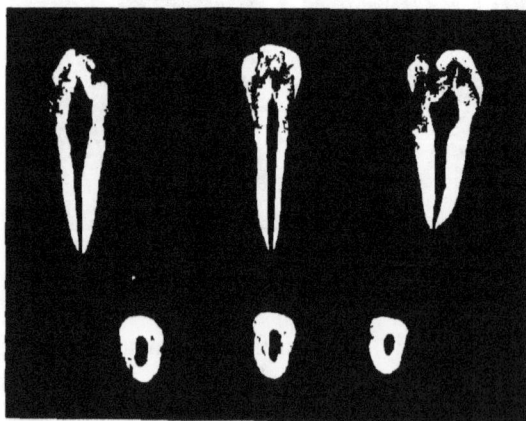

FIG. 157.—SECTIONS OF INFERIOR BICUSPIDS.

age size and form of the canal in these teeth is shown by a number of mesiodistal and transverse sections.

Inferior Molars.—The form of the pulp-chambers of the inferior molars corresponds to the general outline of the crown, and the form of the root-canals is similar to the general contour of the roots. The pulp-chambers approach the quadrilateral form; the buccal and lingual sides are somewhat the longest, the mesial next in length, and the distal, usually slightly rounded, is the shortest. The occluding wall is convex rootward, sloping in the direction of the various cusps, each of which is penetrated by a horn. Like the horns of the pulp-chambers in general, the extent to which these penetrate the cusps is influenced by the age, type, and functional activity of the organ. The floor of the cavity is convex in the direction of the occlusal surface, but this convexity is principally from mesial to distal. From the summit of this convexity the floor slopes to the entrances of the canals, the opening into which is inclined to be funnel-shaped rather than abrupt. The lateral walls of the chamber are much inclined to follow the general contour of the crown. The horns of the pulp-chamber are usually more pronounced in the first than in the second molar, and still less clearly defined in the third than in

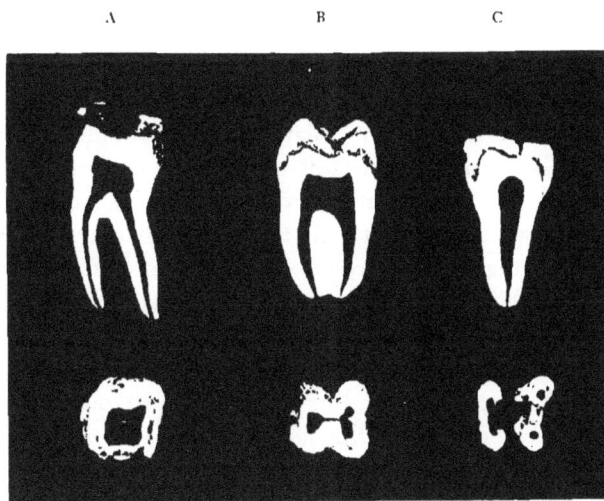

FIG. 158.—SECTIONS OF INFERIOR MOLARS, ENLARGED ABOUT ONE-THIRD.

the second. The roots of the first molar being somewhat further apart than those of the second, the floor of the chamber in the former is slightly more extensive than in the latter. To study the pulp-cavities of these teeth a longitudinal section should be made through the center of the tooth from mesial to distal. Figure 158, A, shows such a dissection through the first molar, and illustrates the average size and form of the chamber and canals at adult age. The canals join the chamber by a funnel-shaped opening, and but little difficulty will be found in effecting an entrance, but to follow them to their apices will be more per-

plexing. The roots of this tooth being much flattened from
mesial to distal, the canals are also flattened in this direction,
but broad from buccal to lingual. The entrances to these canals
may be found at the extreme mesial and distal margins of the
pulp-chamber, and usually extend from the buccal to the lingual
walls of the cavity. It is not uncommon for the mesial canal to
divide soon after leaving the chamber, and continue as two
canals, ending in separate foramina (Fig. 158, B). This condition is seldom present in the distal canal, which is usually
straight from its mouth to the apical foramen. The capacity of
the pulp-chamber is usually a trifle less than that of the first
molar, and the entrances to the canals are somewhat closer
together. In other respects the cavity is similar to that of the
first molar. Figure 158 also shows a number of transverse
sections through an inferior molar, and gives an idea of the size
of the canals at various parts of the roots. In some instances
the roots of the second molar coalesce, in which case a single
root-canal may be present. In the third molar the most common form of the pulp-cavity is one similar to that of the first,
but both the chamber and canals are smaller. Unlike the pulp-
cavity of the corresponding superior tooth, this tooth is not subject to so much variation, although it is sometimes found with a
single root traversed by a single canal, which may be accompanied by a rather large pulp-chamber.

CHAPTER XI.

THE DECIDUOUS TEETH, THEIR ARRANGEMENT, OCCLUSION, ETC.;
THEIR CALCIFICATION, ERUPTION, DECALCIFICATION, SHEDDING
PROCESS, AND AVERAGE MEASUREMENTS; THEIR SURFACES,
GROOVES, FOSSÆ, RIDGES, SULCI, AND PULP-CAVITIES.

THE DECIDUOUS TEETH.

| Central Incisor | Lateral Incisor | Cuspid | First Molar | Second Molar |

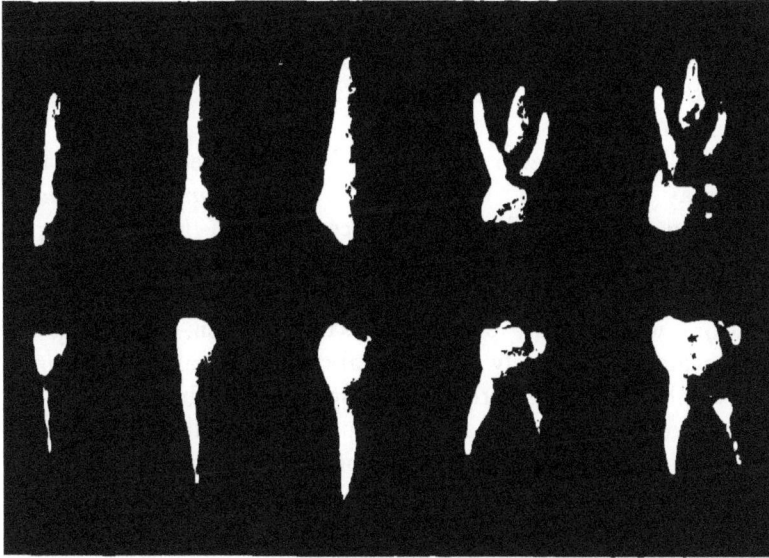

| Central Incisor | Lateral Incisor | Cuspid | First Molar | Second Molar |

Fig. 159.—THE DECIDUOUS TEETH, SUPERIOR AND INFERIOR, FROM THE LEFT SIDE
OF THE MOUTH.

As implied by the term deciduous, these teeth are temporary
in their nature, and, after subserving the purposes of early
childhood, are thrown off by the operation of the economy to
give place to the permanent organs. The shedding process
takes place in the incisors between the seventh and eighth years,

in the molars from the tenth to the eleventh years, and in the cuspids about the twelfth year. This shedding process, however, does not indicate the period at which the degeneracy of the tooth begins, for, in a year or two after the roots are completely formed and the apical foramen established, decalcification begins at the apical ends and continues in the direction of the crown until absorption of the entire root has taken place and the crown is lost from lack of support. Decalcification in the incisors begins between the fourth and fifth years, in the molars from the seventh to the eighth years, and in the cuspids about the ninth year.

The deciduous teeth are *twenty in number*, ten in each jaw, and may be classified as follows: *Four incisors, two cuspids, and four molars.* The incisors, central and lateral, occupy the central portion of the arch, are placed two upon each side of the median line, and are succeeded by the four permanent incisors, which finally occupy the same position. The cuspids are located immediately to the distal of the lateral incisors, and are displaced by the permanent cuspids. The first and second molars come next in the arch, but, unlike the anterior teeth, are followed by permanent successors of another class, the first and second bicuspids, the permanent molars erupting posteriorly to these as the jaw increases in length. In general the deciduous teeth resemble their permanent successors, yet there are a number of minor differences which will require a comparative description. Both the crowns and the roots are much smaller in every direction than those of the permanent teeth, but the diameter of the crowns is proportionately greater than that of the roots, while the roots are proportionately longer. The fact that the roots are smaller in proportion than the crowns is productive of a neck much more constricted. The roots of the deciduous teeth are the same in number as those of the corresponding permanent teeth, the incisors and cuspids being provided with one, superior molars with three, and the inferior molars with two.

THE OCCLUSION OF THE DECIDUOUS TEETH.

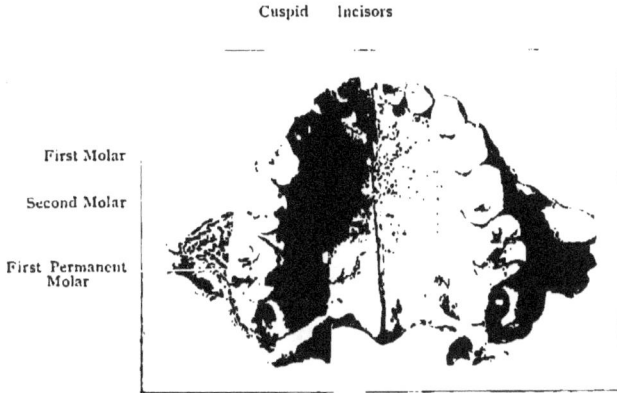

FIG. 160.—THE SUPERIOR DENTAL ARCH ABOUT THE SEVENTH YEAR.

The arrangement of the deciduous teeth in the jaws is similar to that of the permanent organs, the superior teeth describing the segment of a larger circle than the inferior, in consequence of which the superior teeth close over or outside of the inferior. The character of the occlusion in the deciduous teeth is not subject to so much variation as that found in connection with the permanent set, this being accounted for by the more constant form in the crowns of the former. The relations existing between the superior and inferior deciduous teeth when in contact is such that each tooth, with the exception of the inferior central incisor and the superior second molar, occludes with two teeth of the opposite jaw, the superior central incisor being opposed by the entire cutting-edge of the inferior central and the mesial third of the inferior lateral; the superior lateral coming in contact with the remaining two-thirds of the inferior lateral and a portion of the mesial half of the inferior cuspid, this arrangement continuing throughout the series. The foregoing description of the occlusion of the deciduous teeth is applicable to but a small part of their transitory existence. By the time they are fully erupted and have assumed their respective positions in the arch, the increase in the size of the bone is sufficient to create a slight space or diastema between the teeth, which condition is soon followed by a greater separation through a protrusion of the anterior teeth, caused by the growth and approach of the permanent teeth from behind.

The calcification of the deciduous teeth is similar to that of the permanent, the process in the incisors and cuspids beginning along the cutting-edges in three distinct lobes, while in the molars a center of calcification is provided for each cusp (see Development of the Teeth).

THE DECIDUOUS TEETH IN DETAIL.

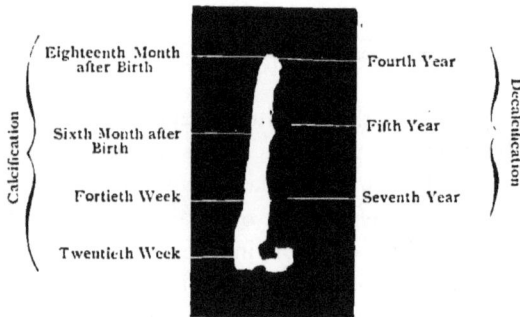

FIG. 161.

SUPERIOR CENTRAL INCISOR.

CALCIFICATION BEGINS, ABOUT THE FOURTH FETAL MONTH.
CALCIFICATION COMPLETED, SEVENTEENTH TO EIGHTEENTH MONTH AFTER BIRTH.
ERUPTS, SIXTH TO EIGHTH MONTH AFTER BIRTH.
DECALCIFICATION BEGINS, ABOUT THE FOURTH YEAR.
SHEDDING PROCESS TAKES PLACE, ABOUT THE SEVENTH YEAR.
AVERAGE LENGTH OF CROWN, .23.
AVERAGE LENGTH OF ROOT, .39.
AVERAGE LENGTH OVER ALL, .62.

This tooth, as well as all of the deciduous teeth, presents for examination numerous surfaces, margins, and angles, these being the same in name and location as those of the permanent teeth.

The Labial Surface of the Crown (Fig. 162).—This surface is smooth and generally convex, but with an inclination to flatness near the incisive margin. The mesial margin is slightly convex in the direction of the length of the tooth, and rounded from labial to palatal. The distal margin is decidedly convex from the cutting-edge to the cervical line, in many instances forming almost a complete semicircle, which is usually at the expense of the distal angle of the crown. The cervical margin is deeply concave in the direction of the root, and the incisive margin is straight over its central portion and rounded or angular at its extremities. The labial grooves are seldom so well defined as those upon the permanent inisors.

The Palatal Surface of the Crown.— In some instances this surface is smooth and concave near the cutting-edge and convex over the cervical portion, with the marginal ridges well defined. In other cases it is concave from the cutting-edge to the cervical ridge, being provided with a longitudinal ridge in the center, a slight depression upon either side, and marginal ridges poorly defined. In the former instance the palatal fossa is

FIG. 162.

present; in the latter it is absent. The *mesial* and *distal* surfaces of the crown are both smooth and convex, the former being inclined to flatness over its cervical third—a condition which is seldom present in the latter. The mesial angle is alone well defined, the cutting-edge passing into the distal surface with a long, gradual sweep, thus in a measure destroying the distal angle. The neck of the tooth is marked by a decided constriction, which is principally produced at the expense of the crown alone. The root of the tooth, when compared with the root of the permanent central incisor, is much longer in proportion to the length of the crown. In some instances it is flattened from mesial to distal, these two sides converging as they pass to the palatal; in others it is flattened from labial to palatal. Generally speaking, it is a straight root. but is occasionally provided

with a slight mesial curve near its apical third, and it is sometimes curved slightly from labial to palatal.

SUPERIOR LATERAL INCISOR.

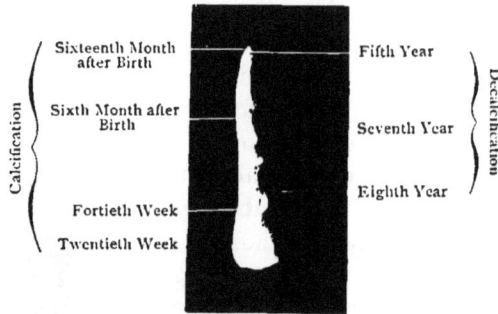

Calcification

Sixteenth Month
after Birth

Sixth Month after
Birth

Fortieth Week

Twentieth Week

Fifth Year

Seventh Year

Eighth Year

Decalcification

FIG. 163.

CALCIFICATION BEGINS, ABOUT THE FOURTH FETAL MONTH.
CALCIFICATION COMPLETED, FOURTEENTH TO SIXTEENTH MONTH AFTER BIRTH.
ERUPTS, SEVENTH TO NINTH MONTH AFTER BIRTH.
DECALCIFICATION BEGINS, ABOUT THE FIFTH YEAR.
SHEDDING PROCESS TAKES PLACE, ABOUT THE EIGHTH YEAR.
AVERAGE LENGTH OF CROWN, .25.
AVERAGE LENGTH OF ROOT, .45.
AVERAGE LENGTH OVER ALL, .70.

FIG. 164.—SUPERIOR
LATERAL INCISOR, LA-
BIAL SURFACE.

The various surfaces of this tooth so
closely resemble those of the central incisor
that a separate description will be unnecessary ; in a general way, however, there are
a few minor points of difference. The tooth
is smaller in every direction excepting in
its length, which is generally equal to and
frequently greater than that of the central
incisor. The diameter of the root is but
little less than that of the central, while
the mesiodistal measurement of the crown
is about one-third less, in consequence of
which the neck of the tooth is not so well defined. The
angles of the crown are more rounded than those of the central
incisors.

SUPERIOR CUSPID.

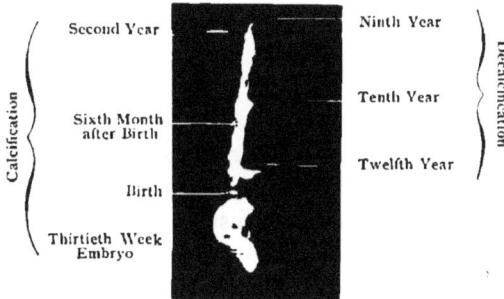

FIG. 165.

CALCIFICATION BEGINS, ABOUT THE FIFTH FETAL MONTH.
  CALCIFICATION COMPLETED, ABOUT TWO YEARS AFTER BIRTH.
    ERUPTS, SEVENTEENTH TO EIGHTEENTH MONTH AFTER BIRTH.
DECALCIFICATION BEGINS, ABOUT THE NINTH YEAR.
  SHEDDING PROCESS TAKES PLACE, ABOUT THE TWELFTH YEAR.
AVERAGE LENGTH OF CROWN, .25.
  AVERAGE LENGTH OF ROOT, .53.
    AVERAGE LENGTH OVER ALL, .78.

Like the permanent cuspid, the general contour of this tooth is that of a double cone, the lines of which are somewhat broken. The greatest mesiodistal extent of the crown is from angle to angle, and this measurement about corresponds to the width of the crown of the central incisor.

The Labial Surface of the Crown (Fig. 166).—This surface is strongly convex from mesial to distal, and slightly so from the cutting-edge to the cervical line. It is bounded by five margins: mesial, distal, cervical, mesial-incisive, and distal-incisive. The mesial and distal margins are rounded and smooth, the cervical well outlined by the cervical line and base of the cervical ridge, while the two incisive margins are formed by the mesial and distal cutting-edges. The labial grooves are thrown well toward the lateral margins, and are usually more distinct than those upon the incisors. The labial ridge is prominent.

FIG. 166.—SUPERIOR CUS-
PID, LABIAL SURFACE.

The Palatal Surface of the Crown.—This surface is generally divided into two portions by the palatal ridge, which extends from the summit of the cusp to the base of the cervical ridge. On either side of this ridge are the palatal grooves, but which appear more in the form of small fossæ. The marginal ridges are fairly well defined.

The Mesial and Distal Surfaces of the Crown.—The extent of these two surfaces is frequently much interfered with by the slope of the mesial and distal cutting-edges, which may be so long that the angles of the crown are forced well toward the cervical line, in some instances almost obliterating these two surfaces. When the cutting-edges are shorter, these surfaces present a marked general convexity. While the summit of the cusp will always be found to be in a direct line with the long axis of the tooth, there is in nearly every instance a difference in the length of the cutting-edges, and, unlike the cutting-edges of the permanent cuspid, the mesial is usually the longest. The neck of the tooth is much constricted and the root straight and conic.

THE SUPERIOR MOLARS.

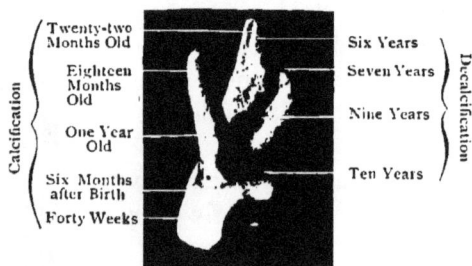

Fig. 167.

The Superior First Molar.

CALCIFICATION BEGINS, ABOUT THE FIFTH FETAL MONTH.
  CALCIFICATION COMPLETED, EIGHTEENTH TO TWENTIETH MONTH AFTER BIRTH.
    ERUPTS, FOURTEENTH TO FIFTEENTH MONTH AFTER BIRTH.
DECALCIFICATION BEGINS, SIXTH TO SEVENTH YEAR.
    SHEDDING PROCESS TAKES PLACE, ABOUT THE TENTH YEAR.
AVERAGE LENGTH OF CROWN, .20.
  AVERAGE LENGTH OF ROOT, .39.
    AVERAGE LENGTH OVER ALL, .59.

The contour and lobal construction of the crown of this tooth is peculiar to itself, being dissimilar to any other class of teeth in the mouth. Calcification takes place from three centers, two for the buccal and one for the palatal half of the crown. The general form of the crown may best be studied by an examination of the occlusal surface.

The Occlusal Surface of the Crown.—The outlines represented are those of an irregular quadrilateral, of which the buccal and mesial sides are the longest. The angles of the quadrilateral are somewhat variable, the mesiobuccal being acute, the mesiopalatal obtuse, while the two distal angles are rounded right angles. The surface is surmounted by three cusps, a mesiobuccal, a distobuccal, and a palatal. These various cusps are separated from one another by three developmental grooves—the mesial, the distal, and the buccal. The marginal ridges are sharp and well defined, this being particularly true of the buccal and palatal, which resemble cutting-edges. The mesial-marginal ridge begins at the mesiobuccal angle, and, after making a long distal curve, ends in the mesial incline of the palatal cusp. The center of the surface is deeply and irregularly concave, producing the central fossa, and descending from the various ridges and cusps surrounding it are numerous supplemental grooves and ridges. The various developmental grooves are not inclined to cross the marginal ridges, although in some instances one or two may be found to do so.

The Buccal Surface of the Crown (Fig. 167).—This surface is generally smooth and convex, with an excessively developed cervical ridge, which is particularly prominent at its mesial extremity. The buccal groove is in the form of a slight depression, and the buccal ridges, common to all molars, are scarcely to be observed. The mesial, occlusal, and cervical margins are distinctly outlined, while the distal is obliterated by the gradual passing of this surface into the distal surface.

The Palatal Surface of the Crown.—This surface is circular in outline, decidedly convex and smooth, and is seldom broken by grooves or ridges. It is most prominent near the center, from which point it slopes in every direction. The cervical ridge is not so pronounced as that of the buccal surface, but there is a sudden rounding of the surface in a cervical direction to meet the palatal root.

The Mesial Surface of the Crown.—This surface is probably more extensive than any of the others ; it is inclined to flatness, with a slight conic convexity over its occlusal third, and a slight concavity near the cervix. The buccopalatal measurement of the surface is nearly twice as great as that from the occlusal margin to the cervical line. It is much more prominent near the occlusal margin, so that a V-shaped space usually exists between it and the distal surface of the cuspid.

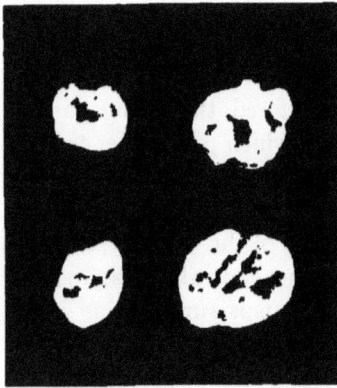

FIG. 168.—OCCLUSAL SURFACES OF THE DECIDUOUS MOLARS.

The Distal Surface of the Crown.—The extent of this surface is much less than that of the mesial ; it presents a general convexity, and is seldom broken by grooves or ridges, although occasionally the distal groove crosses its occlusal margin. Like the deciduous teeth previously described, the neck of the tooth is marked by a decided and abrupt constriction, this form appearing to arise from the heavy enamel folds which are present near the cervical line, rather than from any marked constriction in the base of the roots themselves.

The roots of the tooth are three in number—a mesio-buccal, a distobuccal, and a palatal ; of these, the latter is usually the largest and longest. The two buccal roots are much flattened from mesial to distal, while the palatal is compressed in the opposite direction. The apical ends of the roots are much separated from one another, the triangle which these points form being almost twice the size of the triangle formed by the base of the roots. The apical ends are usually provided with a central curve.

Superior Second Molar.

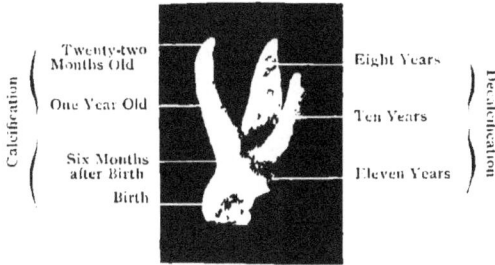

FIG. 169.

CALCIFICATION BEGINS, BETWEEN THE FIFTH AND SIXTH FETAL MONTHS.
CALCIFICATION COMPLETED, TWENTIETH TO TWENTY-SECOND MONTH AFTER BIRTH.
ERUPTED, EIGHTEENTH TO TWENTY-FOURTH MONTH AFTER BIRTH.
DECALCIFICATION BEGINS, SEVENTH TO EIGHTH YEAR.
SHEDDING PROCESS TAKES PLACE, ELEVENTH TO TWELFTH YEAR.
AVERAGE LENGTH OF CROWN, .22.
AVERAGE LENGTH OF ROOT, .46.
AVERAGE LENGTH OVER ALL, .68.

The most remarkable feature about the crown of this tooth is its close resemblance to the crown of the superior first permanent molar. The various surfaces are almost identical, the developmental process, and consequently the cusp-formation, is the same, the marginal and other ridges common to the occlusal surface correspond, and both the central and distal fossæ are present, together with the various developmental grooves. A description of the crown will, therefore, be unnecessary; suffice it to say that it is much smaller in every direction and is somewhat more constricted at the neck. The roots are the same in name and number as those of the first permanent molar, but

FIG. 170.—SUPERIOR SECOND MOLAR, BUCCAL SURFACE.

they are more widely separated at their apical extremities. In general form they are smaller than those of the superior first deciduous molar.

18

THE INFERIOR DECIDUOUS TEETH.

A description in detail of the inferior incisors and cuspids would practically be a repetition of that given of the corresponding superior teeth, and for that reason will be passed with a limited reference to each. The inferior molars being in many respects unlike the superior, they will require a separate description.

Inferior Central Incisor (Fig. 171).

CALCIFICATION BEGINS, ABOUT THE FOURTH FETAL MONTH.
  CALCIFICATION COMPLETED, SIXTEENTH TO EIGHTEENTH MONTH AFTER BIRTH.
    ERUPTED, SIXTH TO EIGHTH MONTH AFTER BIRTH.
DECALCIFICATION BEGINS, ABOUT THE FOURTH YEAR.
  SHEDDING PROCESS TAKES PLACE, ABOUT THE SEVENTH YEAR.
AVERAGE LENGTH OF CROWN, .19.
    AVERAGE LENGTH OF ROOT, .35.
      AVERAGE LENGTH OVER ALL, .54.

FIG. 171.—IN-
FERIOR CENTRAL
INCISOR, LABIAL
SURFACE.

This is the smallest of the inferior teeth, in this respect being at variance to the superior central, which is larger than the lateral. The mesio-distal diameter of the crown is but little less than that from the cutting-edge to the cervical line. The mesial and distal angles are similar, both being pointed and square. The cervical ridge is quite pronounced and the neck much constricted.

The **root** is usually straight and tapers gradually from base to apex. It is broader on the labial than on the lingual side, and the mesial and distal sides are but little flattened.

Inferior Lateral Incisor (Fig. 172).

CALCIFICATION BEGINS, ABOUT THE FOURTH FETAL MONTH.
  CALCIFICATION COMPLETED, TWELFTH TO FOURTEENTH MONTH AFTER BIRTH.
    ERUPTED, SEVENTH TO NINTH MONTH AFTER BIRTH.
DECALCIFICATION BEGINS, ABOUT THE FIFTH YEAR.
  SHEDDING PROCESS TAKES PLACE, ABOUT THE EIGHTH YEAR.
AVERAGE LENGTH OF CROWN, .19.
    AVERAGE LENGTH OF ROOT, .39.
      AVERAGE LENGTH OVER ALL, .58.

This tooth is larger than the central incisor, and closely resembles the superior lateral both in size and form. The crown is

more rounded in its nature than that of the central, forming a
greater general convexity to the labial surface, and less concavity to the lingual. The mesial angle of the
crown is fairly well defined, while the distal is
usually much rounded by a long, circular sweep
of the cutting-edge to meet the distal surface.
The mesial surface of the crown is flattened and
somewhat prominent at the angle, while the distal surface is strongly convex. The labial grooves
are but slightly visible, while the corresponding
lingual grooves are quite pronounced. The neck
of the tooth is even more marked than that
of the inferior central incisor. The root is
long and tapering, slightly flattened from mesial
to distal, with a decided longitudinal groove
on both the mesial and distal sides. The labial

FIG. 172. — IN-
FERIOR LATERAL
INCISOR, LABIAL
SURFACE.

and lingual sides are rounded, and there is an inclination to
crookedness, which is usually from mesial to distal.

Inferior Cuspid (Fig. 173).

CALCIFICATION BEGINS, ABOUT THE FIFTH FETAL MONTH.
CALCIFICATION COMPLETED, ABOUT TWO YEARS AFTER BIRTH.
ERUPTED, SEVENTEENTH TO EIGHTEENTH MONTH AFTER BIRTH.
DECALCIFICATION BEGINS ABOUT THE NINTH YEAR.
SHEDDING PROCESS TAKES PLACE, ABOUT THE TWELFTH YEAR.
AVERAGE LENGTH OF CROWN, .23.
AVERAGE LENGTH OF ROOT, .45.
AVERAGE LENGTH OVER ALL, .68.

The principal variations between this tooth
and the superior cuspid are observed in the diminished mesiodistal measurement of the crown,
together with it being somewhat less angular
in outline. The ridges and grooves common
to the various surfaces are not so marked as
those of the superior cuspid, resulting in a
smoothly formed crown throughout. The root
is larger in proportion to the size of the crown
than that of its superior opponent, thus producing a neck much less constricted. It is
usually straight, or provided with a slight distal

FIG. 173.—IN-
FERIOR CUSPID
LABIAL SURFACE.

inclination near its apical extremity, and much flattened from
mesial to distal, these two sides converging to the lingual,
forming a rounded triangular outline.

Inferior First Molar (Fig. 174).

CALCIFICATION BEGINS, ABOUT THE FIFTH FETAL MONTH.
CALCIFICATION COMPLETED, EIGHTEENTH TO TWENTIETH MONTH AFTER BIRTH.
ERUPTED, FOURTEENTH TO FIFTEENTH MONTH AFTER BIRTH.
DECALCIFICATION BEGINS, SIXTH TO SEVENTH YEAR.
SHEDDING PROCESS BEGINS, ABOUT THE TENTH YEAR.
AVERAGE LENGTH OF CROWN, .24.
AVERAGE LENGTH OF ROOT, .38.
AVERAGE LENGTH OVER ALL, .62.

Upon making an examination of the occlusal surface of this
tooth it will be observed that the crown is made up of four
irregularly formed lobes, separated from one another by four
well-defined grooves. Each lobe is pro-
vided with a cusp, more or less prominently
developed. Between the various cusps are
two fossæ—one occupying the distal two-
thirds of the surface (the distal fossa) and
the other the remaining or mesial third (the
mesial fossa). The outline of this surface,
which represents the contour of the crown in
general, is that of an oblong square, with its
angles more or less rounded, and having a
slight variation in its parallel lines. Each
lobe denotes a separate center of calcification, and the four
grooves the lines of union between the various parts.

FIG. 174.--INFERIOR
FIRST MOLAR, BUCCAL
SURFACE.

The mesiobuccal lobe is somewhat irregular in contour,
and is frequently the largest of the four. It assists in forming
the mesiobuccal angle of the crown and the greater part of the
mesial fossa. Descending from this cusp to the lingual is a
pronounced triangular ridge, which is made continuous by
uniting with a similar ridge from the corresponding lingual
cusp. By this union a transverse ridge is established, separating
the mesial from the distal fossa. The central boundary of this
lobe is formed by the mesial groove, which arises from the distal
fossa, passes over the transverse ridge to the mesial fossa, from

which it continues to the lingual, and by the buccal groove, which branches off from the mesial somewhat to the distal of the transverse ridge, passing over the buccal-marginal ridge to the buccal surface.

The Distobuccal Lobe.—This cusp is generally smaller than the mesiobuccal, and is more pointed and more regular in outline. It assists in forming the distobuccal angle of the crown, and by its central incline forms about one-third of the distal fossa. Its boundaries are formed by the buccal, mesial, and distal grooves, the latter beginning in the distal fossa, and passing over the distal-marginal ridge to the distal surface.

The Mesiolingual Lobe.—In the recently erupted tooth the summit of this cusp is long and pointed, and frequently remains the most pronounced of the four. It is triangular in outline, and, as above referred to, furnishes a triangular ridge, which, by uniting with a like ridge from the mesiobuccal cusp, forms the transverse ridge. By its central incline it assists in forming the mesial fossa. Its boundaries are formed by the mesial groove and the lingual groove, the latter arising near the center of the distal fossa, passing to, and sometimes crossing, the lingual-marginal ridge.

The Distolingual Lobe.—This is usually the smallest of the four. It is inclined to be rounded, rather than angular, and in some instances is poorly developed. It assists in forming the distolingual angle of the crown, as well as a portion of the distal fossa. Its central boundaries are formed by the lingual and distal grooves.

The marginal ridges of the surface are abruptly but irregularly formed, ascending and descending the various cusps in a manner similar to those previously described.

The Buccal Surface of the Crown (Fig. 174).—This surface is smooth and generally convex, with a mesiodistal measurement about twice as great as that from the cervical line to the occlusal margin. The surface is most prominent over its cervical third, forming a well-rounded and bold cervical ridge, a feature strongly characteristic of this tooth. The distal center of the surface is broken by the buccal groove, which usually ends near the center in a shallow depression or pit.

The **Lingual Surface of the Crown.**—This surface is much less extensive than the buccal. It is smooth and convex throughout, and is broken near its distal center by the lingual groove, which gradually disappears as it passes rootward. The cervical ridge is not so prominent as that of the buccal surface.

The **Mesial and Distal Surfaces of the Crown.**—These are slightly convex in every direction, the former passing, by a gradual sweep, into the lingual surface, destroying the angularity of the crown at this point, while the latter passes more abruptly, forming an acute angle. These surfaces are both prominent near the occlusal margin, making the point of contact with adjoining teeth near that surface. The bold cervical ridge of the buccal surface is discontinued or greatly diminished upon these surfaces, both of which are inclined to pass very gradually into the base of the roots.

The **roots** of this tooth are the same in name, position, and number as those of the inferior permanent molars. They are much flattened from mesial to distal, the center of their flattened sides being further compressed by a deep longitudinal groove, which extends from the base to the apex of each root. In passing from the base of the roots to their apices they become more widely separated, until these extremities are much wider apart, proportionately, than those of the permanent molars.

Inferior Second Molar (Fig. 175).

CALCIFICATION BEGINS, BETWEEN THE FIFTH AND SIXTH FETAL MONTHS.
    CALCIFICATION COMPLETED, TWENTIETH TO TWENTY-SECOND MONTH AFTER BIRTH.
    ERUPTED, EIGHTEENTH TO TWENTY-FOURTH MONTH AFTER BIRTH.
DECALCIFICATION BEGINS, SEVENTH TO EIGHTH YEAR.
    SHEDDING PROCESS TAKES PLACE, ELEVENTH TO TWELFTH YEAR.
AVERAGE LENGTH OF CROWN, .21.
    AVERAGE LENGTH OF ROOT, .44.
    AVERAGE LENGTH OVER ALL, .65.

The anatomy of this tooth being almost identical to that of the inferior first permanent molar, it will be unnecessary to enter into a description in detail. The lobes, and consequently the grooves, are the same in position, name, and number, and a similar developmental process is recorded. The tooth is not characterized by a prominent cervical ridge, such as is found upon the inferior first molar, the crown passing very gradually into the root-base with a neck moderately constricted.

THE PULP-CHAMBERS AND CANALS OF THE DECIDUOUS TEETH.

A few general remarks in reference to these cavities, in connection with the information to be derived from the accompanying chart and its annexed description, will sufficiently instruct the reader, without the necessity of treating each tooth individually. The pulp-chambers and canals of these teeth, like those of the permanent organs, assume the form of the external contour

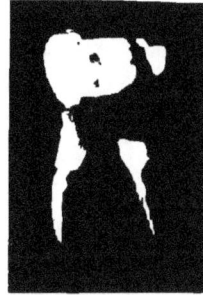

FIG. 175.—INFERIOR SECOND MOLAR, BUCCAL SURFACE.

of the tooth, the crown of the tooth being provided with a central cavity, the pulp-chamber, partaking of outlines closely resembling those of the crown, while the root is traversed by the pulp-canal, likewise conforming to the shape of the root. One very important distinction between the pulp-chambers and canals of the deciduous teeth and those of the permanent organs is that the former are proportionately larger. It must also be noted that the apical foramina in these teeth are so transitory in their nature that there remains but a very brief period during which the canals may be said to be fully formed. It will be recalled that in a very short time after the roots have become completely calcified, decalcification begins, and this process of degeneracy, beginning at the apical extremities of the roots, very early destroys the foramina, which have in a measure served as a protection to the surrounding parts during operations upon the canals.

With the canals proportionately larger than those which occupy the roots of the permanent teeth, the foramina during their very limited existence are also much larger, and much more readily penetrated. With these ever-changing conditions in the pulp-

canals of the deciduous teeth, it is of importance that a definite
knowledge of what takes place should be acquired, and it is for
this purpose that the accompanying chart has been prepared
(Fig. 176).

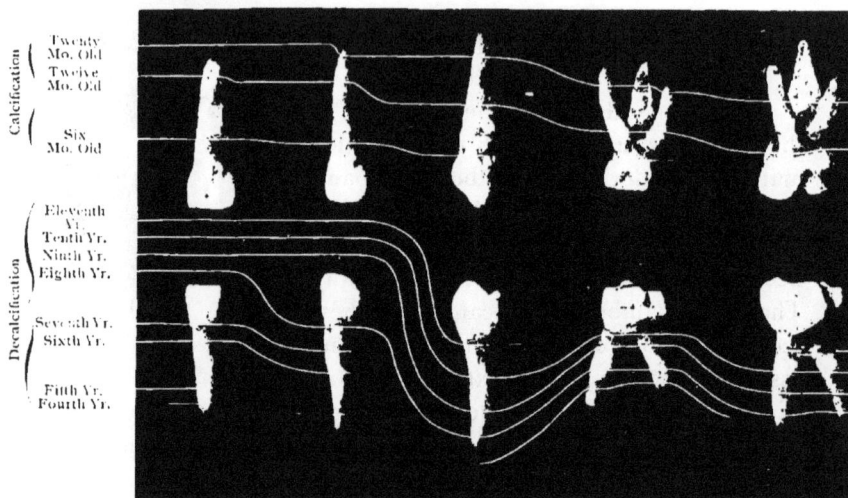

FIG. 176.—CHART SHOWING CALCIFICATION AND DECALCIFICATION OF THE DECIDUOUS TEETH AND THE CONSEQUENT LENGTH OF THE PULP CANAL AT VARIOUS PERIODS.—(After Peirce.)

CHAPTER XII.

DEVELOPMENT OF THE TEETH.—THE DENTAL GERMS, ENAMEL
ORGAN, AND DENTIN ORGAN; THE DENTAL FOLLICLE; CALCI-
FICATION AND ERUPTION.

DEVELOPMENT OF THE TEETH.

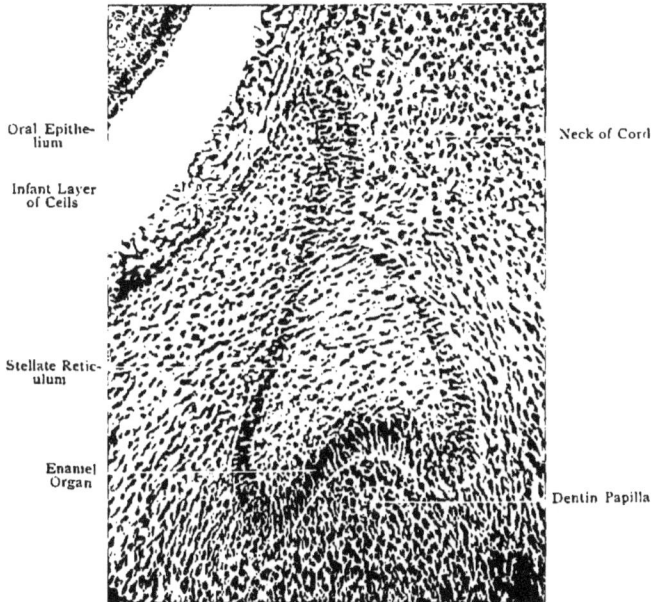

FIG. 177.—DEVELOPING TOOTH-GERM. × 300.

It is not the intention of the author to enter into an exhaustive
treatise of the primary stage of this phase of dental anatomy,
preferring to treat the general subject from a macroscopic rather
than a microscopic standpoint. In order that the student may
obtain a general idea of the structural changes which take place
at a very early period, and which eventually result in the forma-

281

tion of the teeth, the genesis of the subject will be briefly
referred to. Preparation for the development of the teeth
takes place as early as the middle of the second fetal month,
this preparatory alteration in the tissues beginning before the
process of ossification in the bony structures which finally
surround and give support to the organs. At this early period
there will be found following the line of the future alveolar
ridge a slight heaping up of the surface epithelium, while immediately beneath this proliferation of cells there appears a dipping
in of the deep epithelial layer in the direction of the future
alveolar walls. This epithelial inflection is known as the
epithelial band, or *tooth-band*. The structure is not, as might
be supposed, a special inflection for each tooth-germ, but is
continuous from one end of the future jaw to the other. It
must be remembered that at this time the outline of the jaw
has not been established, and the tooth-band, although not generally considered as essential to the developmental process, is
principally instrumental in directing the position of the dental
organs. The position and form of this epithelial band may best
be studied in vertical transverse section. When first making its
appearance it is somewhat broad and shallow, but as it passes
more deeply into the parts it partakes of the outline of the
letter V with its open end directed toward the surface. In
penetrating the subjacent tissue, the free extremity of the band
is inclined to the lingual or palatal, its external surface is slightly
convex, and its internal surface correspondingly concave. Histologically considered, the tooth-band is composed of elements
similar to those which serve to make up the subepithelial layer
of the oral mucous membrane. After the tooth-band has
assumed certain proportions, there appears on its inner or concave surface a thin membranous plate, which is likewise a
continuous structure, extending the full length of the epithelial
band.

This lamina does not spring from the free margin of the tooth-
band, but is given off at a point about midway between this
border and the base of the band. The structure of this secondary
band is so similar to that of the primary one that it should be
considered as an inflexion from it rather than a new structure.

We find then between the seventh and eighth week, the maxillary regions giving place to two bow-shaped bands (one for each jaw), each of which is preparing to throw out from its secondary lamina ten little buds, which soon develop into the germs for the twenty deciduous teeth. When these buds make their appearance they are simple, rounded bodies, placed somewhat closely together, but they do not long retain this simple form. The first change which takes place is one in which they appear to lengthen out into slender cords, the extremities of which soon begin to extend laterally, and the primitive enamel organ is

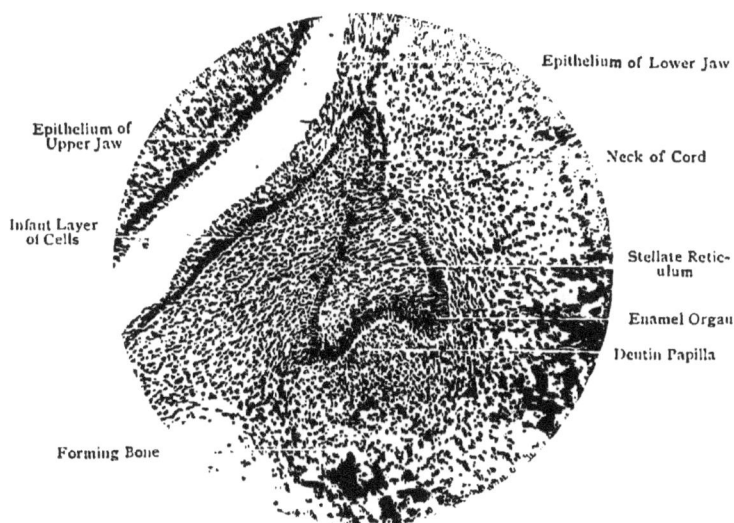

FIG. 178.—DEVELOPING TOOTH-GERM, TWELFTH WEEK. × 40.

formed. Accompanying this lateral extension of the periphery, a bell-shaped outline is assumed, which phenomena is rapidly increased by a specialization of the surrounding cellular tissue forcing into the concavity.

This bell-shaped proliferation of cells, given off directly from the tooth-band, to which it continues for a time attached, together with the specialized submucous tissue crowding into its concavity, constitute the tooth-germs, the former being the *enamel organ*, and the latter the *dentin organ*. It will, therefore, be seen that the enamel is dependent upon the oral epithelium for its development, while the dentin springs from an entirely different source—the submucous tissues of the jaw. The enamel organ rapidly undergoes a cellular transformation : its concavity is increased, and the bell-shaped outline more strongly defined. Accompanying this change in form it gradually recedes from the surface, and its connection with the tooth-band becomes less secure. The submucous cells, which have been rapidly filling in the concavity of the enamel organ, are also preparing to take upon themselves a special function, that of the formation of the dentin. Up to this period (tenth week) the enamel and dentin germs are not definitely separated from the surrounding cellular structure, but now a gradual transformation takes place, whereby the tooth-germs become enveloped in a sac-like covering—the *dental follicle.*

Enamel Organ (Figs. 177 and 178).—This portion of the tooth-germ, as previously stated, is derived from the concave or lingual surface of the tooth-band, which in turn is derived from the surface epithelium. From the free extremity of its slender cord-like attachment it spreads out and forms a hood-like covering to the dentin germ. The surface of the organ contiguous to the dentin germ, or dentin papilla, as it is frequently called, is concave in the direction of the oral surface, being thickest over the center of its concavity, thinning down as its periphery is approached. Externally, the enamel organ is covered by an epithelial layer, which is reflected upon its inner surface or that in contact with the dentin papilla. These two layers are named, according to their location, the external and internal epithelium of the enamel organ. Placed between these two layers, and constituting the bulk of the organ, are numerous stellate bodies which penetrate a layer of rounded cells, the *stratum intermedium,* and finally reach the internal epithelial layer known as the *enamel cells.* It is from this internal layer of epithelial cells that the enamel is calcified, and they are, therefore, the essential cells of the enamel organ. As its name implies, the function of the enamel organ is principally that of enamel calcification, but in the opinion of many writers its primary activity is that of molding the tooth-form as represented by the dentin papilla, and it is not until this latter organ has assumed the form and extent of the future tooth-crown that dentin calcification begins.

Dentin Organ (Fig. 178).—This part of the tooth-germ, formed from the submucous tissues of the primitive jaw, occupies the concavity of the enamel organ, and at an early period begins to assume the form of the future tooth-crown. Thus, primarily, the papillæ for the incisors will have their cutting-edges outlined by three small lobes, each of which represents a separate point of calcification, while the papillæ

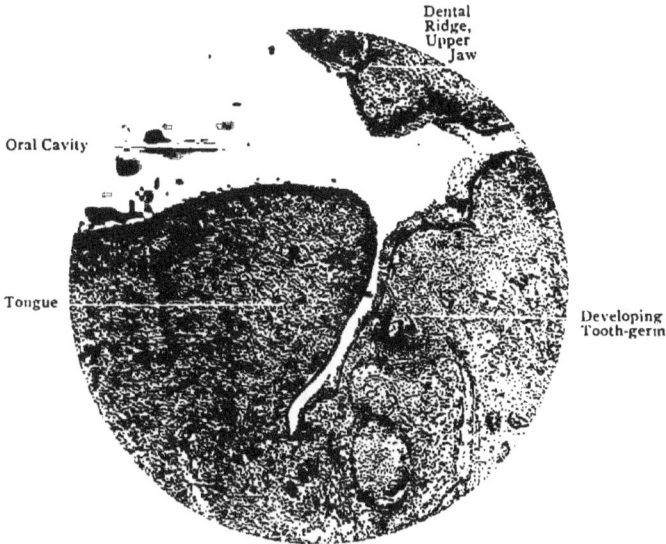

FIG. 179.—SECTION THROUGH THE FLOOR OF THE MOUTH OF TWELFTH WEEK HUMAN EMBRYO. X 30.

for the molars will be molded according to the number of cusps of the future tooth, a small tubercle making its appearance for each cusp.   In its inception the dentin papilla is composed of cellular tissue identical with that of the surrounding parts.   The growth of the papilla is in the direction of the surface; at the same time the enamel organ forces itself more deeply into the substance of the parts, not only overhanging the coronal extremity of the papilla, but extending about and inclosing its lateral walls.   Accompanying the growth of the papilla is a rapid change in its structure, becoming more vascular throughout, and its peripheral cells, differentiating, form the essential dentin cells—*the odontoblasts.* This layer of cells is in close relation to the enamel cells of the enamel organ, the combined activity of the two finally resulting in the calcification of the tooth-crown. The dentin papilla, which eventually becomes the tooth-pulp, decreases in size as calcification proceeds in the dentin, all additions to the calcifying surface taking place from within; while the enamel organ may be said to increase in size, the calcific action in the enamel progressing from within outward.

Reference has been made to the fact that the enamel organs for the deciduous teeth are given off from the tooth-band at a point somewhat distant from its free margin, so that the tooth-band is continued beyond the primitive enamel germ, this free margin of the band afterward generating the enamel organ for the succedaneous tooth. As the twelve permanent molars are not succedaneous teeth, some other means must be provided for their development.

Opinions of various writers upon this subject are somewhat conflicting. The theory is advanced by some that as the jaw increases in length the tooth-band and lamina primarily provided for the deciduous teeth are extended backward, first giving off a bud for the first permanent molar, at a somewhat later period, and with the increase in the growth of the jaw an additional bud is generated for the second molar, the third molar being provided for in a like manner. Another theory, and one generally accepted as correct, is that the cords for the permanent molars spring individually and directly from the subepithelium. There may be found an exception to this in the case of the first permanent molar, which sometimes appears to have its origin from the distal follicular wall of the second deciduous molar. Whatever theory be accepted in regard to the genesis of these permanent organs, the process of development after the appearance of the primary bulb or enamel germ is identical with that of the deciduous teeth.

The Dental Follicle, or Tooth-sac.—During the early life of the tooth-germ, both the enamel organ and the dentin

papilla are differentiated from the surrounding parts by dissimilarity of structure only, but as development proceeds, a more
definite separation appears between the tooth-generating organs
and the general tissues of the primitive jaw, this separating
medium being known as the dental follicle. The term "follicle"
is only one of a number applied to these parts, "dental sacculus," "tooth-sac," and other appellations being employed with
equal significance. By some writers it is customary to apply the
term "follicle" up to the period of complete closure, the term
"sac" or "sacculus" being employed after that time. There

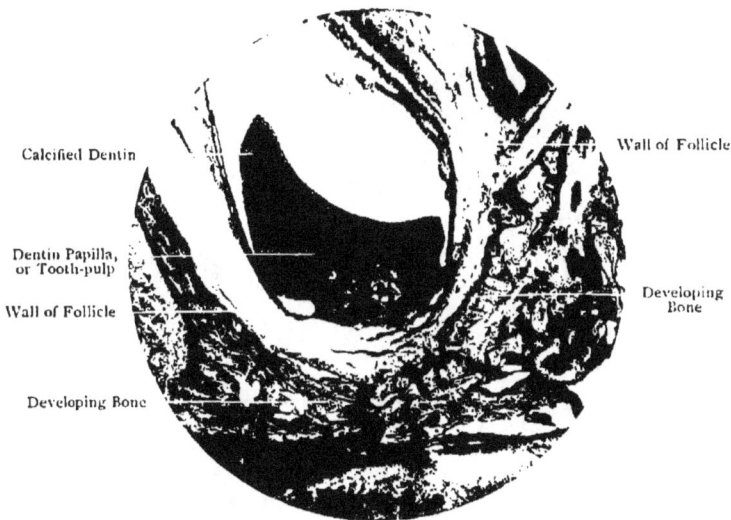

Calcified Dentin

Wall of Follicle

Dentin Papilla,
or Tooth-pulp

Developing
Bone

Wall of Follicle

Developing Bone

FIG. 180.—DEVELOPING TOOTH ABOUT THE FOURTH FŒTAL MONTH. APPEARANCE OF
THE TOOTH-FOLLICLE.

appears, however, to be little foundation for such a distinction,
the terms being synonymous. Again, the follicle is frequently
referred to as meaning the sac and its contents, but this usage
is a misapplication of the term. There appears to be no well-
founded reason why the follicular wall or sac should not be
referred to as such, regardless of the formative organs within.
As to the development of the tooth-follicle, it appears to be a
generally accepted theory that at a very early period there is
developed from the base of the papilla cells which, differenti-

ating, form the walls of the follicle. By this growth of cells the periphery of the papilla is first surrounded, and this step is soon followed by an extension of the cellular structure in the direction of the surface epithelium, to the deep layer of which the cells become firmly attached, and in so doing inclose the enamel organ, which hangs like a hood over the extremities of the papilla. The tissue thus formed from the base of the dentin germ is continuous with and similar in its origin to the pulp-substance. The primitive tooth-germ, during the formation of the follicular wall, is found swinging in a membranous pocket,

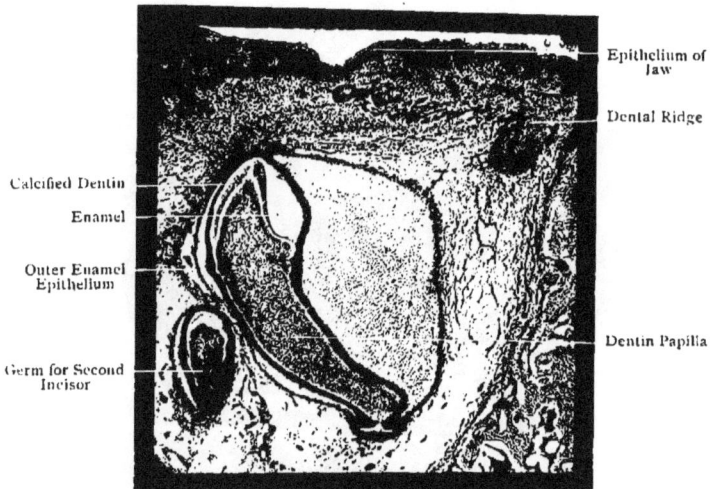

FIG. 181.—DEVELOPMENT OF DECIDUOUS INCISOR, FROM HUMAN FETUS.—
(*After Geise.*)

being supported by the epithelial band, which, in turn, is attached to the oral epithelium; but as the walls increase and completely inclose the germs, which is accomplished about the fourth fetal month, the epithelial band is broken and the second or saccular stage of tooth-development is reached. The walls of the follicles are made up of two layers; the outer layer is dense and firm, and finally becomes the dental perios-teum; the inner layer being thin, frail, and in the recent state somewhat transparent, and at an advanced period assisting in the formation of the cementum.

Having thus briefly described the primary stage of tooth-development, the careful study of which can only be pursued with the aid of the microscope, we will now pass to the secondary or saccular stage. By the introduction of a number of illustrations, prepared from original dissections by the author, this phase of the subject will be readily comprehended.

Before continuing the subject of tooth-development, it will be eminently proper to briefly describe those parts directly concerned in the process. At a very early period of fetal life we find preparations are being made for the development of the

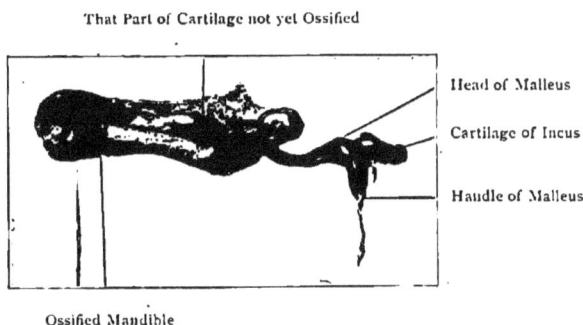

FIG. 182.—DEVELOPING MANDIBLE, THREE-MONTH FETUS.

maxillary bones. That these are about the first bones to be called into functional activity accounts in a measure for their very early development. The osteoblastic activity in the intercellular substance destined to become the inferior maxilla begins about the middle of the second month of fetal life, while at a somewhat later period a similar action takes place in the region of the superior maxilla. A detailed description of the body of these bones having been given on another page, it will not be repeated here, but to that portion which gives lodgment to the tooth-germs, and which in a measure is controlled by their presence, some attention must be given, and for this purpose the mandible will principally be used.

Figure 182 represents the lingual face of the lower jaw after

19

removal from a three months' fetus. Attached to it is the remaining portion of Meckel's cartilage, which by this time is
much wasted.   It will be recalled that this cartilaginous
band appears in the mandibular processes before the beginning of the second fetal month, being formed in two distinct
halves, the free ends of which finally unite at the median line,
forming a continuous support or framework, about which ossification takes place.   At a corresponding period, and in a similar
manner, two like processes are thrown out for the superior
maxillæ, but, unlike Meckel's cartilage, these do not unite at the
median line, but stop short of this, the space thus resulting
being provided for by two additional processes, which shoot
down from the region of the fore-
head and provide for the development of the intermaxillary bones.
About the middle of the second
month a center of ossification appears in the neighborhood of the
future mental foramen, quickly followed by others at the symphysis
and at the angle.   These secondary
centers soon unite with the primary
one, and by the end of the second
fetal month the osseous contour of
the primitive jaw is established.
While ossification takes place in the
membrane surrounding Meckel's

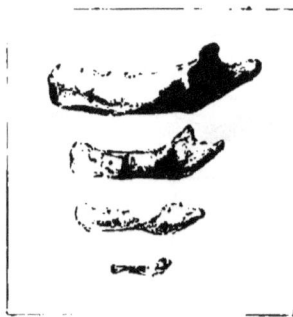

Fig. 183. — Evolution of the
Mandible from the Third Fetal
Month to Birth, Two-thirds
Actual Size.

cartilage, the cartilage itself does not appear to be directly
concerned in the process, and by the sixth or seventh fetal
month the mandibular portion completely disappears, while that
portion near the tympanum is ossified into the malleus.   That
portion of the bone which forms above Meckel's cartilage and
the inferior dental nerve is that which finally gives support to
the tooth-germs.   This cartilage is not confined to the human
species, but is the common heritage of reptiles, rodents, birds,
and fishes, in all of which it gives support to the developing
lower jaw.

Figure 183 represents the evolution of the mandible from the middle of the third fetal month to the time of birth. It will be observed that during this interval there is a gradual increase in the size of the bone, but little alteration in its contour. By a constant and gradual osseous deposit about the distal extremity of the bone its length is increased to accommodate the additional teeth as they make their appearance. While the external form of the bone shows but slight variation during this period, the internal structure, or that wherein the tooth-germs lie, is undergoing a complete transformation.

Figure 184 is illustrative of these changes; beginning with a simple groove, or gutter, into which the tooth-follicles hang, the follicles exerting a controlling influence over its form. Next comes the appearance of septa between the anterior follicles, which at this period are somewhat irregularly placed in the arch, followed in a few weeks by a well-defined partition between the cuspids and molars, until finally, at birth, each follicle is inclosed in its individual crypt, with the single exeption of the second molar,

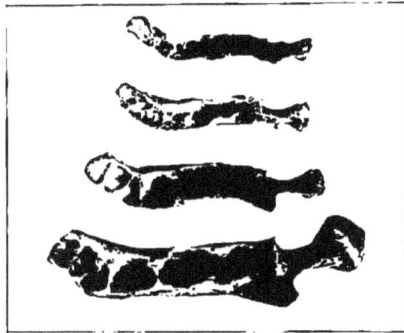

FIG. 184.—EVOLUTION OF THE MANDIBLE FROM THIRD FETAL MONTH TO THIRD MONTH AFTER BIRTH.

in which the distal septum, or that which is to separate it from the first permanent molar, has not yet made its appearance. As the tooth-follicles increase in size, by the development of the teeth within, they become more perfectly inclosed in the bony vaults, the sides of the alveolar walls arching over and almost completely inclosing the developing teeth. Figure 185 shows the lower jaw of a seven-months-old child, embodying the condition above referred to. No sooner have the crypts grown to this extent, than the resorptive action produced by, or provided for, the advancing crowns speedily results in their downfall, to be again built up with the evolution of the permanent teeth.

About the first visible sign of preparation for the development of the teeth, other than that made apparent by dissection, may be observed as early as the beginning of the third fetal month, when, upon opening the cavity of the mouth and looking

Mucous Membrane and Periosteum Lifted Up

FIG. 185.—INFERIOR MAXILLA OF SEVEN-MONTHS-OLD CHILD.

upon the palate, a well-defined infolding of the epithelial eminence will be seen. In figure 186 this condition is shown by a dissection through the oral cavity of a four months' fetus, the outer fold being that of the cheeks and lips, while within is the hard palate and primitive alveolar ridge. The mouth at this

Labiodental Space   Labial Fold

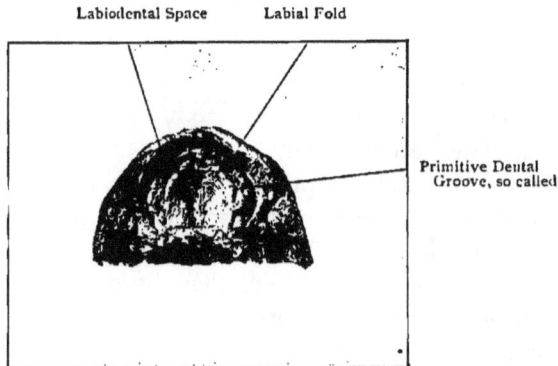

Primitive Dental Groove, so called

FIG. 186.—SECTION THROUGH THE MOUTH OF FOUR-MONTH FETUS.

period has passed the rudimentary state, the transverse plates which contribute to the formation of the hard palate having approached each other until the oral and nasal cavities, heretofore existing as a single buccal cavity, have become separate and distinct. The infolding of the oral epithelium, as outlined on the summit of the primitive alveolar ridge,—the primitive dental furrow, so called,—marks the position of the tooth-band, from which are given off the incipient tooth-bulbs. For the purpose of further investigation, a dissection of these parts was made and the maxillary bones removed, after which they were divested of their fibrous covering, including the periosteum. That portion which overlies the palatal processes was readily lifted in one sheet, while that upon the facial surface was separated at the median line and stripped independently of the other (Fig. 187). The removal of these tissues is readily accomplished until the margins of the partly formed alveoli are reached. Here the periosteum dips down into the various crypts, and serves as a lining membrane for them, and probably contributes fibers to the outer layer of the follicular walls. After advancing thus far, the detached tissues may be grasped, and by careful manipulation the tooth-follicles containing the formative organs removed from their respective vaults and turned over for examination.

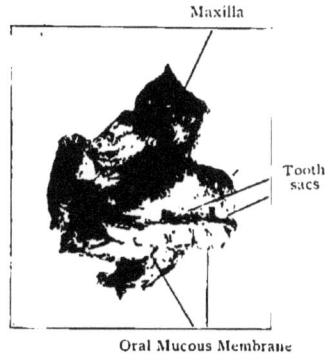

Fig. 187. — SUPERIOR FETAL MAXILLA, SHOWING MANNER OF DISSECTION TO EXPOSE TOOTH-SACS.

Figure 188 shows the result of such a dissection. On the left, the palatal plates and alveolar walls of the divested bones may be observed ; on the right, is the fibrous covering, which has been turned completely over after removal from the bones, having firmly attached to it the ten tooth-follicles for the deciduous teeth. If this dissection be made without the precaution of lifting the periosteum, the follicles would not cling to the oral membrane with sufficient tenacity to permit of their ready removal. It may be of some interest to note that at this early period the position of the follicles containing the germs for the lateral incisors is that which the tooth is forced to occupy up to and frequently beyond the eruptive period, being crowded within the tooth-line by the central and cuspid follicles, in consequence of which the lateral crypts are thrown well into the palatal plates.

Figure 189 represents the sacs broken down, exposing to view the dentin papillæ, or those structures destined to become

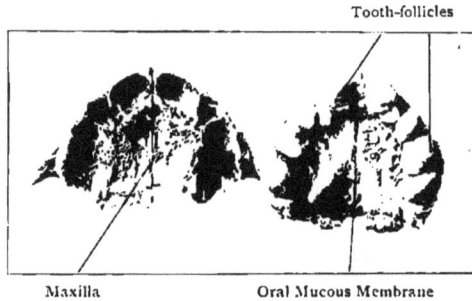

Fig. 188.—Dissection upon Superior Maxilla, Fourth Fetal Month, Exposing Tooth-follicles.

the tooth-pulps. The position which these occupy in the illustration is exactly the reverse from that which they assume when in position in the follicle, being thus reversed that a better idea of their shape may be obtained. Prior to the twelfth or thirteenth week of fetal life this incipient bulb or papilla is with-

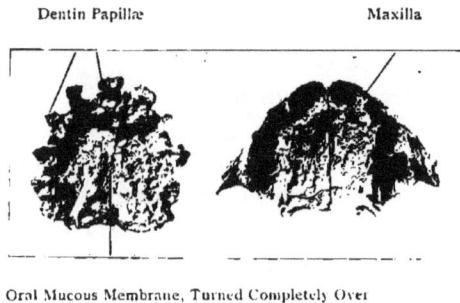

Fig. 189.

out definite form; but by the latter period, each papilla begins to assume the contour of the future tooth-crown, as faintly out-lined in the illustration, those of the incisors presenting the angular form of the future cutting-edge, the cuspids that of the

single cone, while the coronal extremities of the molars are
represented by outlines corresponding to the future cusps and
marginal ridges. Besides the dentin papilla, there is contained
within the follicular walls the organ which later on is productive
of the enamel, but which up to this time has been actively

Oral Mucous Membrane and Periosteum Dissected rom Mandible

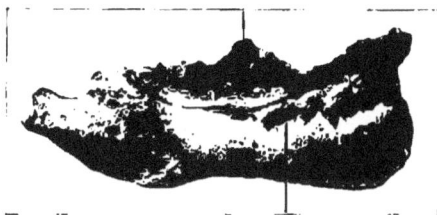

Attachment of Follicle to Oral Membrane (Enlarged One-third)

FIG. 190.—MANNER OF DISSECTION TO EXPOSE TOOTH-SACS.

engaged in molding the tooth-form as outlined by the papilla.
To return to the tooth-follicle, the dissection in this instance
being upon the lower jaw of a four months' fetus. Figure 190
shows the mandible removed and the dissection carried to the

Tooth-follicles   Oral Membrane

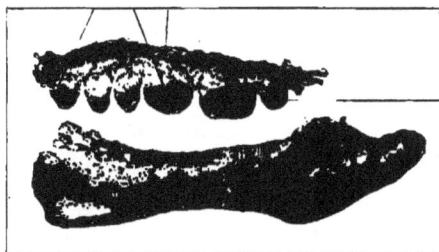

Follicle for First
Permanent Molar

FIG. 191.—TOOTH-FOLLICLES REMOVED FROM MANDIBLE, FOURTH FETAL MONTH.

point at which the follicles may be lifted from their bony encase-
ments, this being accomplished by an incision along the base of
the bone, followed by a stripping of the membrane first from
the facial and then from the lingual side of the bone. When
these two flaps reach the margin of the crypts, they are firmly

grasped and the follicles removed from their sockets, as illustrated in figure 191.

At the beginning of the saccular stage of development the form of the future tooth-crown is well outlined by the dentin papilla, which in figure 192 is brought into view by a dissection of the walls of the follicles shown in figure 191 without breaking the attachment existing between the two. As in the case of the follicle, much confusion of terms in regard to this structure exists, the "dentin bulb," the "pulp," "dentin germ," and the "papilla" being used *ad libitum*, without apparent regard for the structural changes which are continuously affecting the organ. It will, therefore, be proper to simplify this conglomeration of

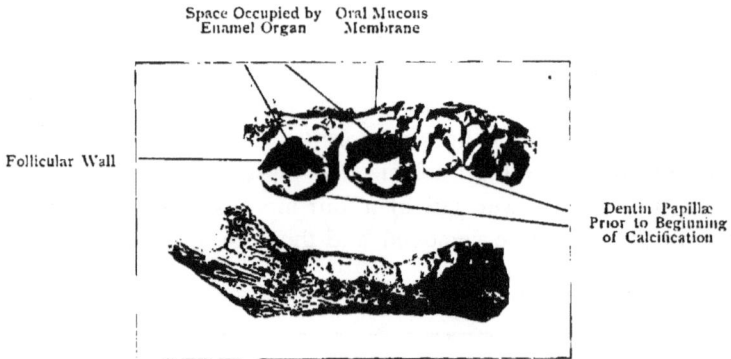

FIG. 192.—TOOTH-FOLLICLES SHOWN IN FIGURE 191 OPENED.

terms by a classification appropriate to the various stages of development. The term "dentin papilla" will best describe this part of the tooth-germ up to the time of beginning of calcification, subsequent to which the term "pulp" should be employed. As previously stated, the enamel organ, until this period, has been principally devoting its energies to the molding of the tooth-form, and it is not until this model, as represented in the papilla, is complete that the process of calcification begins. About the fourth fetal month preparations for the calcification of the deciduous teeth are begun by the development of the odontoblastic cells for the dentin, which first make their appearance on the periphery of the dentin papilla, the summits of the various cusps in the molars and the future cutting-edges of the incisors being first affected. This phenomenon is soon followed by the appearance of the amelioblastic cells for the enamel, which establish themselves in the internal epithelial layer of the enamel organ.

Figure 193 is prepared from a dissection made in a manner similar to that shown in figure 190, but at a period about a month later, being from the superior maxilla of a five months' fetus. The dissection shows the extent of calcification at this period, which process also defines the position of the odontoblastic cells upon the extremity of the papillæ. In the incisors (one of which was lost in the preparation of the specimen) the

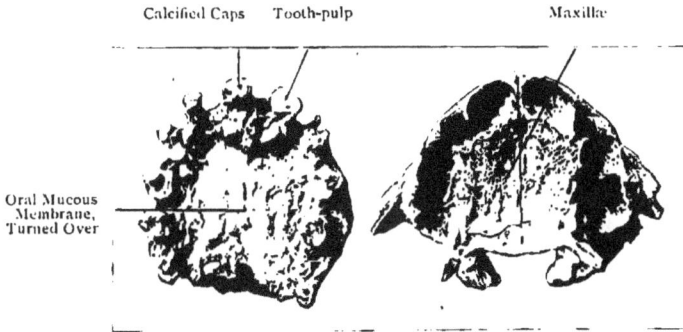

FIG. 193.

dentin may be seen capping the cutting-edges. The cuspids in this subject have not yet begun to calcify, although it is not unusual to find the cusp of this tooth receiving its lime-salts at this early period. In the molars the summits of the various cusps, as well as a portion of the various ridges descending therefrom, are undergoing the change produced by the impregnation of the lime-salts. It is quite probable that these delicate caps are at this time composed of dentin alone, the calcoglobulin which precedes the enamel calcification forming somewhat later. From this time forward the pulp undergoes a gradual transformation as to size and form, and there is likewise a change in its cellular construction on those parts adjacent to the calcific action. While the outline of the pulp is gradually changing, its

original form is permanently recorded upon the periphery of the
dentin cap, which, when once formed, is immutable, all additions
taking place from within.

Figure 194 illustrates the result of a dissection upon the lower
jaw of the same subject, disclosing practically the same condi-

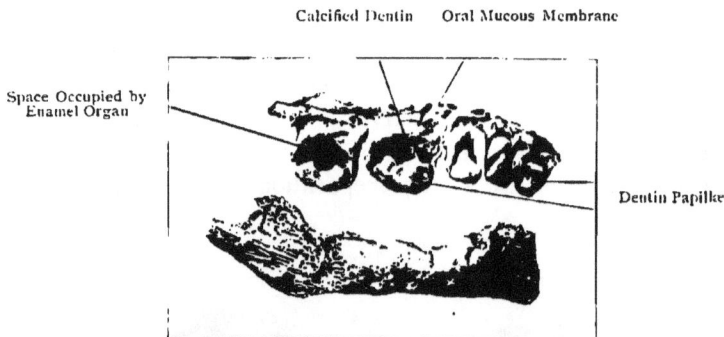

FIG. 194.—TOOTH-FOLLICLES OPENED, EXPOSING DENTIN PAPILLÆ AND BEGINNING OF
CALCIFICATION, FIFTH FETAL MONTH.

tions, with the exception of the sac containing the developing
cuspid, which was found with a slightly calcified cap of dentin.
This slight variation between the development of the superior
and inferior teeth is one that is present in nearly every instance,
the latter being somewhat in advance of the former.

In figure 195 the tooth-pulps, with their primitive cappings

FIG. 195.

of dentin, have been removed from the follicles, and a better
opportunity of studying the relations between the two parts is
presented.   By the conversion of the coronal extremities of the
dentin papilla into odontoblasts, and their active calcification,
some positive union between the two parts might be expected.

On the contrary, the dentin caps are readily removed, leaving the pulp beneath without the slightest rupture, so that we find calcification is not a secretory or excretory metamorphosis, but that the change takes place within the substance of the papilla itself, whereby it is altered from an organic to an inorganic substance.

The next dissection was one upon the mandible of a six months' fetus. Figure 196 shows the tooth-follicles removed from the partially formed bony crypts in which they have been incased. At an early period of fetal life, and at a time prior to the completion of the tooth-follicles, there is deposited beneath the tooth-germs a thin layer of bone, which at once begins to assume the form of the partially developed follicular walls. As the growth of the follicle proceeds, there is a corre-

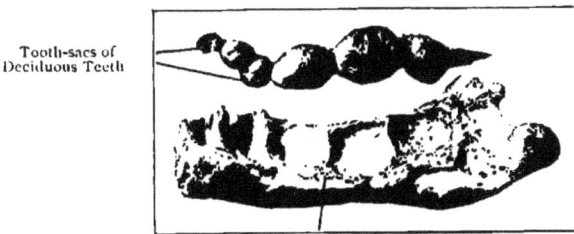

Tooth-sacs of
Deciduous Teeth

Lingual Surface of Mandible.

FIG. 196.

sponding increase in the osseous deposit, the alveolar walls extending about and accommodating themselves to the membranous sacs. Thus we find in this portion of the maxillary bones a feature peculiar to itself—that of a continuous transformation from its earliest inception to the adult period, first developing about the temporary tooth-sacs and completely incasing them, which is speedily followed by complete resorption of the walls, again followed by a rebuilding during the evolution of the permanent teeth, and again swept away with the loss of these organs. Figure 197 illustrates the opposite side of the same jaw, with its outer or facial plate removed, together with the intervening septa. The follicles are opened and the extent of calcification at this period (six months) made apparent. The pulps and calcified caps are approximately as found when dissected, save a slight settling of all the parts. The incisors have calcified to about one-third their full coronal length; the unicusped contour of the cuspid has been established, as shown by the deposit of the lime-salts upon its summit, and the upbuilding of the mesial and distal cutting-edges. The first

Follicular Wall   Calcified Caps   Oral Mucous Membrane

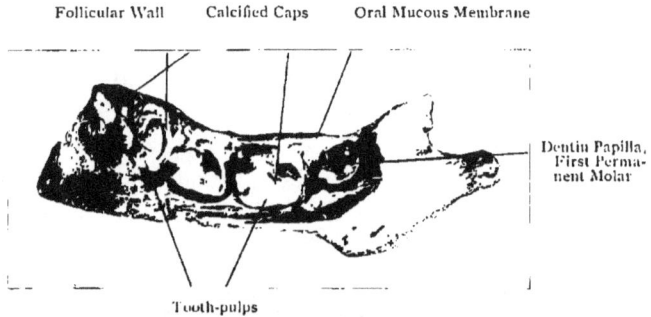

Dentin Papilla, First Permanent Molar

Tooth-pulps

FIG. 197.

molar has about completed its occlusal surface, and, while the cusps of the second molar are nearing completion, there is a lack of union in the central and distal fossæ. Immediately posterior to the second deciduous molar, the sac containing the formative organs for the first permanent molar is shown opened, exposing to view the dental papilla, which at this early

FIG. 198.

period has assumed the form of the future tooth-crown, and calcification is about to begin. Figure 198 illustrates the extent of calcification in the deciduous teeth at the sixth fetal month. As the growth of the teeth proceeds, it will be observed that the angularity which originally accompanied the calcifying caps of the molars is gradually disappearing, as is also the tritubercular form of the incisors and cuspids, this change being brought about by the deposition of enamel to the parts.

Figures 199 and 200 show the calcified caps removed from the pulps, and so arranged that both their external and internal anatomy may be studied. Prior to this time there has been but little alteration in the form of the dentin pulp, only a gradual decrease in its size being noted; but now we find it being divested of many of its angles, particularly those which origi-

FIG. 199.

nally served as a basal form for the coronal extremities of the future tooth. With the disappearance of these the concavity within the cap is slowly assuming the form of the future pulp-chamber.

That a better understanding of the saccular stage of tooth-development might be had, a transverse section was made through the molar follicles, as shown in figure 201. In this the

FIG. 200.

attachment of the follicular walls to the deep epithelial layer is visible, while within the walls is the enamel organ, the calcified dentin, and the tooth-pulp. As previously stated, the enamel organ is seen suspended above and forming a hood-like investment to the calcifying structure. This organ not only overhangs the occlusal surface of the tooth-crown, but completely envelops the sides of the calcified cap and dentin pulp. Previous to the beginning of calcification the enamel organ is in close proximity to the dentin papilla, the original form of the latter being represented by the calcified dentin.

We have now arrived at that period of fetal existence when it is possible to study the macroscopic development of the first

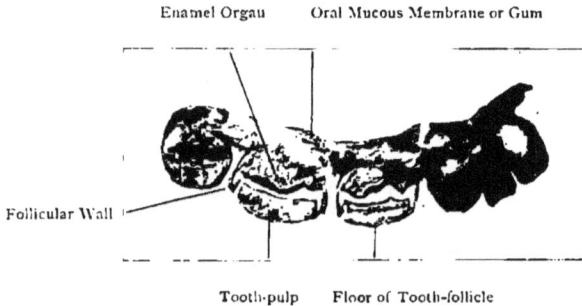

Enamel Organ　　　Oral Mucous Membrane or Gum

Follicular Wall

Tooth-pulp　　　Floor of Tooth-follicle

FIG. 201.—DISSECTION SHOWING PULP, CALCIFIED CAP, AND ENAMEL ORGAN.

permanent molar. Figure 202 represents a section of the lower jaw of a six months' fetus, and displays not only the sacs of the deciduous teeth, but also that of the first permanent molar. In most respects the evolution of this tooth is similar to that of the temporary organs, having its origin from the deep epithelial layer, either directly or by continuation of the tooth-band backward. Preparations for its growth are begun as early as the third fetal month, at which time the enamel organ is given off, and thereafter the developmental process is identical with that of the deciduous teeth.

Oral Mucous Membrane

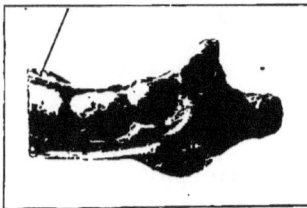

FIG. 202.

There is one structure, however, intimately connected with the development of the permanent teeth not found in connection with the deciduous organs—the *gubernaculum*, or leading cord.

Figure 203 represents another section of a six months' fetal mandible, with the dentin papilla for the first permanent molar turned out from the follicle after being rolled from its bony incasement. Attached to the apex of the tooth-sac (which

has been turned back) and leading from it to the epithelium of
the jaw, is the gubernaculum. This fibrous structure was at
one time thought to be directly concerned in the development
of the tooth. Although this is denied at present, little is
said in regard to its function, but it undoubtedly serves the
purpose of directing the tooth to that position which it should
occupy in the jaw, and where the least resistance to its eruption
is formed by the foramen which the cord has established. Each
of the permanent teeth is provided with a similar membranous
cord, an illustration and more complete description of which will
follow later on.

We have now arrived at a period when the subject under

Oral Mucous Membrane   Gubernaculum

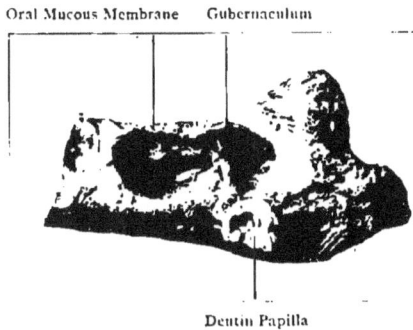

Dentin Papilla

FIG. 203.

consideration naturally becomes of deeper interest. I refer to
that time when the being changes from a complex dependent
condition to one of self-providing independence. Previous to
the time of birth the teeth appear to be but little disturbed by
certain morbid conditions which might be present in the parent,
and from their earliest inception up to this period their development proceeds with but little interruption and with much regularity. Figure 204 illustrates the condition of the deciduous teeth
at birth ; the central incisors are calcified externally to the cervical line, the lateral incisors to a point corresponding to the
summit of the palatocervical ridge ; the cuspids have advanced
somewhat beyond the angles of the crown, while the molars have
their crowns calcified to about one-half their completed length.

With all this progress as represented by the external contour
of the tooth-crowns, the internal form appears to be somewhat
slow in assuming the shape of the future pulp-chamber. From
the beginning of the saccular stage of development up to the
time of birth there is but little increase in the diameter of the
tooth-sac, but there occurs a gradual increase in its length.

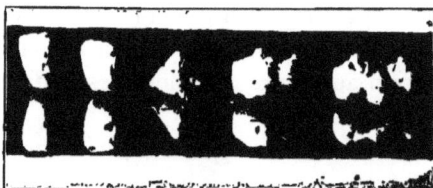

FIG. 204.—DECIDUOUS TEETH AT BIRTH. (Reflected picture.)

Figure 205 shows the mandible from a child one week old, with
the greater part of the external or facial surface of the bone
removed, exposing not only the sacs containing the developing
deciduous teeth, but also that of the first permanent molar.
The relation of the sacs to the inferior dental canal is apparent,
as well as the firm attachment of the follicular walls to the oral
membrane. In figure 206 the tooth-sacs have been dissected

Oral Mucous Membrane

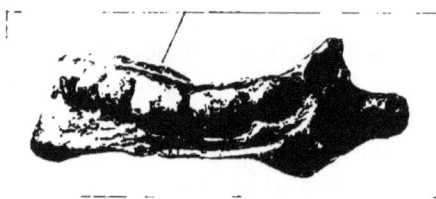

FIG. 205.

and the pulps removed from the calcified caps, presenting an
additional illustration of the amount of dentin deposit at this age.
To further illustrate the size and form of the pulp as compared
with the calcified cap at birth, a transverse incision was made
through the left superior maxilla, at a point corresponding to
the base of the pulp, as shown in figure 207. The calcified

parts remain in position resting against the remaining portion
of the enamel organ, while the pulps are dislodged and may be
observed resting upon the incised surface. In this illustration

Calcified Caps

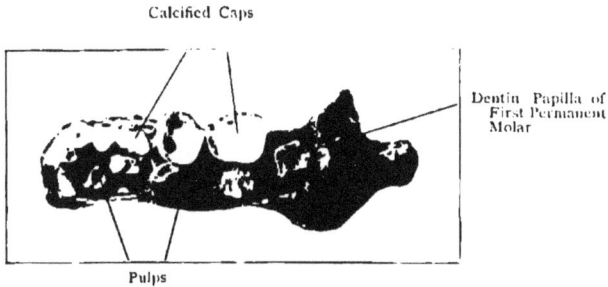

Dentin Papilla of
First Permanent
Molar

Pulps

Fig. 206.

the dentin papilla for the first permanent molar is also seen,
being supported by the walls of the follicle, which in turn are
attached to the oral epithelium by the gubernaculum. It may

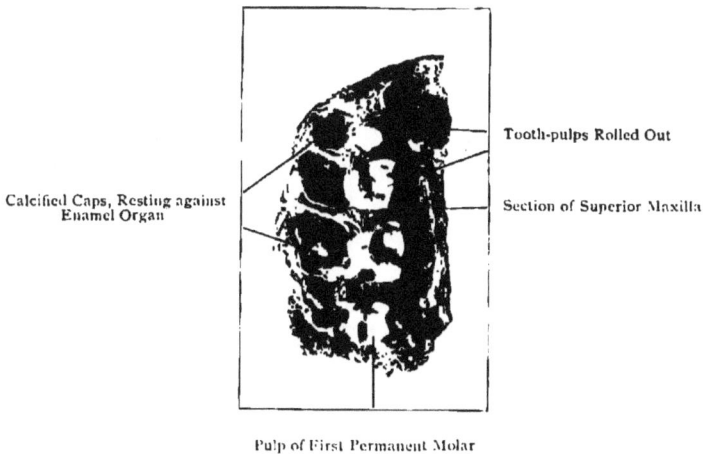

Tooth-pulps Rolled Out

Calcified Caps, Resting against
Enamel Organ

Section of Superior Maxilla

Pulp of First Permanent Molar

Fig. 207.

also be noted that those parts of the pulp corresponding to the
cusps and marginal ridges show a decided convergence of the
surface toward the center.

20

Reference has been made to the formation of the follicular walls by a differentiation of cells, which at an early period are given off from the base of the papillæ, and to the continuity of the two structures. Figure 208 was prepared for the purpose of showing these intimate relations. The sides of the sacs

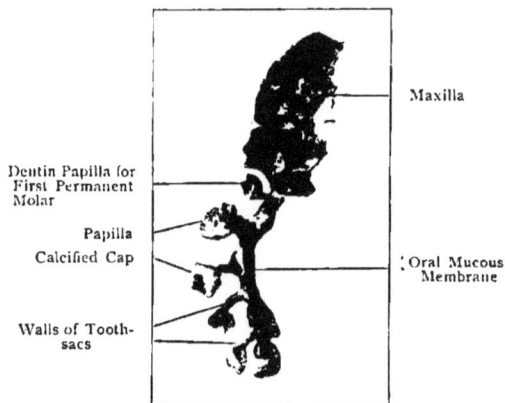

Dentin Papilla for
First Permanent
Molar

Papilla
Calcified Cap

Walls of Tooth-
sacs

Maxilla

Oral Mucous
Membrane

FIG. 208.

were opened and turned back ; the calcified tooth-caps with pulps in position were grasped and given several revolutions, thus twisting the remaining portion of the walls, the floor of which is seen as a continuous structure given off from the pulp and connecting it with the epithelium of the jaw.

PREPARATIONS FOR THE DEVELOPMENT OF THE PERMANENT TEETH.

A little before the time of birth sufficient advance has been made in the development of the permanent teeth to permit a study of their relations with the temporary organs. The early preparation for the growth of these teeth was for a long time a subject of much controversy, some writers advancing the theory that the buds for the permanent teeth were produced or given off from the sacs of the temporary teeth ; others contending that the cords were derived from the remnants of the primitive cords immediately after their rupture. The theory now generally accepted is that the cord is given off from the primitive cord at a point in close proximity to its attachment to the deciduous enamel organ. This can only apply to those teeth which are succedaneous, and, therefore, does not include the permanent molars, as heretofore stated. Whatever theory be accepted in regard to the genesis of the permanent teeth, there can be no mistake in regard to that part of the process which we are permitted to ocularly investigate. I shall, therefore, proceed to describe the position and contents of the permanent tooth-sacs at birth. Figure 209 presents the result of a vertical dissection through the superior maxillary bones, the

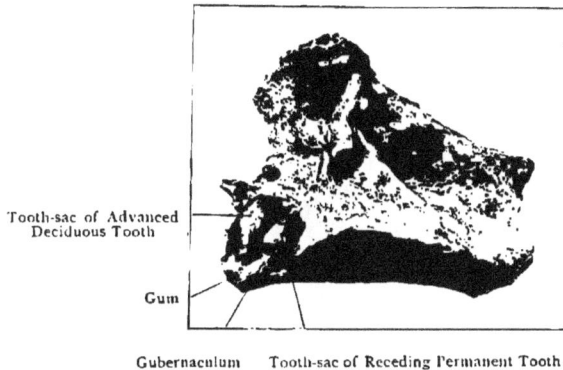

Tooth-sac of Advanced Deciduous Tooth

Gum

Gubernaculum   Tooth-sac of Receding Permanent Tooth

FIG. 209.—SECTION THROUGH SUPERIOR MAXILLÆ, SIXTH MONTH AFTER BIRTH.

inner surface of the right maxilla being exposed to view. Many of the frail processes, particularly those entering into the construction of the nasal cavities, were lost during the preparation of the specimen, the whole purpose of the dissection being to show the sac of the permanent incisor and its relationship to its predecessor. By this time calcification in the deciduous incisor has so far advanced that the contour of the tooth-crown may be plainly outlined through the walls of the sac. Resting against the palatal concavity of the crown of the temporary incisor is the sac containing the formative organs for its permanent successor. This permanent tooth-sac does not long remain in such close proximity to the deciduous tooth-crown, for as the latter advances toward.the surface of the gum the former recedes, and is soon inclosed in a separate crypt, which, were it not for the gubernacular foramen, would completely inclose it. Figure 210 represents a section of the left superior maxilla, introduced at this point for the purpose of showing the position of the foramina for the gubernacula. These may be observed immediately posterior to the incisor and cuspid teeth. At birth these foramina do not exist as such, the partially formed vaults containing the sacs for the permanent teeth appearing as an extension of the temporary crypts in a palatal direction ; but as the temporary teeth advance and the

Foramina for Gubernacula

FIG. 210.

permanent teeth recede, the roof of the crypt is completed and the foramen established by the presence of the gubernaculum. This extension of the temporary crypts is more clearly demonstrated in figure 211, which represents one side of the lower jaw at birth, with the partially calcified deciduous teeth in position in the bone. Suspended above this is the gum, which has been dissected from the bone, having attached to its under surface the sacs for the permanent teeth. Those for the incisors are particularly well defined and their place of lodgment in the bone readily noted. In the case of the cuspid, both the deciduous tooth and the permanent tooth-sacs are in position in the crypt. The cords which support the incisors are somewhat lengthened from the weight of the sacs, the gubernacula not assuming this thread-like form until the permanent sacs have further receded.

Figure 212 shows the result of a dissection upon these

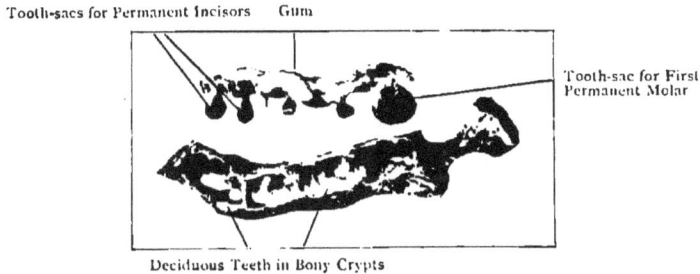

Tooth-sacs for Permanent Incisors   Gum

Tooth-sac for First
Permanent Molar

Deciduous Teeth in Bony Crypts

FIG. 211.

permanent tooth-follicles. The dentin papillæ of the various teeth are seen in a reversed position, with the follicular walls attached to their bases. At this period the papillæ for the permanent incisors may be compared to the tail of a fish, being perfectly transparent over its free extremity, which feature is

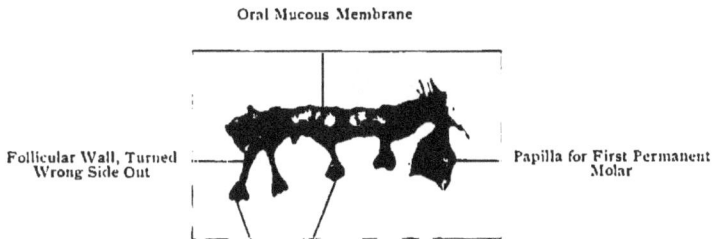

Oral Mucous Membrane

Follicular Wall, Turned
Wrong Side Out

Papilla for First Permanent
Molar

Dentin Papillæ for Permanent Teeth, Upside Down

FIG. 212.

gradually lost as its thickened base is approached. The fish-tail appearance is further represented by the division of the free extremity into three distinct parts, each of which provides a separate point of calcification. The cuspid papilla is missing, and the bicuspids have advanced little beyond the form of the

primitive bulb. The first molar is shown with the full diameter
of the crown represented by the pulpal mass, and the tips of the
cusps are already beginning to take on the calcific action.

Tooth-sacs of Permanent Teeth

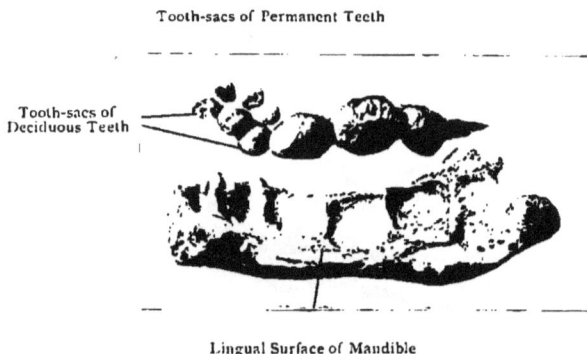

Tooth-sacs of
Deciduous Teeth

Lingual Surface of Mandible

FIG. 213.

Tooth-sacs of Deciduous Teeth

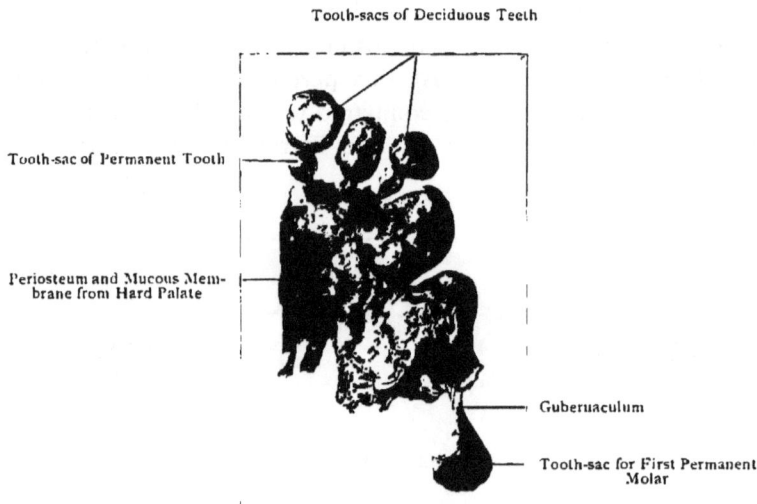

Tooth-sac of Permanent Tooth

Periosteum and Mucous Membrane from Hard Palate

Guberuaculum

Tooth-sac for First Permanent Molar

FIG. 214.—SAME AS FIGURE 213, EXCEPT ON UPPER JAW.

A further illustration of the progress of the development of
the permanent teeth and their relation to the deciduous organs
may be seen in figure 213, the dissection in this instance being

upon the lower jaw of a one-month-old child. The membrane
has been lifted from the bone with the tooth-sacs attached to it.
Immediately posterior to the sacs containing the crowns of the
temporary incisors and cuspids are those for their corresponding
successors, the papillæ of which have already assumed the
tubercular outline of the future cutting-edge. While the permanent tooth-sacs are distinctively independent pouches, there
appears, nevertheless, to be a well-established fibrous connection existing between the outer layer of the two follicular walls.
This fibrous union is gradually broken as the permanent sacs
recede and become incased in their own vaults.

While the permanent tooth-sacs are generally referred to as
"receding," it is a question if this term is fully justified.

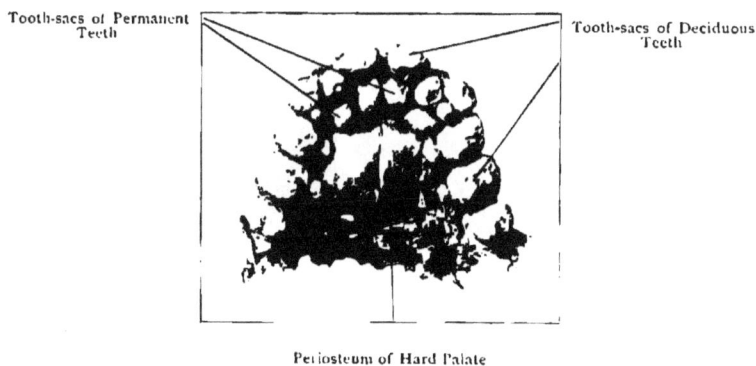

Tooth-sacs of Permanent Teeth

Tooth-sacs of Deciduous Teeth

Periosteum of Hard Palate

FIG. 215.—TOOTH-FOLLICLES FOR DECIDUOUS AND PERMANENT TEETH, THREE MONTHS
AFTER BIRTH.

While the follicles do not remain in close relation with their
predecessors, the change in the relative position of the two is
principally brought about by the advance in the deciduous sacs,
this forward movement being accompanied by a marked growth
of the bone in the direction of the future alveolar ridge, thus
leaving the permanent tooth-sacs well buried in the substance
of the jaw.

Figure 215 represents the result of a dissection upon the
superior maxillæ of a three-months-old child. The mucous membrane covering the hard palate, together with the periosteum,

has been dissected from the bones and turned over for examination. The tooth-follicles for all the deciduous teeth, as well
as those of the succedaneous permanent organs, may be
observed firmly attached to the fibrous tissue. The permanent
incisor sacs at this age are almost equal in size to those of their
predecessors, and the dentin papillæ within possesses a mesiodistal diameter almost equal to those of the calcified temporary
caps. The sacs containing the germs for the permanent cuspids
and bicuspids are somewhat diminutive, but the enamel organs
within are already molding the contour of the future tooth-
crowns upon the dentin papillæ.

By the beginning of the second month after birth calcification
in the crowns of all the deciduous teeth is about complete, and
preparation for the growth of the roots is under way. While

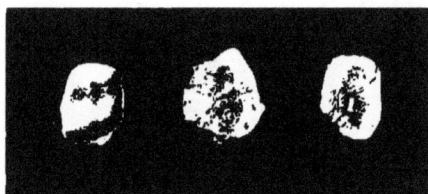

FIG. 216.

at this period the tooth-crowns may be said to be almost completely calcified, this does not apply to the interior of the
crowns, the deposit of dentin internally being a continuous
process, resulting in a gradual reduction in the capacity of the
pulp-cavity. It is also quite probable that the enamel organ is
somewhat active up to the eruptive period, and, if this be true,
the enamel covering of the crown is not complete until this
time. Whatever be the condition in the crowns, the time for
the formation of the roots has arrived, and it is principally
through the activity of the tooth-pulp that they are generated.
We have seen that the contour of the tooth-crown was first
molded upon the dentin papilla ; so it is with the tooth-root : by
a gradual elongation of the sac, accommodations are afforded
the tooth-pulp for a corresponding growth. As the pulp

lengthens out toward the future apex of the root, it is molded
to the root-form, and calcification takes place by the generation
of odontoblastic cells upon the periphery of this organic root-
form.

While the process of root-formation in the single-rooted tooth
may be readily comprehended, the bifurcation or trifurcation
of the molar roots presents a complication which calls for special
reference. Figure 216 will assist in explaining this phenomenon. In the illustration three deciduous molar crowns are
shown, two of which are incased in their tooth-sacs, the third
being stripped of this membrane. The view is directly upon
the base of the tooth-sacs, immediately beneath which is the
base of the pulp. Up to this period the odontoblastic cells have
been generating about the occlusal surface and lateral walls of
the crown only, but now an accumulation of these cells is to be
found upon the base of the pulp, lining up in the position of
the future root-walls. This structural change is faintly outlined in the illustration. By this inward extension of the odontoblastic cells from various points about the margins of the
pulp, and their union near the center of the mass, provision is
made for the calcification of the various roots, which process is
continued separately by an extension and molding of the pulp
into two or more divisions.

Calcification of the Cementum.—While the dentin of the
root is derived from the tooth-pulp, the external covering of the
root (the cementum) is generated from another source. In
every respect cementum is closely allied to bone, and we find
its development provided for in a similar manner. As stated in
another part of this chapter, the tooth-sac is made up of an
outer and an inner layer, both of which are rich in blood-
vessels. These membranous walls continue to invest the roots
of the teeth during their upbuilding. The outer layer of the sac
remains as a permanent structure placed between the root and the
alveolar walls, forming the alveolodental periosteum, while upon
the surface of the inner layer osteoblasts (cementoblasts) are developed, which are speedily converted into bone or cementum. In
this process the tooth-root may be compared to one of the long
bones of the body, and the development of the cementum con-

sidered under the head of Subperiosteal Ossification. The only variation to be observed between this and the subperiosteal development of bone is in the presence of a single Haversian canal (as the pulp-cavity may be considered), and even this difference is sometimes overthrown by small canals running at right angles to the pulp. These small canals are generally found near the apex of the root, at which point the cementum is usually the thickest. Like the enamel cap of the tooth-crown, the cementum is deposited upon the surface of the dentin of the root, thus increasing its diameter.

Eruption of the Teeth.—Up to this time no reference has been made to that process by which the teeth burst forth from their bony incasements, and, penetrating the mucous membrane, make their appearance in the mouth. Attention has been called to the growth of the bone about the tooth-follicles,—first forming beneath them as an open gutter, next surrounding their lateral walls and inclosing each follicle in a separate apartment, and finally each tooth becoming more completely enveloped by an arching-over of the mouth of the bony vault. This condition in the maxillary bones is reached between the seventh and eighth month after birth, and, almost simultaneously with the completed incasement of the teeth by the bone, active resorption begins, that portion of the bone which was last in forming being gradually removed. The cause of the resorption of the bone may readily be attributed to the advancement of the tooth-crown, but the forces which are responsible for this latter phenomenon do not appear to be clearly understood. In a general way, the advancement of the crown may be said to result from the elongation of the root by the addition of dentin to its free extremity. But, when we take into consideration that the cuspid teeth, both deciduous and permanent, have their roots fully or nearly calcified before they begin to advance toward the surface, we at once establish an exception to the generally accepted theory. The eruptive process takes place first in the anterior teeth,[*] and the bone overlying the labial surface is first removed.

This loss of the bony structure is continued until fully one-half

---

[*] See Description of the Teeth in Detail, p. 131.

of the labial surface is uncovered, and, as the crowns continue to advance toward the surface, they assume a more prominent position in the arch, and thus their cutting-edges become bared. The palatal or lingual face of the crypt serves a double purpose, forming not only a covering to the deciduous tooth, but also serving the permanent tooth-sac in the same capacity. This part of the crypt remains unabsorbed, the tooth-crown glides by its margin, and, after penetrating the mucous membrane, makes its appearance in the mouth. Closely following the resorptive process comes a rebuilding of the parts, until, finally, when the tooth is fully erupted, it is firmly supported by the new bone filling in about the base of the root. Accompanying the eruption of the anterior teeth and their establishment in the arch we find an increase in the depth of this portion of the jaw, and, as the molar teeth advance and assume their position, there is a corresponding increase in the depth of the jaw in this locality. At the beginning of the eruptive period the roots of the deciduous teeth are but partially calcified, but as the crowns advance the calcific action at the extremity of the roots is continued, and, in the majority of instances, by the time the crowns are fully erupted the roots are completely formed. During the period of eruption the transitory nature of the alveolar portion of the jaw-bone is made manifest, accommodating itself to the growth of the teeth as well as to their change of position. The free margins of the alveolar walls are taking on new structure, which advances with, and becomes adapted to, the base of the tooth-root. Coincident with this the deeper portion of the alveolar process is formed by a rapid filling-in about the root as the tooth travels onward to assume its final position in the jaw. The eruption of the teeth is usually by pairs, with a slight intermission between each class. The central incisors first make their appearance, followed by the laterals, after which the first molars are erupted. The cuspids usually follow the first molars, and, finally, the second molars take their place in the arch. While this brief description of the eruption of the teeth refers to the deciduous organs only, the process in the permanent teeth is almost identical with this. Further reference to the

eruption of the permanent teeth will be made in connection
with the degeneracy of the temporary set.

To return to the subject of tooth-development, attention is
called to figure 217, prepared from a dissection upon a nine-
months-old child, the illustration representing the hard palate,
or roof of the mouth, at this period. The four incisor teeth
have made their appearance, the labial surfaces of the crowns
being fully exposed, while those facing the palate are but
slightly uncovered. The approach of the remaining deciduous
teeth is plainly indicated by the fullness of the alveolar borders,
and the margins of the crypts are now being removed by
resorption.

FIG. 217.—HARD PALATE FROM A NINE-MONTHS-OLD CHILD, ACTUAL SIZE.

If the mucous membrane should be removed, the crowns of
the advancing teeth would be brought to view after the removal
of the walls of the tooth-sacs, while the approaching cuspids and
second molars yet remain partly covered by an arching over of
the walls of the crypts ; the resorptive process has also begun
in these parts. It would also be observed that, while the walls
of the crypts are molded to the outlines of the tooth-crowns,
there exists a well-defined interspace between the two. During
the growth of the tooth this interspace is filled by the walls of
the tooth-sacs, and even after the teeth have passed the sacular
stage of development, and assumed their positions in the mouth,

there yet remains between the tooth-roots and the alveolar walls a slight space which is occupied by the alveolodental periosteum.

The next dissection made was one upon the superior maxillæ of a two-year-old child (Fig. 218), representing the roof of the mouth of this subject. In this specimen it will be noticed that all of the deciduous teeth are erupted with the exception of the second molars. The palatal surfaces of the incisors are fully uncovered, while in the cuspids the labial surfaces are much more exposed than the palatal. In figure 219 the mucous membrane and sufficient of the bone has been removed to

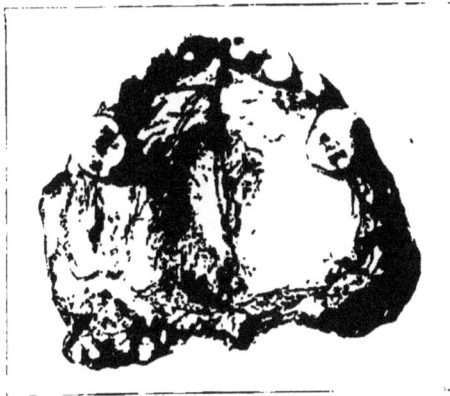

FIG. 218.—HARD PALATE FROM A TWO-YEAR-OLD CHILD, ACTUAL SIZE.

expose the tooth-sacs of the developing permanent teeth. The dissection furnishes no additional information over that obtained from figure 218, excepting that the primitive follicles for the second permanent molars make their appearance at this time.

At this early period the jaw has not sufficiently lengthened to permit of these follicles occupying their future position ; consequently they are found generating immediately over the toothsacs of the first permanent molars. As the first molars advance and the jaw lengthens backward, these follicles will be carried to the distal by the extension of the mucous membrane, to which they are firmly adherent. If these follicles were to be dissected

at this time, the papillæ would be without definite form, showing
that the early function of the enamel organ had not yet begun.
The tooth-sacs containing the permanent lateral incisors are
found immediately beneath the palatal plates, and frequently
during their earlier life they are not even protected by the
bone, being in immediate contact with the mucous membrane.
On account of the imperfect protection frequently afforded these
sacs, the germs are sometimes injured and the teeth fail to make
their appearance.

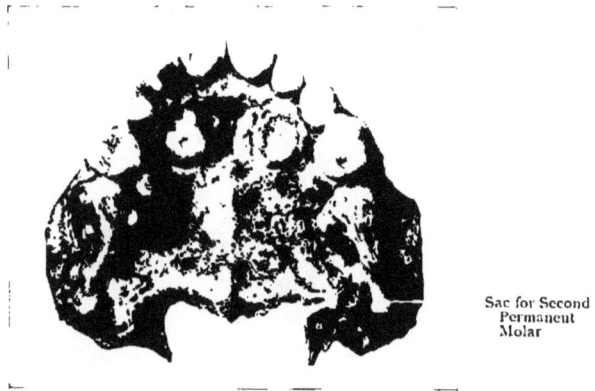

Sac for Second
Permanent
Molar

FIG. 219.—ROOF OF THE MOUTH OF A TWO-YEAR-OLD CHILD.

In figure 220 the walls of the sacs shown in figure 219 have
been opened, and the relations existing between the first and
second dentition at the end of the second year become apparent.
The crowns of the permanent incisors are deeply set in the
substance of the jaw, while the partially calcified crowns of the
permanent laterals are in close proximity to the palatal surface.
The partially formed crowns of the permanent cuspids are still
more deeply seated in the substance of the jaw than those of
the central incisors, and are not visible in the illustration. In
this connection it will be well to again refer to the gubernaculum,
and to its function—that of directing the tooth to its proper
position in the arch. By reference to the illustration the crowns
of the permanent teeth will be observed heading in various
directions, and, while in this instance there appears to be a

general tendency for them to advance and assume their proper positions in the arch, in many cases they will be found directed at right angles to the point at which they should emerge from

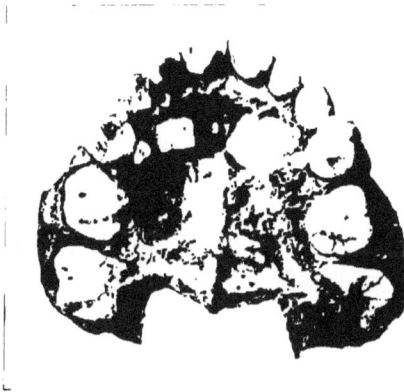

FIG. 220.—DEVELOPMENT OF THE TEETH ABOUT THE SECOND YEAR.

the bone. The gubernaculum, which appears to be nothing other than an elongation of the follicular walls, not only directs

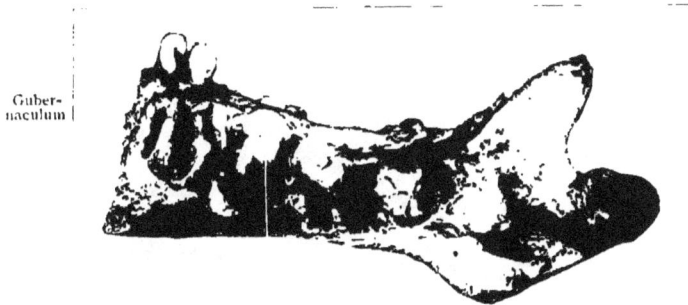

Guber-
naculum

Tooth-sac for First Bicuspid

FIG. 221. — DISSECTION ON LINGUAL FACE OF LOWER JAW, CHILD NINE MONTHS OLD.

the tooth by the tension of its fibers, but the foramen which its presence creates stimulates the resorptive action over the tract to be traveled by the tooth. Figure 221 was prepared for the purpose of better showing the gubernacula, and the manner of connecting the tooth-sacs with the oral mucous membrane. This condition is well shown in the anterior teeth, the tooth-crowns having receded well toward the body of the jaw. The follicle for the first permanent bicuspid may be observed attached to the lingual face of the deciduous molar sac, and the leading cord is not yet an adjunct to the developmental process, but this structure will make its appearance as the follicle recedes from the surface.

FIG. 222.—DECIDUOUS MOLARS WITH TOOTH-SACS FOR PERMANENT BICUSPIDS ATTACHED TO THE GINGIVAL TISSUE.

In figure 222 the deciduous molars have been removed from the jaw and the relations existing between these teeth and the developing permanent molars is shown. The tooth-follicles for the succedaneous teeth are found immediately beneath the gingival margin, and apparently attached to the deep layer of the mucous membrane. This relationship between the per-

Deciduous Molars'   Deciduous Incisors

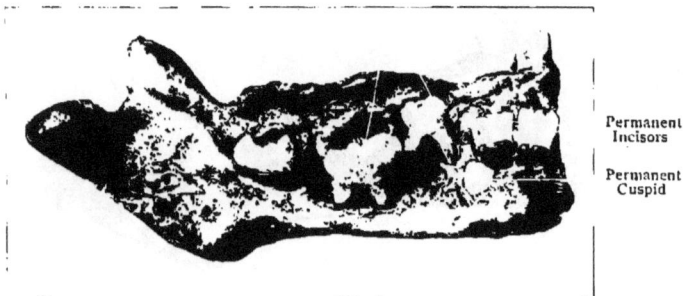

Permanent Incisors

Permanent Cuspid

FIG. 223.—SAME AS FIGURE 221, WITH TOOTH-SACS OPENED SHOWING DEVELOPING TEETH IN THE JAW.

manent and temporary organs is present about the eruptive period, but as the deciduous tooth advances and the permanent tooth-sac recedes, the two organs become more widely separated, and the permanent tooth-sac is connected to the surface

only by the elongated follicular fibers which form the gubernaculum.

Decalcification of the Deciduous Teeth.—By the close
of the second year the twenty deciduous teeth have taken their
place in the dental arch, their roots have become fully calcified,
and the apical foramina established; it is only for a short

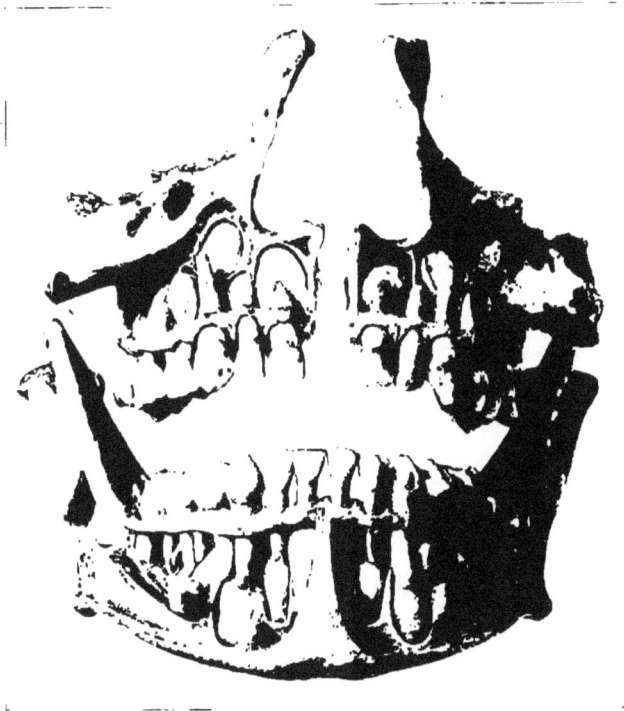

FIG. 224.—DEVELOPMENT OF THE TEETH ABOUT THE SEVENTH YEAR.

period, however, that they remain thus perfect, the process of
decalcification beginning about the fourth year. This resorptive
action begins at the apical extremities of the roots and gradually progresses in the direction of the crowns. Commencing
about the fourth year with the central incisor, decalcification
takes place in the teeth in the order of their eruption, the
lateral incisor following the central, the first molar following the

21

lateral, etc.   By reference to figure 224 an approximate idea
of the progress of decalcification may be obtained, and it will
be observed that about three years elapse from the beginning
of this rather obscure process to its completion, and the final
casting off or shedding of the tooth-crowns.   In reference to the
causation of this dissolution of the deciduous teeth, but little
appears to be known.   It has been said to result from the
presence and pressure of the advancing permanent teeth, but
there is no question but that it occurs absolutely independent of these organs, decalcification frequently taking place
when from some obscure reason, one or more of the successional
teeth are absent.   During the entire period of root decalcification, the pulp of the tooth, which is also involved in the destruction, retains its vitality, but with the loss of vitality in the pulp
resorption of the root ceases ; so that the gradual removal of
the root-substance must be considered as a purely physiologic
action.

Figure 224 shows a dissection upon the jaws of a seven-year-
old child, by a careful study of which, a fair knowledge of the
extent of resorption in the deciduous teeth at this period may
be obtained.

Advance of the Permanent Teeth.—By referring to figure
224, the relations existing between the deciduous and the permanent teeth at about the seventh year may be noted.   While
the crowns of the deciduous teeth remain in position, a part of
the space formerly occupied by their roots is taken up by the
advancing crowns of the permanent teeth, the latter being
calcified but little beyond their cervical lines.   Between the
seventh and eighth years the crowns of the deciduous incisors
are cast off, and gradually the crowns of the permanent incisors
force their way through the gum, the arch by this time having
sufficiently increased in size to accommodate the additional
width possessed by them.   Previous to this time, or about the
sixth year, by a backward extension of the jaws, the first permanent molars have erupted, assuming a position in the arch
immediately posterior to the second deciduous molars.   Between
the tenth and eleventh years the crowns of the deciduous molars
are lost, and the permanent bicuspids advance to take their

places. Usually by the twelfth year there has been sufficient increase in the length of the jaws to permit of an additional tooth, and the second permanent molar gradually takes its position immediately posterior to the first. Between the twelfth and thirteenth year the deciduous cuspids are lost by decalcification of their roots, and they are succeeded by the permanent cuspids. We therefore find, by the fifteenth year, fourteen

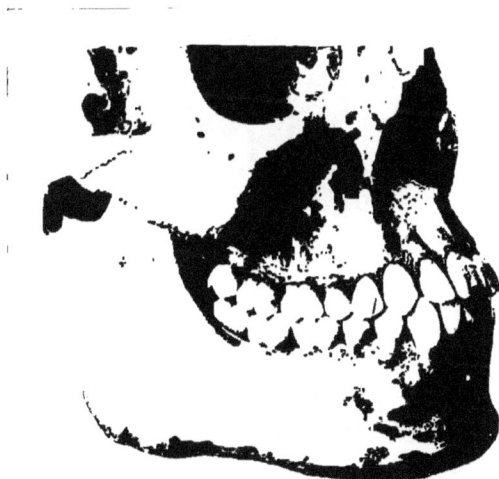

FIG. 225.—THE COMPLETED DENTITION.

fully developed teeth occupying the dental arch of each jaw, the full number, thirty-two, or sixteen in each jaw, not being present until the eruption of the third molar, which, like the other teeth of this class, is compelled to await accommodations for it by a further increase in the length of the maxillary bones. This tooth usually takes its place between the eighteenth and twenty-first years, and thus completes the dentition (Fig. 225).

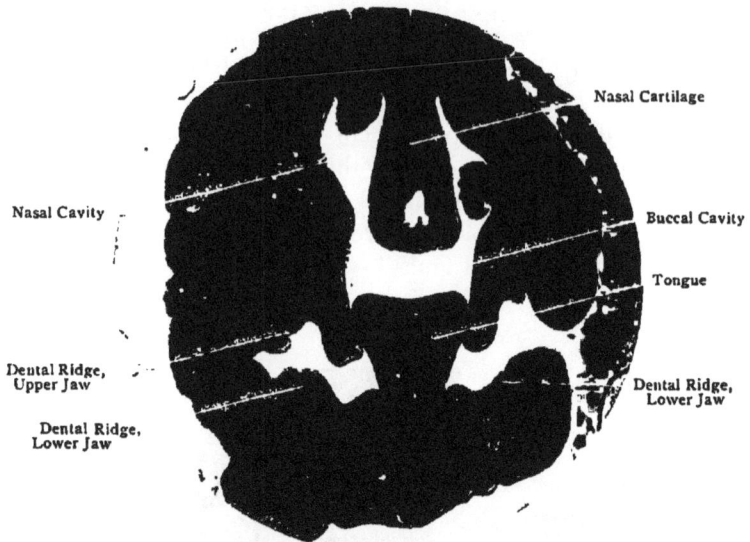

FIG. 226.—VERTICAL TRANSVERSE SECTION THROUGH HEAD OF HUMAN EMBRYO, ABOUT THE TENTH WEEK. × 30.

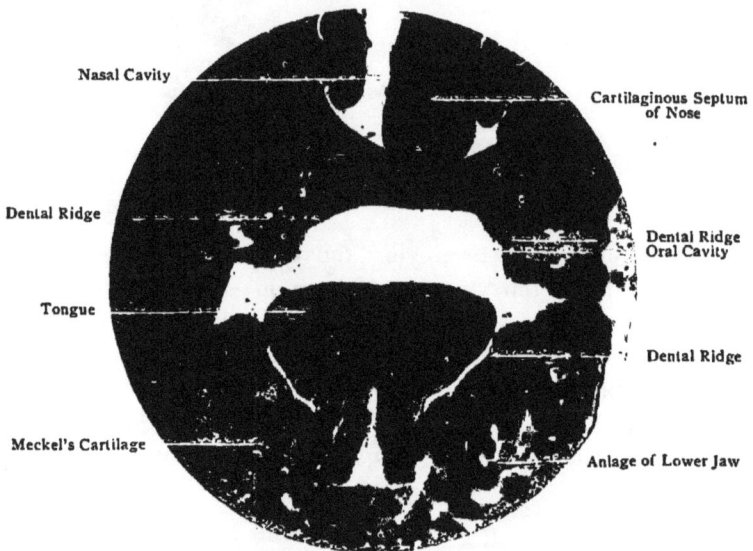

FIG. 227.—VERTICAL TRANSVERSE SECTION THROUGH HEAD OF HUMAN EMBRYO, ABOUT THE TWELFTH WEEK, SHOWING THE SINGLE BUCCAL CAVITY TRANSFORMED INTO THE ORAL AND NASAL CAVITIES. × 30.

PART II.—HISTOLOGY.

CHAPTER I.

GENERAL HISTOLOGY: THE TISSUES OF THE BODY: EPITHELIAL TISSUES; CONNECTIVE TISSUES; MUSCULAR TISSUES; NERVOUS TISSUES.

Part first of this work has been devoted to a gross description of the mouth, its structures, the macroscopic arrangement of the tissues composing the various parts, their relations to one another as made manifest by dissection, etc. The remaining chapters will be given up to the study of the intimate structure of the oral tissues according to their form and organization. The necessity and value of thus continuing the subject may be presented as follows: If we take any of the structures or organs of the mouth,—such as the mucous membrane, the mucous glands, or the muscular fibers common to various parts,— and submit them to a minute dissection, we will find that the component parts may be separated into smaller and smaller portions. Take, for example, the muscles of the tongue : these, to the unaided eye, are distinctly fibrous in appearance, and these fibers may be readily separated from one another by removal of the interposed substance. In like manner the oral mucous membrane is composed of a dense fibrous connective tissue, which may be separated and examined in a similar way. But the possibilities of investigation by this method are limited, as the parts thus dissected are similar to one another, and are soon reduced to microscopic proportions, and when thus examined certain definite anatomic forms present themselves which are no longer capable of being divided. The reduction of any structure to such definite proportions without artificial mutilation or disarrangement of its parts is productive of what is termed an *anatomic element.* It is only in very rare instances that a tissue is composed of a single anatomic element, but

usually two or more kinds are mingled together, and the tissue
thus formed derives its texture, quality, and appearance from
the variety and number of anatomic elements present. Again,
taking the muscular tissues as an example, we find them composed principally of one anatomic element — muscular fiber,
arranged in parallel bundles. In close relation with the fibers
are numerous nerve-filaments and capillary vessels, the whole
being surrounded by a thin layer of connective tissue. Numerous bundles of this nature unite to form larger bundles, with
larger nerve-filaments and blood-vessels. In the hard structures
the same conditions are present, so that a definite understanding
may be obtained by a minute examination only.

Beyond the tissues and the anatomic elements is the *elementary organism*—the *cell*.

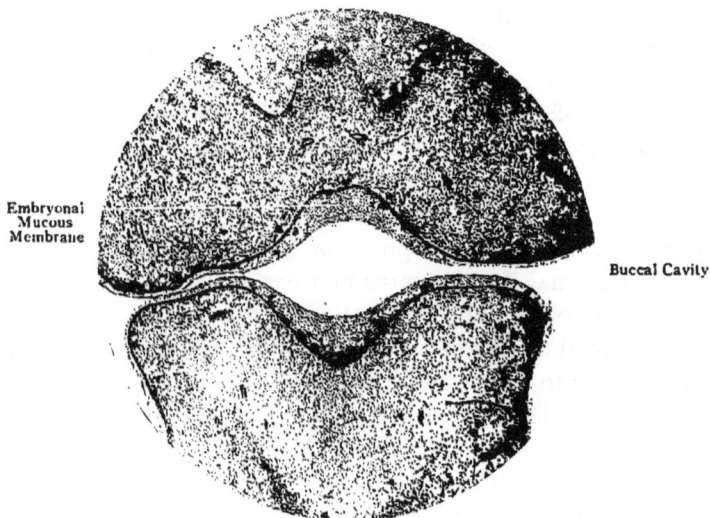

Embryonal
Mucous
Membrane

Buccal Cavity

FIG. 228.—VERTICAL TRANSVERSE SECTION THROUGH HEAD OF HUMAN EMBRYO, ABOUT
THE SIXTH WEEK, SHOWING SINGLE BUCCAL CAVITY. × 30.

A **cell** is a structural element, primarily oval or spheroid in
form, capable of self-nourishment, development, and reproduc-
tion. It is composed of two essential parts—the cell-substance,
consisting principally of a soft albuminous matter, the *protoplasm*, and generally a well-defined central portion, the necessary

part of a typical cell—the *nucleus*. The *nucleus* is usually round and situated near the center of the cell-substance. In many instances there is within the nucleus a small spheric body, the *nucleolus*, which, although not having its function definitely established, undoubtedly is subordinate to the life-activity of the cell. At the beginning of development the cells composing a part are all of the same general form, and similar in other respects ; as development proceeds, the cells arrange themselves in germ-layers, and coincident to this change in relationship cease to resemble one another and become differentiated. By this aggregation of similar cells *tissues* are formed ; and by a combination of different tissues, a structure of definite form and function, an *organ*, is generated.

The tissues of the body are divided into four grand divisions—namely, *epithelial tissue, connective tissue, muscular tissue,* and *nervous tissue*. During early life these tissues are composed of similar elements—of cells ; but as growth proceeds, a twofold change takes place, in one of which the cells produce a special deposit, which accumulates between them and forms the *intercellular substance ;* in the other the tissues of one kind become mixed with those of another.

The Epithelial Tissues.—The cells composing the epithelial tissues are known as epithelial cells ; they are definite in outline, and consist of the cell-substance, protoplasm, and a nucleus. Two forms are most frequently met with—the flattened or squamous, and the columnar or prismatic. The *squamous* or pavement cells are flattened and scaly; in the former the nucleus is round, while in the latter it is flattened. The *columnar* cells present a greater variety in outline, being conic, club-shaped, or spindle-shaped, sometimes short and sometimes long. The nucleus is inclined to the form of the cell-substance, usually being more or less oval or oblong. In size the epithelial cells are quite variable, this difference occurring in cells of the same part as well as in those of distinct parts. By the arrangement of the cells the character of the epithelium is established ; thus, if they are disposed in single layer, a single-layered epithelium is produced ; or if stratified, forming several superposed layers, a stratified epithelium results. The consistency of the epithelial cells is such that they readily succumb to the pressure of neighboring cells, resulting in much variation in their outline. They are cemented to one another by an exceedingly thin layer of an albuminous cement-substance, being in contact by their flattened sides or by irregularly formed processes, the result of force applied by adjacent cells.

A third class of epithelial cells are those columnar cells which are beset upon their free surfaces with numerous minute processes which during life are constantly vibrating ; these are known as *ciliated* cells. Continuous masses of epithelial cells, "epithelia," are found covering the surface of the skin, and lining internal canals and organs. They are found in abundance, covering all parts of the mucous membrane of the mouth, and lining the numerous ducts which empty into that cavity. In some instances the epithelia are composed of a single stratum, in others of a number of strata, and the following varieties are to be observed : simple squamous, simple columnar, and simple ciliated epithelium, when a single stratum is present ; and stratified squamous, stratified columnar, and stratified ciliated epithelium, when several strata are present. The epithelium of the mouth is principally of the stratified squamous variety. Another classification is that known as glandular epithelium, composing the numerous glands and ducts (see Glands of the Mouth).

The Connective Tissues.—In general this term is applied to all those tissues which support and connect other tissues. While the cells are numerous, they are proportionately less numerous than in the epithelial tissues. In the connective tissues the intercellular substance predominates where it is prominently developed and variously differentiated. The intercellular substance is characteristic of the connective-tissue group, and is especially interested in its function. By a variation in the quantity, nature, and disposition of the intercellular substance, three classifications are formed—namely, *fibrous connective tissue, cartilage,* and *bone* (including dentin). Each of these groups is subdivided into several varieties, but in all instances the intercellular substance is to be distinguished from the cells.

Fibrous Connective Tissue.—This tissue is present in the skin and mucous membrane, in the intermuscular tissues, in tendons, in fascia and aponeuroses, and in the tissues connecting various organs. It is composed of a meshwork of fine fibers of two kinds. The first, which makes up the greater part of the tissue, is formed of very fine, white, structureless fibers arranged closely in bundles and bands crossing and intersecting in all directions. The second variety, or the yellow, elastic fiber, has a much sharper and darker outline, not arranged in bundles, but is intimately mingled with the white fibers by twisting around and among its filaments. These are known as the elementary connective-tissue fibers. The size of the connective-tissue bundles depends upon the number of elementary fibers present, and by a variation in the arrangement of the bundles a variety in the character of the fibroconnective tissue is produced in different localities. When the fibrous connective tissue is formed into an unbroken mass, as in mucous membrane, the minute bundles are collected into smaller or larger groups (the *trabeculæ*), and these are in turn associated into groups. In the skin and mucous and serous membranes, the trabeculæ of the connective-tissue bundles are separated, and, by crossing and recrossing one another, form a dense, fan-like structure. In other tissues, as the tendons and fascia, the bundles are arranged in parallel layers. In the submucous tissues the connective-tissue fibers are loosely woven, the fibers crossing and intermingling, with the intervening spaces unusually large, resulting in a loose, flabby tissue—*areolar tissue.* Three varieties of fibrous connective tissue are distinguished—namely, mucous connective tissue, fibrillar connective tissue, and reticular connective tissue. Mucous connective tissue consists of an abundance of undifferentiated intercellular substance in which are a few bundles of fine fibrils and a number of round or oval cells. Fibrillar connective tissue, or areolar tissue, consists of an abundance of intercellular substance differentiated into connective-tissue fibers and cells. Reticular connective tissue, as its name implies, is a network of slender bundles of connective tissue associated with flattened, nucleated cells.

The fibrous connective-tissue cells are few in number, of several varieties, and variously shaped, being flattened, stellate, or apparently distorted by pressure from surrounding cells or fibrous bundles. In the mucous membrane the cells are oblong and somewhat flattened, having many branches which reach out and, uniting with like processes from neighboring cells, form a network. Other connective-tissue cells are comparatively larger, oval or rounded in form, granular in appearance, rich in protoplasm, and are known as *plasma-cells*. The body of connective-tissue cells, besides containing a nucleus, frequently contain pigment-granules; these are known as *pigment-cells*. These are seldom found in mucous or serous membranes, being principally confined to the integument. Fat-globules may also be found in fibrous connective tissue, and when of considerable size unite and form a rounded cell, called a *fat-cell*. Numerous fat-cells uniting, and well supplied with blood-vessels and nerves, form *adipose* tissue, or fat. Fat-cells are frequently found in areolar tissue as well. When fibrous connective tissue is immediately contiguous to epithelium, it becomes somewhat modified and a new membrane is formed, called the *basement membrane*, or *membrana propria*. This membrane is a thin, transparent, structureless layer, and, when in connection with those mucous membranes provided with a layer of vascular fibrocellular tissue, may appear as the formative substance out of which successive layers of epithelial cells are generated. In the ducts and glands—for example, the salivary glands—the basement membrane forms the proper walls of the tubes, and the cells here generated, and corresponding to the epithelial cells of the coarser mucous membranes, are known as *gland-cells*, rather than epithelial cells. This, however, is a distinction without a perceptible difference, the location and function as secreting cells being alike in each.

Cartilage.—Cartilage is a semi-opaque, non-vascular tissue, white in color, and composed of a matrix containing nucleated cells. The matrix is somewhat elastic and rather dense. The cells are simple in form, being spheric or slightly inclined to angularity. The variation in the character of cartilage is due rather to the difference in the character of the matrix than to the cellular structure, the principal variation in the cells being in their size. The cells lie in the spaces or lacunæ of the matrix, which they completely fill. Investing the free surface of most cartilaginous tissue (articular cartilage excepted) is a thin but tough and firm fibrous membrane—the *perichondrium*. This membrane is well supplied with blood-vessels and nerves, and is essential to the growth and maintenance of the cartilage. There are three varieties of cartilage—namely, hyaline cartilage, elastic cartilage, and fibrocartilage.

Hyaline cartilage is of a faint pearly-blue color, slightly transparent, and is found investing the articular ends of the bones—for example, the condyles of the mandible; also forming the costal and nasal cartilages, as well as those of the trachea, bronchi, and a part of the larynx. Hyaline cartilage is distinguished by a granular or homogeneous matrix. The cells, which contain a nucleus with nucleoli, are usually grouped together in patches, and are somewhat irregular in outline, appearing flattened near the free surface of the tissue in which they are placed, and inclined to be perpendicular to the surface in the more deeply-seated portions. The matrix is dimly granular in appearance, resembling ground glass, and receiving its name from this fact. That part of the cartilage close to the perichondrium is supplied with cells much smaller than those occupying the lacunæ in the substance of the mass, and the growth of the cartilage is most active in this part. Lining each lacuna is a delicate membrane (the capsule), which primarily is but partly filled out, but as the cell or cells increase in size, this membrane is carried to the walls of the lacuna. Articular hyaline cartilage is non-vascular, being nourished by the blood-vessels of the bone beneath.

Elastic cartilage is of a dull-yellow color, and is sometimes called yellow cartilage. It is not present in the mouth, but occurs in the external ear, in the epiglottis, and in part of the larynx. Its structural composition is quite similar to hyaline cartilage, but may be distinguished from it by a network of fine elastic fibers which penetrate the matrix. The cells are rounded or oval, containing nuclei and nucleoli.

Fibrocartilage is yellowish or milky white in color, and is much more widely distributed throughout the body than the elastic variety. It is present in the temporomandibular articulation. Like those previously described, it is composed of cells and a matrix, the latter being made up of fibrous connective tissue arranged in bundles, and for this reason it is scarcely deserving the name of cartilage, only that in other portions continuous with it cartilage-cells may be found in abundance. Between the strata of the fibrous bundles are numerous nucleated cells, which are oval and more or less flattened, and each enveloped in a delicate capsule.

Cartilage is further classified into two divisions,—*temporary* and *permanent*,—the former term being applied to that kind of cartilage which in the fetus and in youth is destined to be con verted into bone (for example, Meckel's cartilage); the latter class including all those cartilages which are generated as such, and continue to serve in that capacity. Temporary cartilage closely resembles the hyaline variety, being formed of a matrix in the lacunæ of which the cells are located. These cells, however, are not grouped together as in hyaline cartilage, but are more uniformly distributed throughout the matrix.

Bone.—Bone is mainly composed of tricalcium phosphate and cartilage. The matrix of osseous tissue has a distinguishing feature produced by the blending of organic and inorganic substances, resulting in hardness, solidity, and elasticity. The combination of organic and inorganic elements in bone is of such a nature that either part may be removed without destroying the other. The matrix is composed of the salts of lime, especially calcium phosphate, and of slender fibrils united by a cement-substance into bundles of various sizes. The cement-substance is chiefly composed of insoluble lime-salts, principally carbonates and phosphates. These two kinds of structure are found to be present in different parts of the same bone, forming a dense or compact and a spongy or cancellated tissue. The former occur in the shaft of long bones and in the outer layer of flat or irregularly formed bones. Cancellated bone-substance occurs in the extremities of the long bones and in the interior of flat and irregular bones. The irregularly formed maxillary bones give place to both kinds of bony structure; the external layer of the superior maxillæ and the body and rami of the inferior maxilla are composed of compact tissue, while the interior of these bones and the condyloid processes of the mandible are spongy or cancellated in their nature. When examined by the microscope the bony substance is found occupied by numerous little spindle-shaped spaces—the *lacunæ*. Branching out from these in various directions are minute canals—the *canaliculi*—which anastomose with similar canals from neighboring lacunæ. In the maxillary bones no other canals than these may be visible, but if a transverse

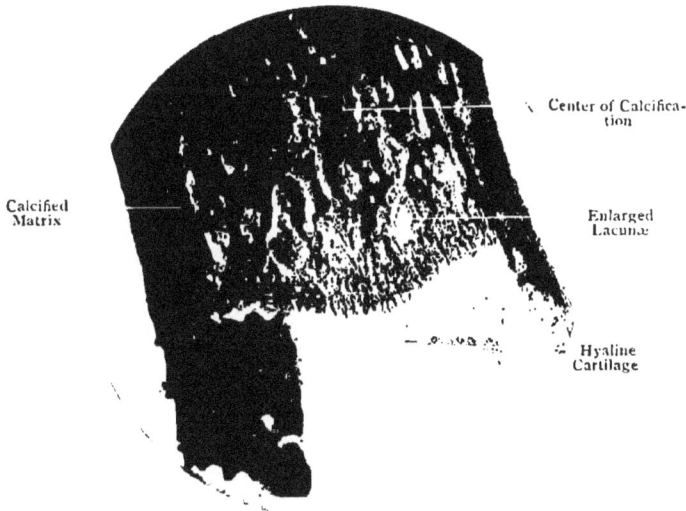

FIG. 229.—DEVELOPING BONE. × 40.

section be cut through one of the long bones, an additional space makes its appearance (Fig. 230).

These spaces are known as the *Haversian* canals. They are circular in outline and appear as a center for a small, circular district mapped out by concentric layers, the lacunæ and canaliculi following the same concentric plan, and through each other communicating with the Haversian canals. The general direction of the Haversian canals is longitudinal with the long axis of the long bones, and in the flat or irregular-shaped bones they are somewhat irregular in formation and ramify in various directions. In the osseous matrix each lacuna contains a bone-cell. These are nucleated, protoplasmic cells. In developing bone, these cells, which do not completely fill the lacunæ, are connected by numerous branches or processes passing through the canaliculi ; in older bone very few processes are observed on bone-cells.

There are two processes by which bone may be prepared for histological examination, either by one method which results in the destruction of the organic elements, or by another which removes the inorganic elements. In the former process the organic matter is removed by simply drying the structure, after

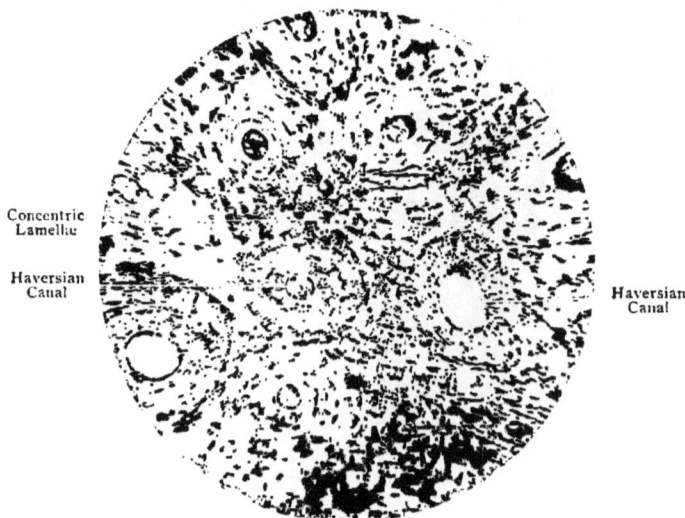

FIG. 230.—TRANSVERSE SECTION THROUGH SHAFT OF LONG BONE. × 30.

which thin sections may be prepared and carefully examined under the microscope, when the Haversian canals, lacunæ, and canaliculi will be seen forming a complete concentric net-work. In the latter method the inorganic substance is removed by immersing a fresh bone in dilute picric acid, $C_6H_2(NO_2)_3OH$, which readily decalcifies it, and when properly prepared sections are placed under the microscope the organic contents of the lacunæ and canaliculi alone are visible.

The concentric laminæ of bone is riveted together by numerous delicate rods or processes named *Sharpey's fibers*, these delicate fibers passing through the laminæ to perform this office.

Periosteum and Bone-marrow.—The interstices of spongy bone are filled with a soft mass,—the bone-marrow,—and the external surface of the bone is covered by a fibrous membrane —the periosteum. This membrane is absent where bones are joined to each other by ligament or cartilage, and over articular surfaces. The periosteum is a compact connective-tissue membrane. It consists of two layers: an *outer*, fibrous layer rich in

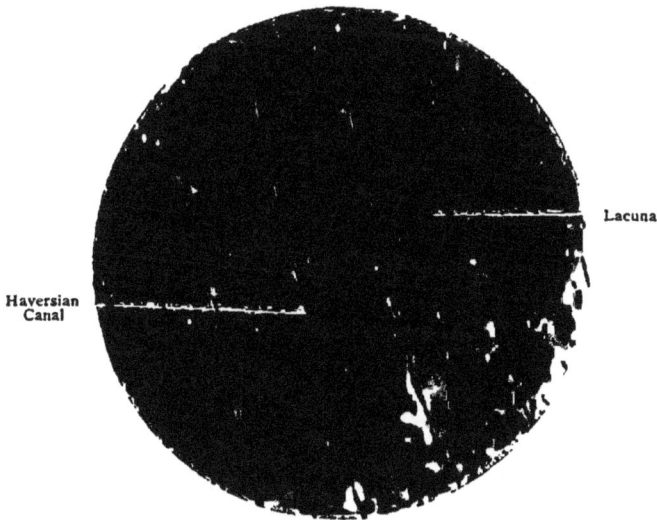

FIG. 231.—LONGITUDINAL SECTION OF LONG BONE. × 30.

blood-vessels, which forms the connection with adjacent structures; an *inner* or osteogenetic layer containing few blood-vessels, loose in texture, but rich in elastic fibers and spheric connective-tissue cells, with oval nuclei. These are the formative cells of bone and are called *osteoblasts*. These cells appear in the lower strata of the inner layer, or the layer in contact with the bone, and are especially numerous during the period of development. Through the blood-vessels of the bone the marrow, internally, is placed in communication with the periosteum external; small branches given off from the numerous arteries and veins of the periosteum enter the Haversian canals, upon which they pass to the canaliculi, thus communicating with the blood-vessels of the marrow. In like manner numerous nerves enter the substance of bone, first passing into the Haversian canals, after which they become closely associated with the minute blood-vessels and are distributed to the periosteum and bone-marrow. The bone-marrow, besides filling the interstices of the spongy substance, is also found occupying the central cavity of long bones, and in the larger Haversian canals. The marrow is of two varieties, distinguished by its color, being either red or yellow. The red marrow is found in the flat bones (including the maxillæ), the vertebra, and ribs, while the yellow marrow occurs in the long bones of the extremities. Red marrow is composed of a delicate connective-tissue network supporting, besides the marrow-cells, a few fat-cells and giant-cells. In the long bones the yellow marrow is surrounded by a connective-tissue membrane lining the medullary canals. Marrow-cells and giant-cells are present in abundance. Marrow is very vascular and contains many osteoblasts.

Dentin.—This structure, which in many particulars closely resembles bone, will be fully considered in connection with the histology of the tissues of the teeth.

Muscular Tissues.—The muscle-fibers, as the structural elements of the muscular tissues, are divided, according to the arrangement of their bundles, into two classes—non-striated and striated.

Non-striated, Smooth, or Involuntary Muscular Tissue. —This tissue consists of contractile fiber-cells, which are elongated, spindle-shaped, and cylindric, with exceedingly elongated extremities, which become shorter and thicker through contraction. They are quite variable in length ($\frac{1}{70}$ to $\frac{1}{150}$ of an inch), and are composed of a pale, homogeneous-looking protoplasm, each inclosing an elongated or rod-shaped nucleus, which is flattened if the cell is so formed. The muscular fibers are firmly bound together by a cement-substance, forming fasciculi, which in turn are collected into strata or membranes, which may be disposed parallel, or crossing and recrossing, forming an intricate network. The connective-tissue septa provide a passageway for the larger blood-vessels, while the capillaries penetrate the fasciculi, forming a complicated network with oblong meshes. Involuntary muscular tissue is not found in the mouth except in the ducts of the salivary glands.

Striated or Voluntary Muscular Tissue.—Striated muscular tissue is composed of long, cylindric fibers, which are regularly transversely striated. In most instances their extremities are attached to bones by means of tendons, as, for example, the cheek- and lip-muscles. The fibers are grouped together by fibrous connective tissue into various-sized bundles, forming

FIG. 232.—TRANSVERSE SECTION OF STRIATED OR VOLUNTARY MUSCULAR TISSUE. × 40.

fasciculi. There is much variation in the length of the fibers composing the fasciculi in different muscles. In most instances the fasciculi which serve to make up the bundles of a single muscle continue parallel with one another throughout their length. Surrounding the muscular fasciculi, and forming a covering or sheath for the individual bundles, is a layer of connective tissue called the *perimysium*, and passing from this into the substance of the bundle is a delicate connective tissue (the *endomysium*), which separates the individual fibers from one another. The

22

former structure carries the larger blood-vessels and nerve-fibers, while the latter supports the capillaries.

Each muscular bundle may again be divided into smaller bundles, which in turn are ensheathed in a similar manner and further divisible, so continuing until the primitive fasciculi, or so-called muscular fiber, is reached. Striped muscular fiber consists of a structureless, elastic sheath (the *sarcolemma*), which structure represents the cell-membrane, and closely invests a number of filaments or fibrils. Besides the fibrillæ, there is contained within this fine, structureless, transparent membrane the

FIG. 233.—STRIATED OR VOLUNTARY MUSCULAR TISSUE. × 40.

sarcoplasm, a faintly granular substance resembling protoplasm, but not identical with it. This substance serves in the capacity of a matrix for the fibrillæ. The fibrillæ are arranged parallel to one another, being supported by the sarcoplasm. It will thus be seen that each fiber of a striated muscle comprises the sarcolemma, the muscle-nuclei, the fibrillæ, and, finally, the sarcoplasm, filling all the interstices, first between the fibrillæ of each muscle-column, between the columns of each group, and between the groups themselves. The disposition of the sarcoplasm may be most favorably studied by a cross-section through the fibers, appearing as a delicate but clear network, within the meshes of which are the muscle-columns. Striated muscular fibers are usually spindle-shaped, tapering off and becoming thinner toward their extremities. In rare instances they are branched at their ends. This condition is present in the tongue, the extremities of the fibers passing transversely into the oral mucous membrane, where they become further subdivided. The striated or voluntary muscles make up the muscular tissues of the lips, cheeks, tongue, and soft palate.

Nervous Tissues. — Until within recent times it has been stated that the nervous tissue consists of two histologic elements known as nerve-cell and nerve-fiber ; that these two elements differed not only in their mode of origin, but in their structure and physiologic endowments. At the present time it is believed that the entire nervous system consists of an infinite number of definite independent morphologic units, which, though having a common origin and a similarity of structure, have, nevertheless, different functions in different parts of the body. This neurologic

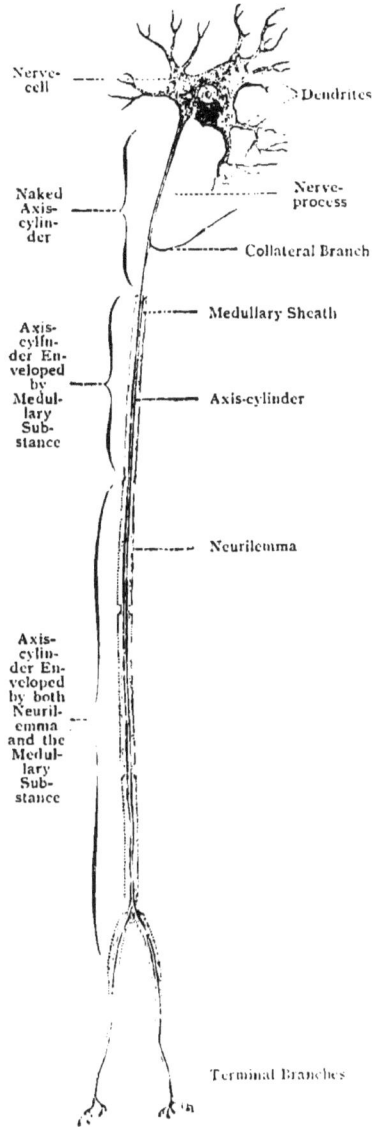

FIG. 234.—DIAGRAM OF A NEURON.—
(From Stöhr's "Histology.")

unit has been termed the *neuron*, and, as represented schemati
cally in figure 234, may be said to consist of: First, the nerve-
cell, or neurocyte; second, nerve-process, or axon; third, the
end-tufts, or terminal branches. Each of these three main
portions of the neuron presents a variety of secondary features
which are related to their functional activities.

The Nerve-cell, or Neurocyte.—The nerve-cells are found in
the cortex of the brain, in the interior of the spinal cord, in the
various ganglia of the cerebrospinal and sympathetic nervous
systems, and in the organs of special sense. All neurocytes are
the modified descendants of independent oval or pear-shaped

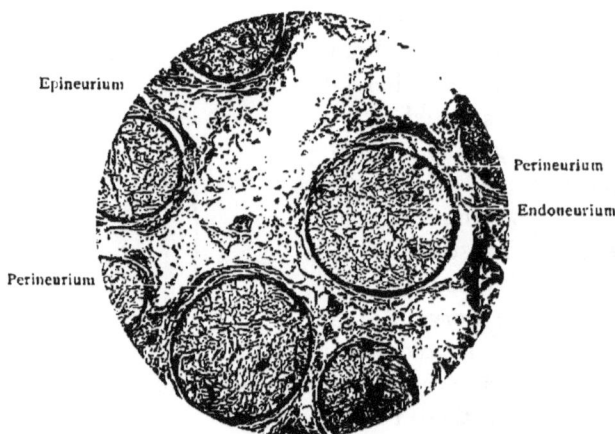

FIG. 235.—TRANSVERSE SECTION, BUNDLES OF NERVE-FIBERS, HUMAN MEDIAN NERVE.
× 30.

cells (the *neuroblasts*), which originate from the epithelial cells
which form the medullary tube. The neurocyte is at first
smooth, devoid of processes, and endowed with ameboid movement. In the course of development the cells project a greater
or less number of processes and assume a variety of shapes
and sizes, in accordance with variations in functions; thus, the
cells may be spheroid, pyramidal, spindle-shaped, stellate, etc.
The body of the cell consists of a protoplasmic basis, more or
less granular, containing a well-defined nucleus and nucleolus.
A centrosoma has also been found in the nerve-cell in many

situations. There is no evidence, however, of the existence of
a cell-membrane. From the body of the neurocyte there arises
one or more protoplasmic processes, which, passing outward in
various directions, divide and subdivide into a greater or less
number of branches, which are collectively known as *dendrites*
or *dendrons*. The ultimate subdivisions and terminations of a
dendrite, though forming an intricate feltwork, always end free,
and never anastomosing with one another. Arising from the
cell-body, the dendrites resemble in appearance and structure
the cell-protoplasm, or cytoplasm. In the cortex of the cerebrum

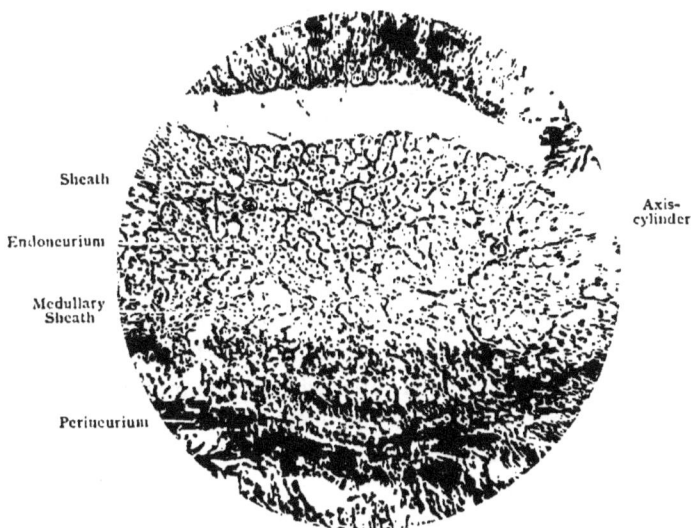

FIG. 236.—PORTION OF TRANSVERSE SECTION OF HUMAN MEDIAN NERVE. × 200.

and in the cortex of the cerebellum the dendrites are characterized by short, lateral projections known as lateral buds or gem-
mules, which impart to the dendrite a feathery appearance.

The Axon, or Nerve-process.—The axon is the first outgrowth
of the protoplasm of the neuroblast, but with the development
of the neurocyte it becomes so differentiated from the dendrites
that it can be readily distinguished from them. It usually arises
from a cone-shaped projection of the cell-body, though occasionally it arises from a dendrite itself. It is characterized by a
short, regular outline and a hyaline appearance. The majority

of the cells, especially in the mammalia, possess but one axon, though in the developing ganglion-cells of the spinal nerves two distinct axons are present. In their subsequent development the two axons appear to blend together to form but a single axon, which, at a short distance from the cell, again divides into two branches, which pursue opposite directions, one passing directly into the spinal cord, the other toward the periphery. The axon may continue as an individual structure for an indefinite distance, varying from a few millimeters to 100 cm. In the former instance the axon, at a distance of a few millimeters from the cell, breaks up into a number of branches, which form an intricate feltwork in the neighborhood of the cell. This type of cell is not widely distributed, being confined largely to the cerebellum. In its course the axon, more especially in the central nervous system, a number of side-branches or collaterals are given off, which do not differ from the axon itself, either in structure or appearance. The axon of the peripheral nerves, especially the spinal nerves, are devoid of collaterals throughout their extent, except, perhaps, in the immediate neighborhood of the cell. The more or less elongated axon becomes inclosed at a short distance from the cell with a thick layer of fatty material, forming a medulla or myelin, inclosed by a delicate cellular sheath (the neurilemma), and thus constitutes what is commonly known as a medullated nerve-fiber. In the central nervous system the neurilemma is frequently wanting. In the sympathetic system the myelin is wanting, though the axon is inclosed by a delicate sheath resembling the neurilemma, thus constituting a non-medullated nerve-fiber. The collateral branches are provided with similar investments.

The End-tufts, or Arborizations.—Each axon, as it approaches its final termination, breaks up into a number of branches, which vary in complexity and appearance in different regions. They are always free from any medullary investment, and appear to be formed by the splitting of the axon into a number of fine filaments, which remain independent of one another. In peripheral organs, as muscles, glands, and blood-vessels, the tufts are in direct organic connection. In the central nervous system the end-tufts are in more or less intimate relation with the dendrites of other neurons.

Nerves.—Nerves are to be regarded, therefore, as groups of axons, with their medullary investments connecting the peripheral organ with the central nervous system.

The nerves are arranged in two great systems—the cerebrospinal and the sympathetic. In the cerebrospinal nerves the conducting media—the nerve-fibers—are arranged in parallel or interlacing bundles, and these are further grouped into nerve-branches or nerve-trunks. The bundles are connected by intervening fibrous connective tissue (the *epineurium*), and through this tissue the principal blood-vessels ramify to supply the nerve-trunks, together with a plexus of lymphatics and numerous fat-cells and plasma-cells.

The size of the nerve-bundles, or *funiculi*, is regulated according to the size and number of nerve-fibers which they contain. Investing each funiculus, or primary bundle of nerve-fibers, is a connective-tissue sheath—the *perineurium*. The fibers composing this sheath are arranged in lamellæ, being separated from one another by lymph-spaces variable in size, through which communication is afforded the lymphatics of the epineurium. Within the bundles the nerve-fibers are held together by fibrous connective-tissue—the *endoneurium*. We find, then, the epineurium holding together and enveloping the several funiculi of the nerve-trunk, the perineurium investing each funiculus, or primary bundle of nerve-fibers, and the endoneurium extending among and around the individual fibers. Nerve-fibers are divided into two classes, which classification is dependent upon the presence or absence of a medullary sheath or covering into the *medullated* or *white*, and the *non-medullated* or *gray*. The *medullary sheath*, or white substance of Schwann, is a bright, fatty substance (the myelin) surrounding the axon, or *axis-cylinder*, the conducting or central part of a nerve-fiber. Between the medullary sheath and the axis-cylinder there is present a small amount of albuminous fluid. Closely surrounding the medullary sheath, and forming the outer boundary of the nerve-fiber, is the *neurilemma*, or *sheath of Schwann*. Between this delicate, structureless membrane and the medulla there are placed at intervals oblong nuclei, surrounded by protoplasm ; these are the *nerve-corpuscles*. Besides the division of nerve-fiber into medullated and non-medullated, each division is susceptible of further subdivision, dependent upon the presence or absence of the neurilemma. Non-medullated nerve-fibers without a neurilemma are composed of an axis-cylinder only; they are cylindric or band-like in form, transparent, and show faint, longitudinal striations. Non-medullated nerve-fibers with a neurilemma are composed of an axis-cylinder surrounded by a neurilemma, and are homogeneous throughout their extent.

Medullated nerve-fibers are those which are partly, but never entirely, invested by a medullary sheath. They may or may not possess a neurilemma; in the former instance they consist of an axis-cylinder and a medullary sheath only. The axis-cylinder, or essential part of nerve-fiber, is cylindric or band-like, occasionally exhibiting a delicate, longitudinal striation, which appearance is due to its being composed of a primitive fibrillæ.

The nerve-cells or ganglia-cells are found in the ganglia as well as along the course of the nerves. They are composed of granular or faintly striated protoplasm, inclosing a characteristic nucleus within which is a nucleolus. They differ greatly in form as well as in size, the spheric, spindle-shaped, and irregularly stellate forms being the most common. In the latter numerous processes are given off, forming the stellate outlines. The cells are variously named, according to the number of processes. If one process is present, the cell is termed a unipolar cell; if two, a bipolar; and if a number of processes exist, they are named multipolar. The processes are of two varieties—the axis-cylinder process and the branched, protoplasmic process. The various forms are most readily distinguished in the multipolar cells. The axis-cylinder process is readily characterized by its hyaline appearance and unbroken outline. The protoplasmic processes are thicker, granular, and striated.

LITERATURE.

Stöhr, "Text-book of Histology," 1896.
Sudduth, "American System of Dentistry," vol. 1, 1886.
Klein, "Elements of Histology," 1889.
Holden, "Human Osteology."
Stricker, "Human and Comparative Histology," vol. 1, 1870.
Morris, "Human Anatomy," 1896.
Lee, "The Microtomist's Vade Mecum," 1896.

THE MUCOUS MEMBRANE OF THE MOUTH; OF THE LIPS; OF THE CHEEKS; OF THE GUMS; OF THE ROOF OF THE MOUTH, HARD AND SOFT PALATE; OF THE FLOOR OF THE MOUTH; THE TONGUE.

HISTOLOGY OF THE TISSUES OF THE MOUTH.

Mucous Membrane of the Mouth.—The mucous membrane lining the cavity of the mouth consists of two parts—the epithelium and the tunica propria ; beneath the latter, and forming the deeper part of the mucous membrane, is the submucosa.

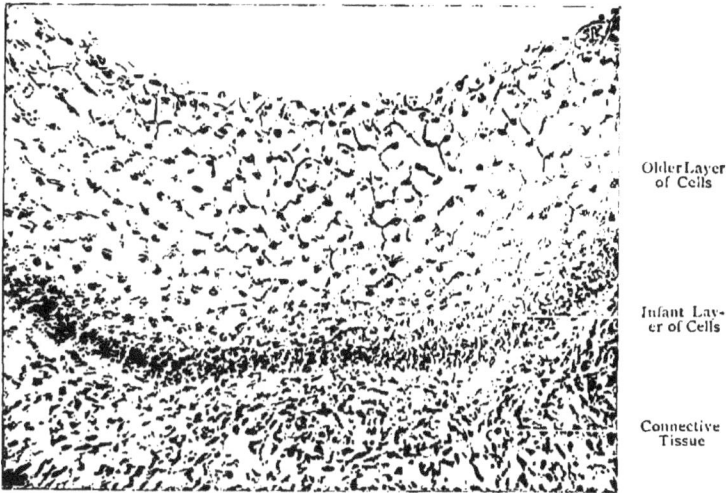

Older Layer
of Cells

Infant Layer of Cells

Connective Tissue

FIG. 237.—VERTICAL SECTION, MUCOUS MEMBRANE OF THE MOUTH, HUMAN EMBRYO. × 150.

The epithelium of the mouth is a thick, stratified, squamous epithelium, the most superficial cells being scale-like or horn-like. The cells are arranged similar to those in the epiderm, are columnar in form, and contain very little pigment.

The tunica propria is a somewhat dense feltwork of interlacing connective-tissue bundles, interspersed with elastic fibers.
The tunica propria penetrates the epithelium in the form of
cylindric or conic papillæ, which differ in length with the
variation in the thickness of the epithelium. As the mucosa is
usually thickest in the lips, gums, soft palate, and uvula, accord-
ingly the papillæ are of the greatest length in these parts. The
tunica propria passes into the submucosa so gradually that a
positive line of demarcation can not be established.

The submucosa consists of a bundle of fibrous connective
tissue with but few elastic fibers. This structure is somewhat
loose in texture and is loosely attached to the underlying periosteum. Over the major portion of the gums and the entire hard
palate the submucosa is attached to the bones of the mouth
through the medium of their periosteal covering. It is in this
loosely constructed tissue that the glands of the mucous membrane are situated. These are for the most part branched,
tubular, mucous glands. Besides adipose tissue in the form of
groups of fat-cells, striped muscular tissue is present in the submucosa. In some parts of the mouth this tissue forms a con-
spicuous portion—namely, in the sphincter muscle of the lips
(orbicularis oris); also in the soft palate, uvula, and pillars of the
fauces.

The blood-supply to the mucous membrane of the mouth is
principally distributed in two plains, the larger vessels to the submucosa and the capillaries to the tunica propria. The larger
vessels break up and send a dense network of capillaries through
its substance and to the numerous papillæ which extend into the
epithelium. Numerous veins ramify through the superficial part
of the tunica propria. The lymphatics form two networks, the
submucosa giving place to the coarser vessels, while the fine
parts are distributed to the tunica propria.

Nerve-supply to the Mucous Membrane of the Mouth.—In the
submucosa the medullated nerve-fibers form a wide-meshed reticulum, from which numerous primitive fibrillæ pass to the
tunica propria, where they terminate, or continue as non-
medullated nerve-fibers, and penetrate the papillæ of the epithelium, forming networks.

Mucous Membrane of the Lips.—Beginning as a direct continuation of the integument or external covering of the lips, the labial covering, including the integument, may be divided into three parts—namely, a cutaneous portion (best described in this connection), a transitory portion, and a mucomembranous portion.

The *cutaneous* portion, covered by a thin epidermis, consists of a double layer of somewhat flattened epithelium. Immediately beneath this is a thin, cellular, mucous layer, the cells composing it being spheroid in form, and containing nuclei which are proportionately large. Subjacent to this is the cutis, composed of fasciculi of fibers intersecting and closely woven together, the principal fibers passing toward the free border of the mucous membrane covering the contiguous surface of the lip. These fibers are for the most part connective-tissue fibers, intermingled with elastic-tissue fibers. Numerous small, vascular papillæ are found upon the surface of the cutis ; these are cylindric or conic in form, and project for some distance into the rete mucosa—the lower layers of living cells of the epidermis. Equally distributed at various depths in this tissue are numerous hair- and sebaceous follicles. The general direction of the hair-follicles in the upper lip is downward, while those occupying the lower lip is upward. Other than the distinction noted by the difference in color of the parts, the cutaneous portion may be distinguished from the transitional mucous membrane by the absence of hair-follicles and sebaceous glands in the latter.

The *transitional* portion of the mucous membrane of the lips is outlined externally by the outer border of the red portion of the lips, and internally by that prominent part of the labial convexity which comes in contact with the opposing labial fold, leaving the transitional portion exposed to view when the lips are in occlusion. The epithelial layer of this surface does not begin where the hair-follicles cease to exist, a slight interspace appearing upon the cutaneous portion which is devoid of these follicles. At its line of beginning the transitional portion of the labial mucous membrane is quite thin, but rapidly increases in thickness in passing toward the mucomembranous portion. Superficially the cells are much flattened, closely associated with

one another, and devoid of nuclei. The cells of the middle and
deeper layers are oblong or spheric, and provided with irregularly shaped nuclei. The chief fibrous tissues of the transitional
portion, which are thinnest at the point where the hair-follicles
cease to exist, are united into flexiform fasciculi, which are separated at various points to give passage to numerous minute
blood-vessels. The fibrous tissues increase in thickness as the
mucomembranous portion is approached. Numerous thin and
somewhat elongated papillæ are distributed over the surface of
the transitional portion.

The *mucomembranous* portion of the mucous membrane of
the lips includes all that portion covering the labial folds within
the mouth, beginning at the line of occlusion on the contiguous
surface and extending to the gums. The epithelium is much
thicker than that previously described, and presents the characteristic layers common to stratified, squamous epithelium.
Superficially, the cells are flattened and tubular, provided with
nuclei of similar form. In the middle layer the cells are flattened and oblong, followed in the deeper layer by irregularly
formed nucleated cells. A variety of fibers make up the
structure—one class fine in texture and united into fasciculi,
intermingled with elastic fibers, together with another set of
coarse, strongly looped fibers. Whenever the fibers of the
tunica propria assume a definite general direction, they are horizontal, passing from right to left and encircling the oral aperture. The tunica propria is beset with numerous conic papillæ
which project into the epithelium; these are longest where the
epithelium is thickest. The mucous membrane forming the
labial frena is covered by an epithelial layer which is much
thinner than that distributed to other parts of the lips. The
fibers in these are irregularly distributed, and the papillæ are
smaller and not so numerous. The coronary arteries and their
accompanying veins course through the lips near the junction
of the transitional with the mucomembranous portion of the
mucous membrane.

Mucous Membrane of the Cheeks.—The mucous membrane of the cheeks presents but little variation in its structure
from that of the mucomembranous portion of the labial

mucous membrane. The buccal epithelium is the same in structure and thickness as that of the lips, excepting the disposition of cells in the middle layer, where they are greater in number and more closely associated, being somewhat distorted by contact. The papillæ, which project from the mucosa into the epithelium, are somewhat broad at their base, with elongated extremities, the height of which is quite variable, in some instances penetrating well into the epithelium, at others merely entering its deeper layer. At the anterior portion of the cheek, or that in the region of the angle of the mouth, the mucous membrane, by its submucous portion, is in immediate contact with the fibers of the buccinator muscle, and throughout the entire surface of the cheek it is closely associated with this muscle. The membrana propria is most dense immediately beneath the epithelium, but as the buccinator is approached it becomes more loose.

Mucous Membrane of the Gums.—The mucous membrane forming these parts is, on account of the numerous tendinous fasciculi which enter into its construction, extremely dense and tough, these characteristics being more strongly manifest here than in any other portion of the oral mucous membrane. These qualities are especially pronounced about the gingival margins and over the major portion of the alveolar walls, being closely bound down to the bone by direct prolongations of the tendinous fasciculi of the periosteum which penetrate the membrane. As the gingival mucous membrane passes into that of the lips and cheeks it gradually becomes less dense. The epithelium of the mucous membrane of the gums is composed of lamina of tessellated and ribbed cells. The superficial cells are the flattened cells of pavement epithelium; subjacent to this they become thicker and deeply ribbed, while the deepest cells are conic, or cylindric with conic extremities. The tissue composing the tunica propria is composed of flattened fasciculi of connective tissue, the fibers of which run parallel with one another. Numerous elastic fibers are also present. Three sets of fibers are to be distinguished in the mucous membrane of the gums—those which run vertically, those which pass in a horizontal direction, and those which radiate or are distributed fan-like. Of the first named, the fibers extend from above downward; in the second class they pass from right to left parallel with the surface; the third class, including those fibers which are reflected from the alveolodental periosteum, are distributed in fasciculi about the margins of the alveoli.

Mucous Membrane of the Roof of the Mouth.—*Hard Palate.*—The mucous membrane covering the hard palate is, in very many respects, dissimilar to that surrounding the necks of the teeth and forming the palatogingival margins. Like the mucous membrane of the gums, that overlying the hard palate is dense and tough. The papillæ of the tunica propria, which penetrate the epithelium, are not so numerous as those upon the gums. In the posterior third of the hard palate they are somewhat more numerous and generally a little more prominent than those in the anterior portion. In the median raphe and over the rugæ, the papillæ are especially sparingly distributed. The epithelium is of the pavement variety, somewhat thinner in front than behind, the cells being more freely distributed at some points than at others. The mucous membrane of the hard palate is less in thickness anteriorly than posteriorly. The distribution of the fibers is such that they radiate from the alveolar borders toward the center of the palate, the anterior fibers passing obliquely backward, while those from the lateral walls pass parallel with one another to the median line. For the most part the fibers are broad and form a plexus between the epithelium and the submucous tissue. The submucous tissue is sparingly distributed over the central portion of the hard palate, but is somewhat more abundant, containing a few fat-cells laterally.

Soft Palate, Uvula, and Fauces.—Passing backward from the posterior margin of the hard palate, the mucous membrane overlies the fibrous aponeurosis of the soft palate and its median and lateral prolongations—the uvula and pillars of the fauces. The epithelium is of the laminated pavement variety, with the deeper cells larger than those placed superficially. The substance of the mucous membrane is composed of fasciculi of connective tissue, intermingled with a plexus of elastic fibers. The fibers are distributed in three principal directions—from side to side, or horizontally, longitudinally, and obliquely. The oblique fibers are instrumental in forming the submucous tissue of both the soft palate and uvula. Numerous conic papillæ project from the tunica propria into the epithelium; these are larger and more numerous on the uvula than on the soft palate. The tunica propria is somewhat variable in thickness to accommodate the glands, which are more or less numerous, and present in greater numbers in one instance than in another. In general the membrane as a whole is thinnest along the margin of the hard palate, gradually increasing in thickness as the free border is approached. The folds of mucous membrane forming the pillars of the fauces, present no peculiarity differing from that of the soft palate, save a more generous supply of elastic fibers.

Mucous Membrane of the Floor of the Mouth.

The Tongue.—The entire unattached surface of the tongue is

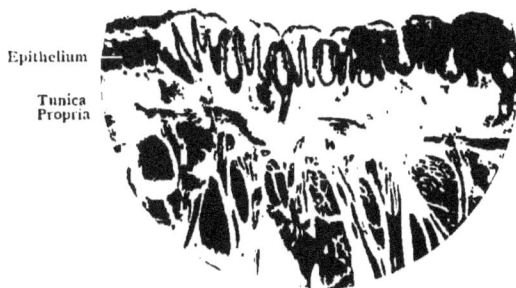

FIG. 238.—LONGITUDINAL SECTION THROUGH MUCOUS MEMBRANE OF THE HUMAN TONGUE. × 20.

covered by a reflection of the mucous membrane of the floor of the mouth. In this organ the general structure of the mucous membrane does not vary from that of other oral mucous membrane, being composed of an epithelium, a tunica propria, and a submucosa. The mucous membrane covering the dorsum of the tongue presents special characteristics, which differ from that of the under surface and the floor of the mouth in general. In the former location the papillary elevations of the tunica propria are conspicuously developed, and with their covering of stratified, scaly epithelium cause the peculiar furred appearance. Three classes of papillæ are distinguished, named, in accordance with their form, filiform papillæ, fungiform papillæ, and circumvallate papillæ.

The filiform papillæ, which are very numerous over the entire dorsum and sides of the tongue, are conic and frequently pro-

FIG. 239.—LONGITUDINAL SECTION OF THE MUCOUS MEMBRANE OF THE HUMAN TONGUE, SHOWING THE FUNGIFORM AND SECONDARY PAPILLÆ. × 80.

longed into numerous horn-like processes, known as secondary papillæ. As elevations from the tunica propria they are composed of well-defined fibrillated tissue, intermingled with numerous elastic fibers. The pavement epithelial cells are found overlapping one another, and provided with processes which project beyond the papillæ.

The fungiform papillæ are also distributed over the entire dorsum and sides of the tongue, but are somewhat less numerous than the filiform variety. They appear as well-defined elevations, and are connected with the tunica propria by a constricted portion or neck. The entire free or rounded surface of these papillæ is beset with secondary papillæ. The epithelium is slightly thinner than that over the filiform papillæ, this being the principal distinguishing feature. The numerous capillaries produce a rich red color, plainly observable through the transparent epithelium. Connective-tissue bundles make up the bulk of these papillæ, few elastic fibers being present.

The circumvallate papillæ are placed on the posterior portion of the dorsum of the tongue, and are few in number (eight to sixteen). They are much larger than those already described, and in general resemble modified fungiform papillæ. They are flattened and broad, and differ from the fungiform by having

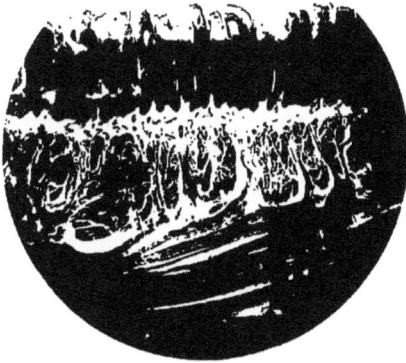

FIG. 240.—SECTION THROUGH EPITHELIUM, NEAR THE TIP OF THE TONGUE. $\times$ 40.

a circular furrow or wall surrounding them. Secondary papillæ are present on the free surface only, the sides, and in some instances the walls surrounding them, being occupied by the end-organs of the special sense of taste—the taste-buds. Other taste-buds are found upon the lateral margins of the tongue posteriorly, nestled in a group of parallel mucous membrane folds—the papillæ foliata. The connective tissue within these papillæ is similar to that in the fungiform papillæ. On other parts of the tongue, or those portions not occupied by these specially constructed papillæ, the epithelium is similar to that in other parts of the mouth. The tunica propria is less in thickness in and about the tip of the tongue, and is intimately connected with the subjacent muscular structure. As the root of the organ is approached, the tunica propria becomes thicker and more dense. The submucosa is especially intimately connected with the underlying parts at the margins and tip of the tongue. The extreme portion of the root of the tongue has its mucous membrane particularly modified by a special aggregation of adenoid tissue—developed lymph-nodules. These are large and readily perceptible to the naked eye. They are provided with a central opening, which dips down into a well-defined vault or crypt, which is lined by a reflection of the stratified oral epithelium.

Blood-supply to the Mucous Membrane of the Mouth. —The oral mucous membrane derives its supply of blood from numerous branches of the external carotid artery—namely, the superior and inferior coronary, buccal, lingual, transverse facial, pterygopalatine, and the alveolar. Entering the submucosa, the minute terminal branches of these arteries are distributed parallel to the surface, and by anastomosis form plexuses from which other minute branches are given off to supply the papillæ of the tunica propria. After coursing through the papillæ the blood is discharged into a similar venous plexus, and thus conveyed from the parts. In a like manner the mucous membrane and papillæ of the tongue is supplied, branches of the lingual artery conveying the blood to the parts. The dorsalis linguæ supplies the mucous membrane of the dorsum of the tongue and pillars of the fauces, while the ranine artery by its minute branches supplies the remaining mucous membrane. Each papilla is entered by two or more arterial terminals, which divide, anastomose, and finally send off capillary branches to the secondary papillæ.

Nerve-supply to the Mucous Membrane of the Mouth. —The distribution of the nerve-fibers to the oral mucous membrane is approximately similar in all parts. The fibers, which are of the medullated variety, are first distributed to the submucosa, forming a wide-meshed reticulum. From this fibers are given off to the tunica propria, terminating in end-bulbs, or, after loosing their medullary sheath, are distributed to the epithelium, where their free extremities lie between the epithelial cells. The nerves of the mucous membrane of the tongue (the glossopharyngeal and lingual branch of the fifth) may have their endings similar to those in other parts of the mouth, or they may be intimately associated with the taste-buds.

LITERATURE.

Legros and Magitot, 1880.

Klein, "Structure of the Oral Lips," 1868.

Sebastian, "Anatomy and Physiology of the Labial Glands," 1842.

Kolliker, "Mikroskopische Anatomie."

Kirke, "Physiology."

Stricker, "Human and Comparative Histology."

Stöhr, "Text-book of Histology," 1896.

CHAPTER III.

GLANDS AND DUCTS OF THE MOUTH; OF THE LIPS; OF THE CHEEKS; OF THE HARD AND SOFT PALATES; OF THE TONGUE.—THE SALIVARY GLANDS.

GLANDS AND DUCTS.

Glands of the Mouth.—The glands of the mouth, like the glands of other parts of the body, are composed almost entirely of epithelium, and may, therefore, be classed with the epithelial tissues. Glands exist in two principal forms—tubular and saccular (alveolar). The former occur either singly or in groups, and are further subdivided into simple tubular and compound tubular glands. A like condition is present in the saccular glands and similar terms are employed to qualify them—simple saccular glands and compound saccular glands.

A simple tubular gland is one composed mainly of a simple, tube-like structure; a *compound tubular gland* is one composed of a number of smaller tubes emptying into a single duct.

A simple saccular gland is one formed by a sacculation of serous or mucous membrane into a single, simple sac, or by branched saccules having an excretory duct (alveolar system); a *compound saccular gland* is composed of a combination of branched saccules.

In the larger glands a sheath is formed by the surrounding connective tissue, from which numerous septa are given off to the interior of the gland, dividing it into compartments varying in size. These are known as *gland-lobules*. The connective-tissue walls of the gland-lobules carry the larger blood-vessels and nerves. Most glands are divided into two essential parts, —the gland-follicle and the excretory duct,—the former being specialized for the secretory function, while the latter, by communicating with the surface, conveys the secreted substance to that point.

The gland-follicles are composed of a layer of gland-cells,

356

usually simple in character, surrounding the follicular walls. External to these is a specially modified connective tissue, forming the basement membrane, or membrana propria. The appearance of the gland-cells and their nuclei is continually changing, being thus influenced by their functional activity.

The excretory ducts consist of a wall of connective tissue and elastic fibers, lined by a columnar epithelium, either simple or stratified. In some instances the arrangement of the excretory

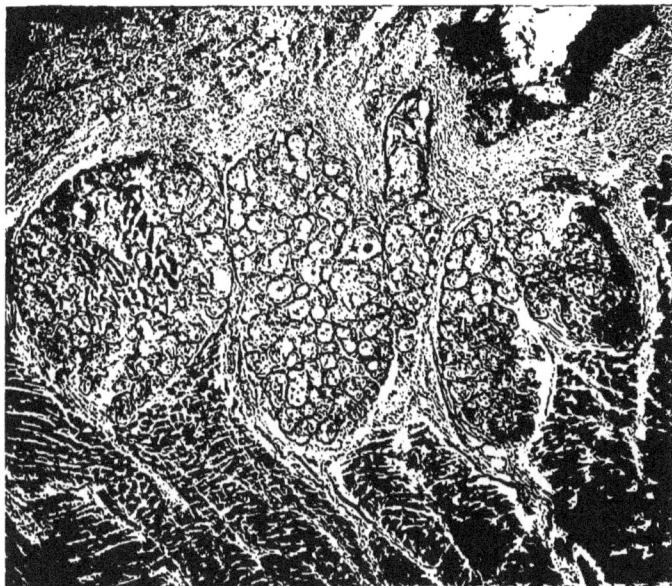

FIG. 241.—SECTION THROUGH THE GLANDULAR TISSUE OF THE TONGUE. × 40.

ducts is much complicated, being divided into secretory tubes, which in turn are subdivided into smaller tubules—intercalated tubes.

The Glands of the Lips (*Labial Glands*).—The glands of the lips are situated in the submucosa, and are first observed immediately within the line of labial occlusion, at which point the thickness of the epithelium becomes somewhat definite and general. These glands are variable in size, but all are sufficiently large to be observed without the aid of the micro-

scope. They are of the compound tubular variety, and communicate with the surface through an excretory duct, which
throughout the greater part of its extent is lined with stratified,
scaly epithelium. In passing from the surface toward the gland-
follicle, the main duct takes a spiral course obliquely through
the tunica propria, and upon reaching the submucosa gives off
numerous branches and twigs which terminate in the individual
acini. The larger branches from the main duct are lined with
stratified squamous epithelium, while the smaller twigs are provided with columnar epithelium. In many instances the main
excretory duct, in its passage through the tunica propria, receives
the principal duct from small accessory ducts. The framework
of the labial glands is formed by the flexiform tissue composed
of fasciculi of the fine connective-tissue fibers belonging to the
submucous layer, together with delicate, coiled, elastic fibers.
This framework gives support to a minute system of capillaries
and small nerve-fibers supplying the acini. The acini are so arranged that those belonging to a large duct are united into a
lobule by the submucous connective-tissue fasciculi, and these in
turn are formed into lobes. By a continuation of the same fasciculi and fibers which limit a lobe, and in the meshes of which
the acini are situated, a sheath to the excretory duct is formed.
Besides the branched, tubular, mucous glands of the lips, there
are occasionally found, at the edges of the lips, sebaceous glands.

    The Glands of the Cheeks (*Buccal Glands, Molar Glands*).
—The glands of the cheek are also situated in the submucous
layer of the mucous membrane. They, like the labial glands,
are of the compound tubular variety, and when microscopically
examined are found to be similar in structure. They are somewhat larger than the labial glands and proportionately less
numerous. The chief duct from each of these glands usually
opens with a narrow mouth on the surface of the oral mucous
membrane, and in its passage through the tunica propria takes
a vertical or oblique direction. In the submucosa the chief duct
branches into two or more smaller ducts, taking up alveoli. As
the buccal glands are somewhat larger than the glands of the
lips, they are composed of a greater number of ducts and
alveoli.

The Glands of the Hard and Soft Palate (*Palatal Glands*).
—The mucous glands of the hard palate are situated in the submucosa and closely associated with the periosteum. They are
compound tubular glands, and in all essential particulars are
similar to the labial and buccal glands. They are quite numerous (200 to 300), isolated in the anterior portion, but are
grouped into a single row or into two rows posteriorly. The
glands are freely distributed in each lateral half, but are absent
at the median line.

In the soft palate the glands are of the same character, somewhat variable in size, the largest being found in the uvula. The
excretory ducts from these glands vary in diameter, in the
nature of their fibrous structure, and in the direction taken in
passing to the surface. Over the surface of the soft palate the
mouths of these ducts are represented by minute orifices slightly
smaller than the body of the duct, but in the uvula the opposite
condition is present, the mouth of the duct being wider than the
body. The course taken by the excretory duct is seldom a
direct one, but after receiving all tributary branches passes
obliquely through the tunica propria, and before entering the
epithelium turns at an abrupt angle, and so continues until the
surface is reached. The ducts are lined by a simple columnar
epithelium, which in some instances is ciliated ; the walls of the
tubes consist of gland-cells and a structureless membrana propria. In some instances the surface epithelium may be reflected
for a short distance and partly serve in the capacity of a lining
to the tubular walls.

The Glands of the Tongue (*Lingual Glands*).—In this organ
two varieties of glands are found, occurring both in the mucous
membrane and in the superficial muscular strata, and being
principally distinguished by the nature of their secretions. The
gland-cells of the one set are mucigenous, secreting mucin ;
these are the mucous glands. The other set is productive of a
serous fluid, thin, watery, and containing albumin ; these are
the serous glands.

The mucous glands of the tongue are found along the lateral
margins and over the root of the organ, being most numerous in
the latter situation. They are of the compound tubular variety,

and in most particulars are identical with the mucous glands of
other parts of the oral cavity.   The ducts are lined with ciliated
columnar epithelium, and the walls of the ducts consist of a
homogeneous membrana propria and gland-cells.   The glands
occupying the root of the tongue are frequently found with
their excretory ducts opening into the follicular crypts.   The
tubules consist of a structureless membrana propria and numerous gland-cells, the latter varying in appearance according to
their functional activity or functional condition.   The crypts of
the follicles constitute reservoirs for the acinous glands, and

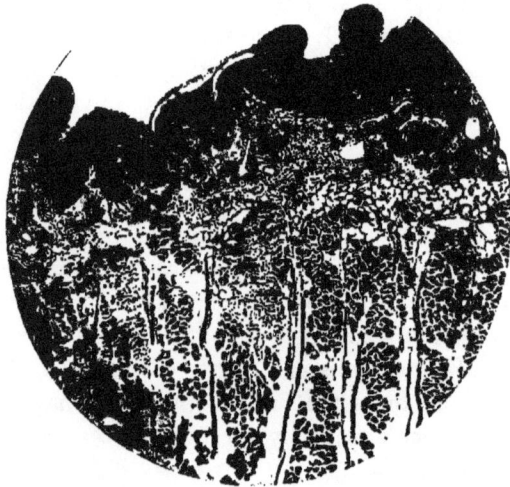

Fig. 242.—Section Through Base of Tongue, Showing Serous Glands.

these receptacles frequently extend for some distance beneath
the surface, receiving at various points the main excretory ducts
from the mucous glands.   These saccular-like reservoirs are
lined by a well-defined capsule surrounded by a fibrous sheath,
internal to which is an epithelial covering, a prolongation of
the common epithelium of the mouth.   Between these two
layers are a number of minute, closed lymph-follicles placed in a
single layer.   The mucous glands on the lateral walls of the
tongue are, for the most part, situated near the middle or

posterior portion. The ducts from these glands usually open directly toward the cheek, but in rare instances they pass obliquely downward and open near the proper floor of the mouth. At the tip of the tongue, buried beneath the mucous membrane and some of the muscular fibers, may be found a pair of mucous glands (Nuhn's) which open by free orifices on the under surface. At the root of the tongue, flat, lenticulated elevations of the mucous membrane are present, beneath which is imbedded conglobate, glandular substance. These show a central orifice leading to a small pit lined with tessellated epithelium.

The serous glands of the tongue are compound tubular glands, and are found in the region of the circumvallate papillæ, closely associated with the taste-buds. The excretory ducts, lined with a simple or stratified columnar epithelium, the latter sometimes ciliated, open near the base of the papilla, or between the papilla and its wall. The tubules are similar to those in the mucous glands, consisting of a delicate, structureless membrana propria and gland-cells. The gland-cells are composed of a frail, transparent protoplasm, containing rounded nuclei.

The Salivary Glands.—The parotid, submaxillary, and sublingual glands each consists of an excretory duct, branching frequently in a tree-like manner into smaller ducts, lined throughout with a layer of epithelial cells. From the smaller ducts terminal branches are given off, which in turn are lined with epithelium. The other portions of the glands are invested by columnar epithelium, and arranged like grapes about the main excretory duct, and consequently belong to the group of racemose glands. The terminal branches or alveoli attached to the smaller excretory ducts are so numerous that they become much compressed from pressure, and the grape-like appearance is more or less destroyed, and but little space is left for interstitial tissue. Each gland is inclosed in a fibrous connective-tissue capsule, and from this numerous septa of fibrous trabeculæ pass to the interior and divide the glandular substance, first into lobes, these being subdivided into lobules, the lobules by further subdivision forming the alveoli. The glandular connective tissue is loose in texture, containing many elastic

fibers and lymphoid cells. Fine bundles of fibrous tissue,
together with branched connective-tissue corpuscles, constitute
the connective-tissue matrix between the alveoli.

The Ducts.—Entering the interior of the gland, the chief duct
divides into a number of large branches, one of which passes
to each lobe, each of these giving off several branches which
connect with the several lobules. Upon close examination the
central tube of each lobule is observed to throw off several
small tubes—the intralobular tubes. Following these are the
intermediate tubules, which continue into the terminal compartments. The chief excretory duct consists of a double layer
of cylindric epithelium and fibro-elastic cartilage. Close beneath the epithelium is a compact membrana propria. The
intralobular tubes are each provided with a distinct lumen.
The walls are composed of a membrana propria lined by a
layer of columnar epithelium, the cells of which contain a central
round nucleus.

The Parotid Gland.—The distinguishing histologic feature
in this gland is found in its excretory duct (Stenson's duct),
which is provided with a membrana propria, especially broad
and compact, placed immediately beneath the epithelium. The
duct is composed of a double layer of cylindric epithelium and
fibrous tissue, intermingled with elastic fibers. The main duct
divides and passes into the intralobular tubes, beyond which
are the intermediate tubules. The intralobular tubes are lined
by columnar cells, while the intermediate tubules are lined by
elongated, spindle-shaped cells. The salivary cells lining the
acini are different in character from those in the submaxillary
and sublingual glands. The parotid gland is a true salivary
gland, and the serous gland-cells composing its epithelial lining
are disposed in a single layer. The cells are columnar or pyramidal in form and composed of a dense protoplasm, containing
a spheric nucleus.

The Submaxillary Gland.—The excretory duct (Wharton's duct), like the main duct of the parotid gland, is composed
of a double layer of columnar epithelium, external to which is
a layer of cellular connective tissue, the whole being surrounded
by a thin stratum of muscular fibers placed longitudinally. The

intralobular tubes are lined by a specialized, elongated, cylindric epithelium, which, in the intermediate tubules, becomes clothed with cubic cells. The acini are lined either with serous gland-cells similar to those lining the acini of the parotid gland or with mucous gland-cells, the former being most constant in their presence. The two kinds of acini are uninterruptedly connected. In most instances there are but a few mucous acini present within the lobule, but occasionally they are found in abundance. The submaxillary is a mixed or mucosalivary gland.

The Sublingual Gland.—The excretory duct (Rivini's duct) is similar in structure to the chief excretory duct of the submaxillary gland. The intralobular tubes are lined with columnar epithelium. The intermediate tubules are not positively known to exist, and it is quite probable that the intralobular tubes pass directly into the terminal compartments. The acini are composed of a membrana propria and gland-cells, both mucous and serous. The former are much more numerous than in the acini of the submaxillary. The membrana propria is composed of stellate connective-tissue cells. This gland is also a mixed or mucosalivary gland.

Blood-vessels and Lymphatics in the Salivary Glands. —The lobules of the salivary glands are richly supplied with blood-vessels. The many arterial branches break up into numerous capillaries, which, forming a dense network, surround the acini, being supported by the interalveolar connective tissue. The lymphatic vessels accompanying the intralobular tubes are in communication with numerous lymph-spaces which exist between the interalveolar connective tissue and the walls of the acini. The substance of the gland is further supplied with blood by numerous plexuses of lymphatics which are carried or supported by the interlobular connective tissue.

Nerve-supply to the Salivary Glands.—The nerve-fibers distributed to the salivary glands are both of the medullated and non-medullated variety, and other nerve-tissue in the form of ganglion-cells is present. The medullated nerve-fibers are abundantly numerous, and are distributed to all parts of the gland. In many respects the fibers are peculiarly constructed. They are extremely delicate, made so by the frail nature of their medullary sheath ; they divide and give off so many branches as to almost give them a feathery fineness. This peculiarity is especially noticeable toward their extremities, where the fibers lie between the alveoli and give off minute branches in all directions. The nerve-fibers are placed in close relation to the tubes and tubules, which they freely encircle ; they perforate the membrana propria and break up into finer subdivisions, from which they are distributed to the exterior of the epithelial cells. In the alveoli two kinds of nerve terminations are found. The primitive fibers branch between the alveoli and are distributed to the membrana propria, upon entering which numerous branches are thrown off which pass to the epithelial cells beneath. The non-medullated fibers, which are much less numerous, are composed of an extremely delicate fasciculi of transparent fibers resembling axis-cylinders, and invested by a sheath of connective-tissue cells containing nuclei. The distribution of these fibers is similar to the medullated fibers, encircling the tubes and penetrating the membrana propria, being similarly distributed to the alveoli.

LITERATURE.

Stöhr, "Text-book of Histology," 1896.

Stricker, "Human and Comparative Histology," vol. I, 1870.

Klein, "Elements of Histology," 1889.

Sebastian. "Recherches anatomique, physiologiques, pathologiques, les Glans Labiales," 1842.

Ward, "On Salivary Glands," Tood's "Cyclopedia of Anatomy and Physiology."

Klein and Vernon, Stricker's "Human and Comparative Anatomy," vol. I, chap. XVI.

Sudduth, "Embryology and Dental Histology," "American System of Dentistry," vol. I, part III.

Brubaker, "Transactions Odontological Society of Pennsylvania," 1889–'95.

Todd and Bowman, "Physiological Anatomy," vol. I.

Szontagh, "Essays on Minute Anatomy of Hard Palate in Man," 1866.

CHAPTER IV.

MUSCULAR TISSUES OF THE MOUTH; OF THE LIPS; OF THE CHEEKS;
OF THE SOFT PALATE; OF THE TONGUE.

MUSCULAR TISSUES OF THE MOUTH.

Muscular Tissues of the Lips.—The minute bundles forming the fasciculi of the oral sphincter muscle—the orbicularis oris—are distributed between the submucosa of the mucomembranous portion and the subcutaneous tissue of the cutaneous portion of the lips. The muscular fibers radiate in three principal directions upon either side of the median line : from the angle of the mouth, toward the median line, and from the fleshy slips of the maxilla and mandible—the musculi incisivi. As the fibers from the angle of the mouth pass to the substance of the lip, they are arranged in a laminated manner. When the median line is reached, one set of fibers terminate somewhat abruptly in the subcutaneous tissue, another set is continued beyond the median line and attached to the cutis of the opposite side, while a third set, without crossing the median line, is attached to the incisive fossæ of the maxilla and mandible. The numerous muscular fibers of the internal, labial, or mucomembranous portion, and the external, facial, or cuticular portion, penetrate the parts and terminate in close proximity to the epithelium or to the base of the papillæ. Delicate, hair-like fibers which are continuous with the sarcolemma slightly penetrate the cutis and membrana propria. A few of the fibers, which may be classed with the terminals of the outrunning muscles from the lips, are arranged in a number of fasciculi in the subcutaneous portion, pass through the fasciculi of the orbicularis oris, reach the submucous tissue, where they cross and recross one another, and finally pass into the membrana propria, where they end in fan-like terminals. The fasciculi of the orbicularis differ somewhat in the upper and lower lips; in the former the bundles are strongly developed

toward the angle of the mouth, while in the latter the median
bundles are the strongest. The labial muscular tissues are of
the transversely striated variety. The fibers are cylindric in
form, having rounded or pointed extremities in the interior, and
broad or flattened ends where they come in contact with the
periosteum. When examined with a high power each fiber
shows alternately broad and narrow striæ, the former being
dim, while the latter is bright in appearance. With a stronger
power both the broad and narrow striæ are seen to be transversely striated.

Muscular Tissues of the Cheeks.—The muscles entering
into the construction of the lateral walls of the mouth have
already been described in part I, page 24, giving the relations
existing between the individual muscles, together with the general disposition of the various fasciculi. Histologically considered, these muscles partake of all the characteristics of
striated or voluntary muscular tissue. In the body of the buccinator and masseter muscles the fibers are cylindric and have
definitely pointed or rounded ends. Near their termini, particularly in the latter muscle, the inner extremities of the terminal fibers are pointed, while the outer ends, or those by which
the attachment is formed, are broad and rather flat.

Muscular Tissues of the Soft Palate.—The disposition of
the striated muscular tissue of the soft palate is extremely complicated. The azygos uvulæ, the only true longitudinal muscle
in the soft palate, has its origin from aponeurosis of the soft
palate and from the nasal spine of the palate-bone, the fibers
passing backward upon either side of the median line. This is
a double muscle, and near its point of origin the two portions
are distinct and separated by a definite space, but upon reaching
the base of the uvula they become closely associated. The
fasciculi do not continue to the apex of the uvula, but immediately beyond the center of its length are thrown out fan-like
toward the sides, terminating in a manner similar to the fibers
of the lips. In passing from before backward a number of
small fasciculi are given off, which reach out laterally and
traverse the glandular lobes, completely surrounding them,
after which they again return to the principal fibers at the

median line. The palatopharyngeus muscle is divisible into two parts, the upper extremities of which lie partly in front and partly behind the levator muscles. The greater number of the fibers of one set, situated in front of the levators, form a curved, flattened aponeurosis. The fibrous border of the hard palate serves as an attachment for the convex border of this portion, while the other border, which is concave, is directed toward the arch of the levators. The fibers of the palatopharyngeus, situated behind the levators, form a number of loose fasciculi interspersed by fat-cells. In passing toward the free border of the soft palate the fibers become much more delicate, and, separating, some course in front and others behind this muscle. In this location the fibers become closely associated with the glands, and either end here or are continued to the submucosa, or even to the membrana propria of the mucous membrane. The fibers of the palatopharyngeus unite with the fibers of the levators, and an arch-like fasciculus is formed by this union which, subdividing, passes in front of the azygos uvulæ to the opposite side. All of these fibers run outward and downward, and unite with the extremities of the other palatal muscles, the fibers of which are somewhat more regularly distributed. Like the muscles of the lips and cheeks, the several fasciculi of the palatal muscles form a delicate plexus, and a quantity of fatty tissue is found between the various fasciculi.

Muscular Tissues of the Tongue.—The tongue is divided into two equal lateral portions by a median septum—the septum linguæ. This central septum, composed of a vertical layer of compact, fibrous, connective tissue, extends the entire length and depth of the lingual median line. Beginning at the hyoid bone, it gradually increases in prominence until the middle of the organ is reached, beyond which point it becomes less pronounced and finally disappears near the tip. The bundles of the muscular tissues are arranged longitudinally, transversely, and vertically. The former lie immediately beneath the mucous membrane, including the superior lingualis above and the inferior lingualis below, together with the greater part of the styloglossus. The superior lingualis extends from the base to

the tip of the organ, and by short fasciculi its fibers are attached
to the overlying tissues.   The fibers of this muscle are placed
between the hyo- and styloglossi muscles of the opposite side,
both of which overlap the fibers of the lingualis near the base
of the tongue.   The inferior lingualis also gives off several
small fasciculi and fibers to the mucous membrane beneath, and
is composed of two bands which reach from the base to the
apex, each being placed between the hyo- and geniohyoglossus
muscles.   The transverse fibers, which are placed between the
superior and inferior lingualis muscles, originate from the septum
linguæ, and form the bulk of the organ.   From their point of origin
these fibers course outward and upward to the sides of the
tongue.   Those fibers which are vertically disposed decussate
with the transverse fibers, and pass from the dorsum toward the
under surface of the tongue, the fibers curving gracefully with
their concavity directed toward the under surface.   In most
instances the ascending vertical fibers, as well as the transverse
fasciculi, pass between those longitudinally disposed and connect with the submucosa.

CHAPTER V.

TISSUES OF THE TEETH—ENAMEL; DENTIN; CEMENTUM; THE
TOOTH-PULP.—THE ALVEOLODENTAL PERIOSTEUM.

TISSUES OF THE TEETH.

Enamel.*—The enamel, which forms a cap-like covering
of varying thickness over the entire crown of the tooth, is a

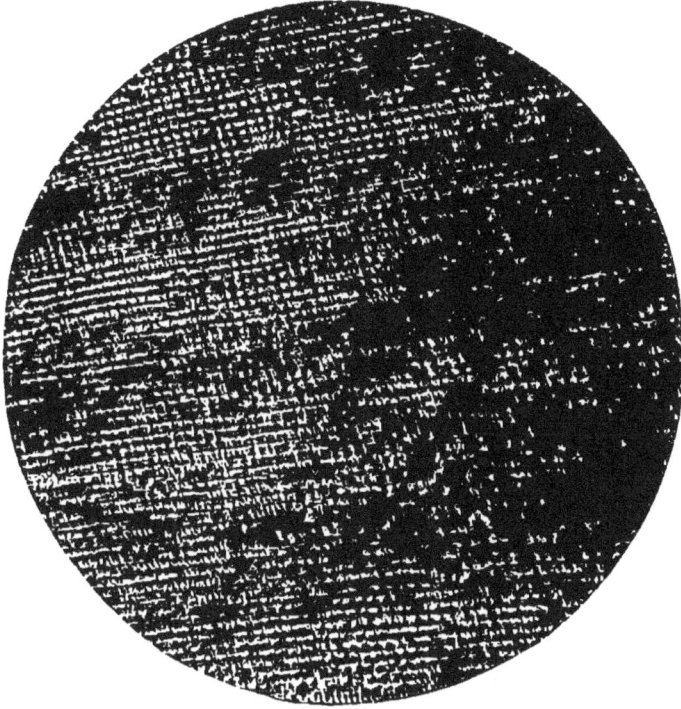

FIG. 243.—SECTION OF ENAMEL FROM HUMAN TOOTH (SPECIMEN BY J. HOWARD MUM-
MERY). × 350.—(*After Williams.*)

vitreous, hyaline substance, containing but little, if any, organic

---

* The author is indebted to Dr. J. L. Williams for a number of illustrations used under this
heading.

matter. The thickness of the enamel cap appears to be strongly influenced by the function of the different parts of the tooth-crown, being thickest over the cutting-edges of the anterior teeth, while in the cuspidate teeth, the entire occlusal surface is provided with the thickest enamel layer. It is about evenly distributed over the lateral walls of the crown, but as the cervical line is approached, its thickness is gradually diminished (Fig. 244). Chemically, enamel is composed of the salts of lime, calcium phosphate predominating. Calcium carbonate, magnesium phosphate, and calcium fluorid are present in smaller quantities. The proportionate quantity of lime-salts in enamel is not fixed, a slight variation in density occurring in the enamel of different individuals. These essential differences are regulated by the proportionate quantity of calcium phosphate and carbonate—a greater amount of the former being productive of additional hardness, while an increase in the latter beyond the minimum amount decreases this quality. As a general rule, the teeth of males contain a greater amount of calcium phosphate than the teeth of females, as shown by the following analysis by von Bibra :

	MAN.	WOMAN.
Calcium phosphate and fluorid,	89.82	81.63
Calcium carbonate,	4.37	8.88
Magnesium phosphate,	1.34	2.55
Other salts,	.88	.97
Cartilage,	3.39	5.97
Fat,	.20	a trace.
Total organic,	3.59	5.97
Total inorganic,	96.41	94.03

In general structure enamel is composed of numerous hexagonal prisms, with a common direction at right angles to the long axis of the tooth. These prisms are known as enamel prisms, enamel fibers, or enamel rods. While the general direction of the fibers is, as previously stated, nearly at right angles to the body of the tooth, they do not pursue a perfectly straight course in passing from the dentin to the surface, but are disposed in a tortuous or wave-like manner. The enamel prisms may be said to sit on end against the surface of the dentin, minute depressions in the latter receiving the extremities of the

rods. The direction of the enamel prisms, as compared to the
body of the tooth-crown, varies according to the part of the
crown which they occupy. Taking the entire crown of the tooth,
they radiate in such a manner from the surface of the dentin
that at the cutting-edge or occlusal surface of the tooth they are
more or less vertical, while over the lateral surfaces they tend
to the horizontal direction. An examination of the prisms
when isolated and decalcified exhibits numerous evenly distributed varicosities, producing a transversely striated appearance to the rods. Tomes has pointed out that the enamel rods

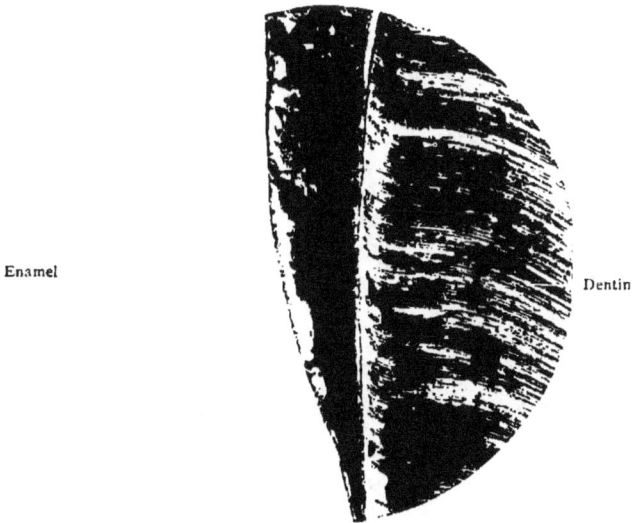

Enamel   Dentin

FIG. 244.—ENAMEL AND DENTIN FROM HUMAN TOOTH, SHOWING GRADUAL REDUCTION
IN THE THICKNESS OF THE FORMER AS THE CERVIX IS APPROACHED. × 30.

are variously disposed in that portion of the enamel most closely
associated with the dentin. On the cusps of the teeth they are
twisted and curved in various directions, while near the surface
on the incisors they are uniform and straight. In general, the
enamel rods, which begin on the surface of the dentin, are continuous through the entire thickness of the enamel. In passing
from the interior to the exterior, the individual rod, to occupy a
proportionate space in all parts, would have to increase in diameter; but this it does not do. In consequence of this arrange-

ment there exist numerous supplemental or peripheral rods, which extend but a short distance from the surface, filling in the interprismatic spaces formed by the longer rods. With the exception of the faint transverse striations, the enamel prisms appear to be structureless. A variety of opinions have been expressed in regard to the cause for the striated appearance of the enamel rods. It was claimed by Hertz to be attributable to a temporary arrest of calcification, but more recent investigation has shown the cause to be the presence of varicosities in the individual fibers, precisely as the varicosities in the individual

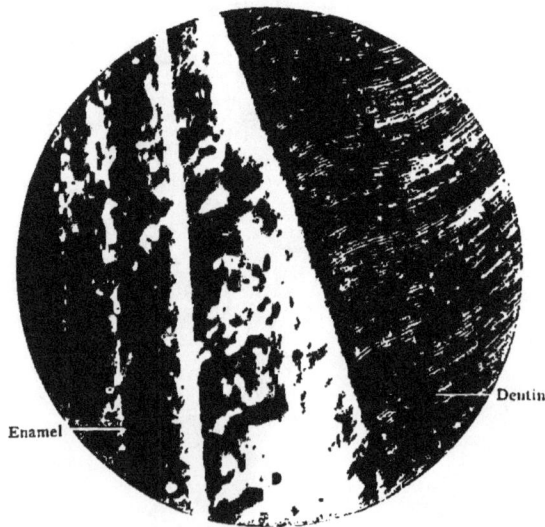

FIG. 245.—COMPARISON IN THE APPEARANCE OF THE ENAMEL AND DENTIN UNDER LOW POWER OF THE MICROSCOPE. × 40.

muscular fibers produce the striated appearance in that tissue. Bodecker asserts that fully developed normal enamel is non-striated, and von Ebner practically makes the same statement, claiming that they are due to the preparation of the specimen, which, usually being mounted in Canada balsam, suffers sufficiently from the slight acid reaction to produce the striated appearance. These statements Dr. J. Leon Williams, in his recent investigations,* emphatically denies, saying that, while in some specimens the varicosities are apparent in some parts, they are decided in others. If this statement be accepted, it would seem to entirely overthrow the theory of von Ebner in regard to the action of the acid, which would be distributed to all parts alike.

According to Williams, the varicosities of one enamel prism are opposite those of the adjoining prisms, and by the coming together of the varicosities the prisms become united by means of processes which they send out. In like manner the varicosi-

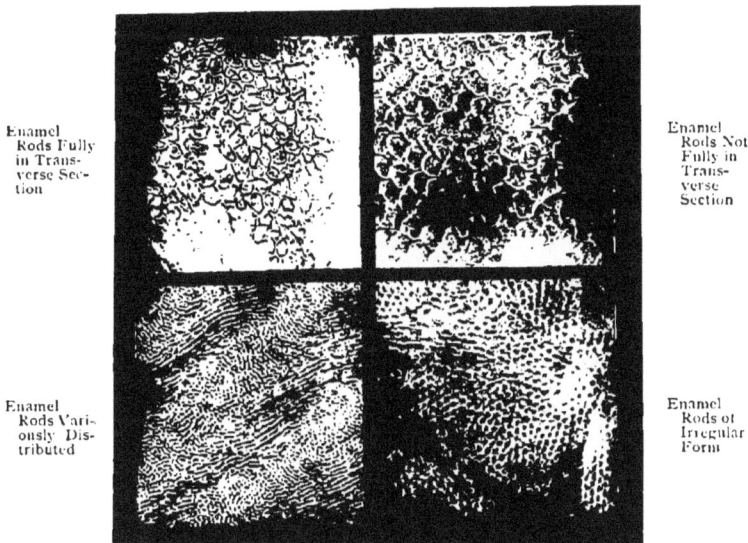

Enamel Rods Fully in Transverse Section

Enamel Rods Not Fully in Transverse Section

Enamel Rods Variously Distributed

Enamel Rods of Irregular Form

FIG. 246.—HUMAN ENAMEL. TRANSVERSE GROUND SECTIONS.—(*After Geise.*)

ties upon the same rod are connected by processes running parallel with the prism. According to von Eber, enamel is traversed by numerous minute canals, and Heitzmann claims to have found organic fibers in its substance. Williams, while admitting the enamel structure to be far more complex than past research has shown, appears to have fully demonstrated that neither canals nor organic fibers are present—in fact, he denies the presence of the least trace of organic matter in this structure. The interprismatic matrix, heretofore considered by most authorities to be an organic structure, now appears, by the thorough methods employed by the last-named gentleman, as a transparent, inorganic substance. By numerous experiments he was enabled to secure a specimen in which the interprismatic spaces of one layer were not backed up by the rods of another layer. In some instances the specimen showed the rods well separated and the interspace closed by a perfectly transparent substance, in the interior of which might be seen connecting processes passing from one rod to another. In ad-

FIG. 247.—THICK SECTION OF ENAMEL OF HUMAN TOOTH, SHOWING BROWN STRIÆ OF RETZIUS. × 40.

dition to the striated appearance formed by the varicosities of the individual prisms of fully developed enamel, other structures of a different character, and upon a much larger scale, are present and known as the "brown striæ of Retzius" (Fig. 247). These markings, readily seen with a low power, are of a brownish color, and run nearly parallel with the surface of the dentin or enamel. Those striæ nearest the surface of the dentin are inclined to follow the contour of that structure, extending in many instances the entire length of the crown. The lines nearest the surface are longest in the region of the cutting-edge, or occlusal surface of the crown, becoming shorter as the neck of the tooth is approached, being directed at an acute angle to the surface of the dentin at that point. A number of theories are advanced to account for the presence of the "brown striæ of Retzius." Tomes suggests that, coinciding as they do to the outer surface of what was at one time the primitive enamel cap, they might be considered as in a measure outlining the stratifications of the primary deposit. Another theory, but one seemingly without foundation, is to the effect

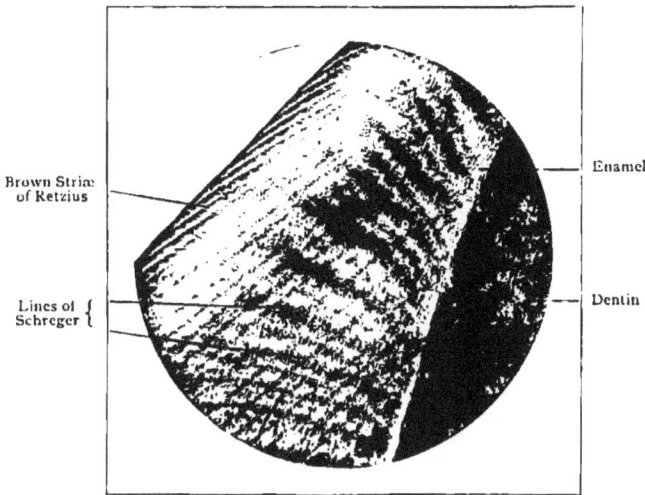

FIG. 248.—ENAMEL AND DENTIN, HUMAN TOOTH.—(*After Geist.*)

that the striæ are produced by an arrest in the calcifying process; while a third theory attributes the cause to a variation in the character of the nourishment taken by the mother during pregnancy. The acceptance of this latter theory would seem to indicate a set diet for all mothers and at stated intervals, the striæ always being present and somewhat regularly distributed throughout the enamel tissue.

Still another set of lines or markings are to be observed upon the surface of sections of enamel, these being known as the "lines of Schreger." Figure 248 shows these lines as they appear upon the surface of the enamel by reflected light, the same being quite invisible by transmitted light. The presence of these lines is due to the various directions assumed by the contiguous groups of enamel rods. Beginning at the surface of the dentin they are well defined, but gradually become less marked as the exterior of the enamel is approached. At the line of union between the enamel and dentin irregularly formed cavities are occasionally observed, into which the dentinal tubes may extend, and in rare instances individual tubes

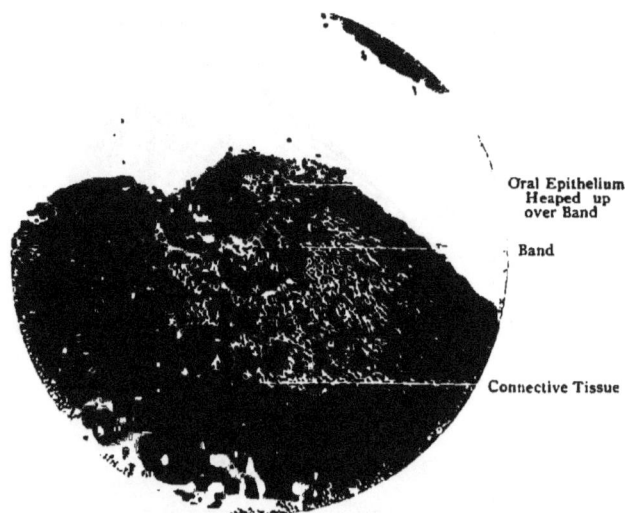

Oral Epithelium Heaped up over Band

Band

Connective Tissue

FIG. 249.—VERTICAL SECTION THROUGH JAW OF HUMAN EMBRYO. FORMATION OF TOOTH-BAND, ABOUT SIXTIETH DAY. × 300.

may pass beyond the boundary-line of the dentin and enter the enamel, but in all probability both of these conditions are pathologic. Such a state of affairs could hardly be considered normal when we take into consideration that the dentin and enamel calcify in opposite directions, and that the outer wall of the former is completed before enamel calcification begins.

Development of Enamel.—Preparations for the development of the enamel begins toward the close of the second fetal month, appearing first as a multiplication of the primitive epithelial cells in the form of a continuous linear projection extending somewhat obliquely into the subjacent connective tissue. From this crest, or tooth-band, the germs for the future enamel organs are given off. These primary dental bulbs, as they are called, number one for each tooth to be generated, and coincidently with their appearance an aggregation of closely associated connective-tissue cells make their appearance in the surrounding submucous tissue. This papilla-like specialization of the submucous tissue is the primitive dentin germ, or dentin papilla. It will thus be observed that the enamel is a product

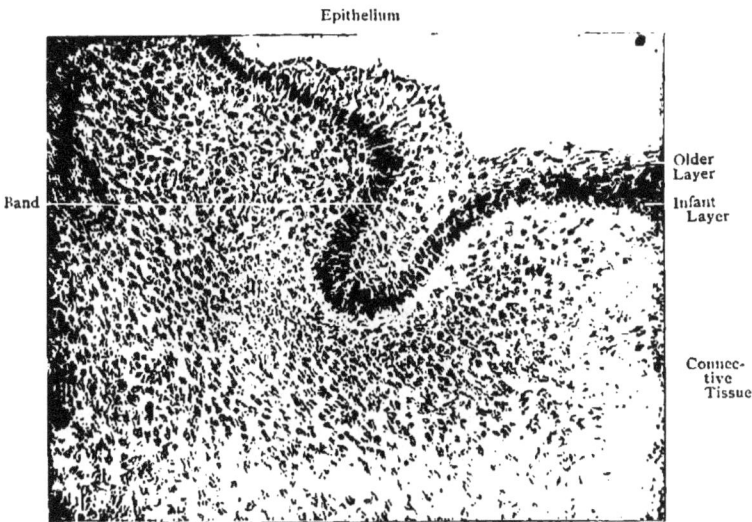

FIG. 250.—VERTICAL SECTION, TOOTH-BAND, HUMAN EMBRYO. TENTH WEEK.
× 300.

of the surface epithelium, while the dentin is generated from the submucous tissues. Soon after the appearance of the club-shaped thickening, or tooth-bulb, by further differentiation its form becomes bell-shaped, with the concavity directed toward the surface. The dentin papilla gradually pushes into the concavity of the forming enamel organ, and at a later period the odontoblastic cells are generated about the periphery of the papilla, closely followed by a surface calcification of the dentin. Soon after the forming of the external layers of dentin the

ameloblastic or enamel-forming cells become active, and a
deposition of enamel prisms takes place upon' the exterior of
the dentin cap.   Before taking up the subject of enamel calcification, brief reference will be made to the further development
of the enamel organ.   As the growth of this organ proceeds,
we find, as the result of a rapid proliferation of the cellular
structure, a marked tendency for the organ to become separated
from the tooth-band.   The peripheral cells are columnar or
prismatic, and remain so, while those in the center, primarily
polygonal, soon become transformed into a radiating network

Dental Ridge

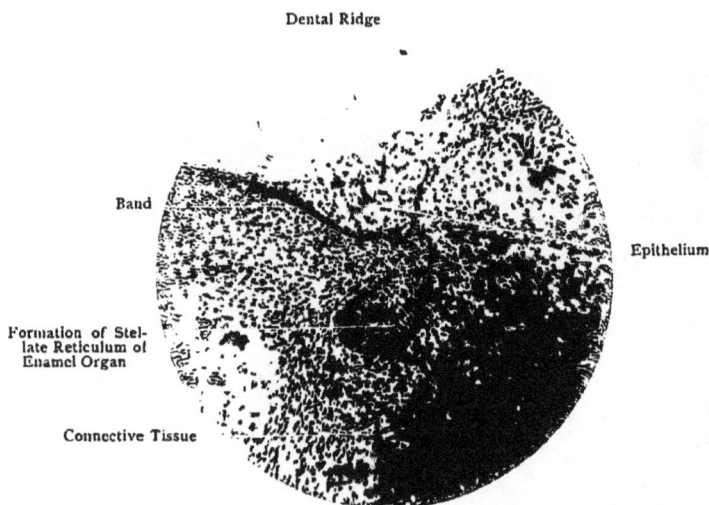

Band

Epithelium

Formation of Stel-
late Reticulum of
Enamel Organ

Connective Tissue

FIG. 251.—SAME AS FIGURE 250.   ABOUT TWELFTH WEEK.   × 200.

or stellate reticulum.   The appearance of a stellate reticulum is
first observed to take place in the cells occupying the central
portion of the enamel-organ, this cellular transformation progressing from the center outward, but ceasing before reaching
the columnar surface cells contiguous to the dentin papilla.
Between the surface, columnar, enamel cells and the stellate
reticulum is a layer of unaltered cells—the "stratum intermedium."   In the earlier stages of the development of the
enamel organ the peripheral cells are alike, being columnar or

prismatic, but almost coincident with the appearance of the
dentin papilla the cells most closely related to it are observed
to become elongated, and form the internal epithelium of the
enamel organ. As the cells forming this internal epithelium
become elongated, their nuclei, instead of occupying the center
of the protoplasmic body, are carried to their extremities. It
will thus be seen that the completed enamel organ consists of
four divisions or layers. Beginning with its convex surface is
an external epithelium or outer tunic, successively followed, in
passing toward the dentin papilla, by a stellate reticulum,

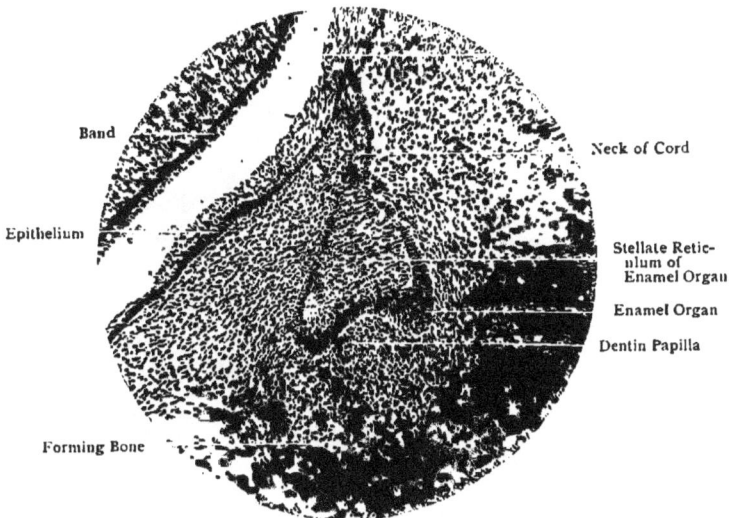

FIG. 252.—VERTICAL SECTION THROUGH TOOTH-BAND, CORD, DENTIN PAPILLA, AND
ENAMEL ORGAN, ABOUT FIFTEENTH WEEK. × 150.

stratum intermedium, and an internal epithelium or inner tunic.
As the growth of the enamel organ proceeds, the tooth-band becomes smaller and smaller in size, until finally a complete rupture
takes place. This rupture, however, does not occur until the
enamel organ has about or fully completed its development, and,
after remaining so long under the influence of the oral epithelium, must be considered, as before stated, an epithelial structure.
It is through the agency of the internal epithelial cells of the

enamel organ, the *enamel cells* or *ameloblasts*, that calcification of the enamel takes place, and that subject will next be considered.

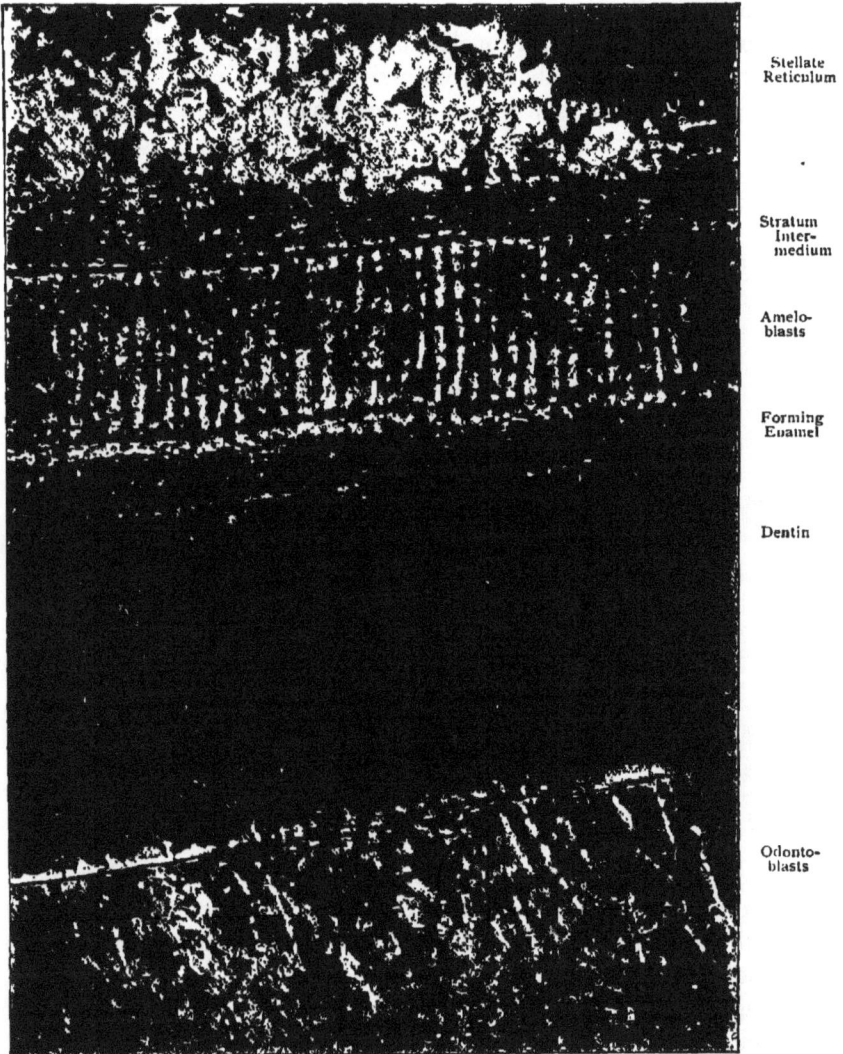

Stellate Reticulum

Stratum Inter- medium

Amelo- blasts

Forming Enamel

Dentin

Odonto- blasts

FIG. 253.—SECTION DEVELOPING TOOTH OF DOG. × 800.—(*After Williams.*)

Amelification (Fig. 253).—Two theories are advanced in regard to the calcification of the enamel. In one it is claimed that the ameloblasts or enamel cells become directly calcified or converted into enamel; in the other, the ameloblasts are simply considered as controlling agents, by secreting or depositing the calcium salts which form the enamel prisms. In the latter theory it is generally believed that the enamel is secreted or shed out from the extremities of the ameloblasts, thus being

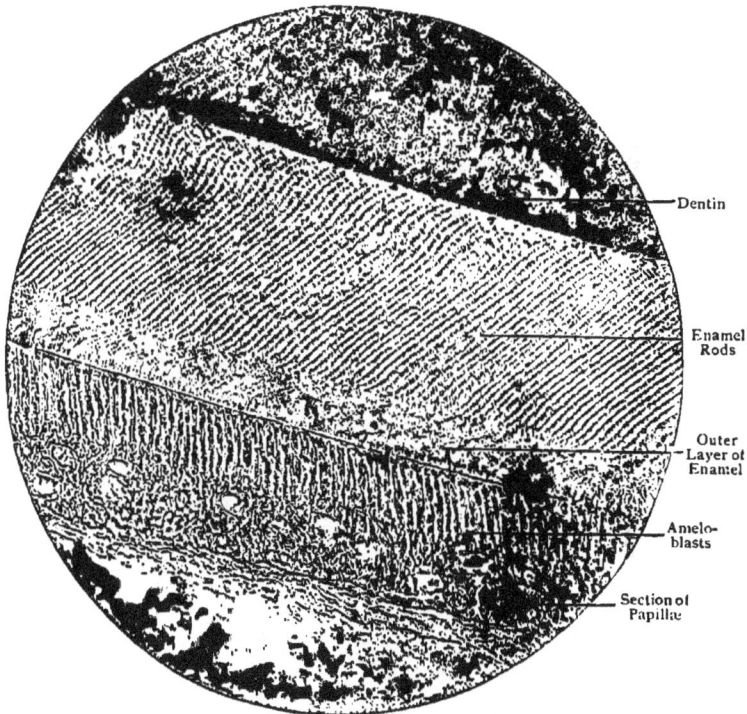

FIG. 254.—SECTION OF INCISOR OF RAT. ✕ 200.—(*After Williams.*)

productive of enamel fibers corresponding in size and position to the secreting cells. In the recent investigations by Williams he takes exception to this latter view, substantiating his opinion by stating that while the ameloblasts of many animals are similar in shape and arrangement, the enamel produced from these similarly arranged cells varies greatly in structure. The same writer also states that, when such a similarity of arrangement exists between the ameloblasts and the enamel prisms, it occurs near the commencement of enamel calcification, and that at a later period the relative position of the ameloblasts and the prisms is always in longitudinal section. Dr. Williams calls attention to the fact of the enamel prisms or rods not ex-

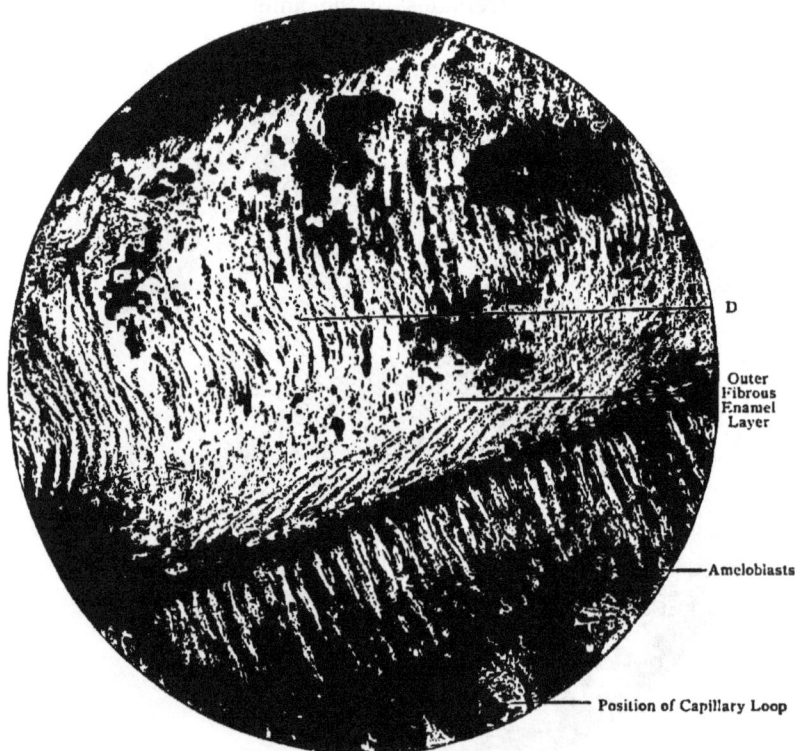

FIG. 255.—SECTION OF INCISOR OF RAT, SHOWING PARTIAL DECALCIFICATION OF ENAMEL. × 600.—(*After Williams.*)

tending through the entire distance between the enamel cells and the calcified dentin (Fig. 254). That part of the structure lying between the ameloblasts and the extremities of the enamel rods is made up of a double set of fibers, some of which are almost at right angles with the long axis of the ameloblasts. In figure 255 D the two sets of fibers previously mentioned are found to join and become closely interwoven.

The ameloblasts are connected with the cells of the stratum intermedium, and more recent investigation goes to prove that this latter structure is directly interested in furnishing to the ameloblasts the proper material for the calcifying process. The stratum intermedium can not, however, take part in the primary enamel calcification, as this process commences before the stratum intermedium is fully developed. This being the case, it is generally supposed that the stellate reticulum furnishes the material for the upbuilding of the first enamel prisms.

Lying between the free extremities of the ameloblasts and the enamel in course of formation is what has been generally considered as a structureless basement membrane or membrana præformativa. The existence, exact location, and structure of this membrane has been and still remains a matter of conjecture. The generally accepted theory appears to be that given above, but Williams refers to it as a layer of newly formed enamel, and does not consider it as a structureless membrane. The structure is to be observed in figure 256, at either extremity of the ameloblastic cells, this writer claiming that the so-called structureless membrane is present at both of these points.

As to the formation of the enamel rods, Dr. Andrews, in 1894, referring to the presence of calcoglobulin in the enamel cells, considered these refractive bodies as calcospherites, which, after being taken up by the ameloblasts, were excreted by them, and after coalescing formed globules of larger size, from which the rods were built up. Williams partly agrees with this statement, but he is of the opinion that the calcospherites coalesce while in the ameloblasts, forming large, spheric bodies, but that the deposit of this substance is in no way productive of building the enamel rods. The theory of Tomes in regard to the forming of enamel rods was that the walls of the ameloblasts themselves became calcified, while the contents of the cells also became solidified, the first forming the interprismatic substance, while the second became the enamel rod. Whatever theory be accepted as to the formation of the enamel rods or prisms, there appears to be no question in regard to the general process of enamel calcification. In the first place, an organic matrix is formed, into which the first-formed layer of enamel is deposited. Gradually the organic matter disappears, leaving behind the inorganic elements closely resembling in appearance the organic matrix, which it has by atomic change supplanted. The question of an organic interprismatic cement-substance is also one upon which various writers disagree. Klein partly believes that such a substance does exist, basing his

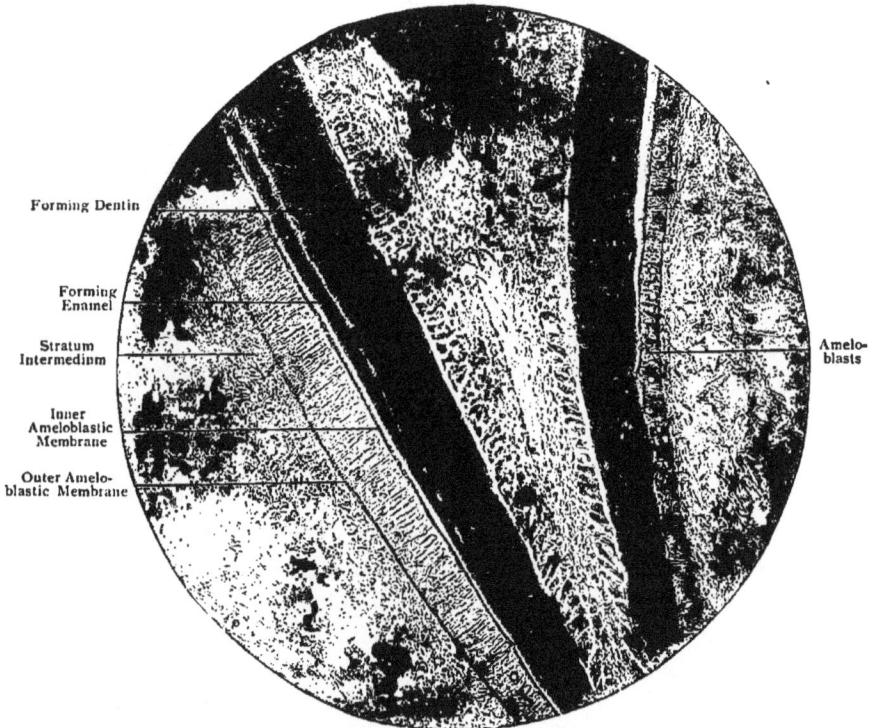

FIG. 256.—SECTION OF DEVELOPING TOOTH OF EMBRYO LAMB. × 150.—(*After Williams.*)

opinions upon the fact that the ameloblastic cells, in common with all epithelial cells, are separated from one another by a homogeneous intercellular substance, and that a certain propor-tion of this organic substance must remain between the enamel prisms after calcification. Dr. Sudduth, by a series of experi-ments made some years ago, appeared at that time to have furnished conclusive proof that an organic, interprismatic cement-substance does not exist between the enamel prisms of fully developed enamel. By the use of a dilute solution of chromic acid, the action of which is the preservation of organic substance, the prisms were liberated, which would not have been the result had they been cemented by an organic cement-substance. By substituting dilute muriatic acid, the action of which

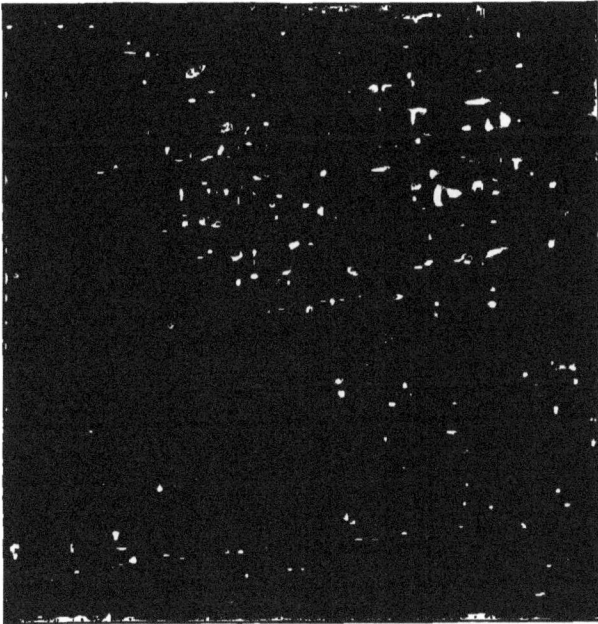

FIG. 257.—SECTION OF ENAMEL FROM HUMAN TOOTH (DARK GROUND ILLUMINATION). × 2000.—(*After Williams.*)

is the destruction of organized tissue, the prisms were not liberated, the acid acting evenly upon the whole mass of enamel, and finally resulting in its complete destruction, not leaving the slightest trace of an organic matrix behind. Dr. Williams furnishes additional evidence in figure 257 of the existence of a transparent, inorganic cement-substance between the enamel rods. This transparent cement-substance, he claims, is formed by the distribution of a translucent liquid substance about the previously formed pattern for the enamel rods. This pattern, generated through the activity of the enamel cells, is composed of a translucent material somewhat more solid than that structure which surrounds it. These two substances calcify together, the latter forming the enamel prisms, while the former makes the cement or interprismatic substance.

Dentin (Fig. 258).—This tissue, which constitutes the principal bulk of the hard part of the tooth, forms a complete cap-like investment over the pulp, from which it is generated. It is

FIG. 258.—SECTION THROUGH CROWN OF HUMAN CUSPID. ✕ 30.

white or slightly yellowish white in color, somewhat elastic, and a trifle harder than bone, which it resembles in many of its characteristics. In a perfectly developed tooth no part of the dentin appears upon the surface, that part within the crown being covered by the enamel, while that of the root is inclosed by the cementum. While the thickness of the dentin varies somewhat over the different parts of the tooth, there is a decided disposition to an equal distribution in every direction. Dentin, unlike enamel, consists of an organic matrix—a reticular tissue of fine fibrils richly impregnated with the salts of calcium, in this resembling the matrix of bone. Traversing the matrix are long, fine canals or tubes (Fig. 259),—the *dentinal tubules,*— which pass from the margins of the pulp toward the surface. Immediately surrounding the dentinal tubules the matrix is especially dense, forming a lining or sheath to the tubes, known as the *dentinal sheaths.* Occupying the lumen of the dentinal tubules are solid elastic fibers—the *dentinal fibers.* In the general structure of the dentin we, therefore, have four parts to ex-

FIG. 259.—SECTION THROUGH ROOT OF HUMAN INCISOR, SHOWING MANY DENTIN TUBULES IN TRANSVERSE SECTION. × 200.

amine : First, the matrix ; second, the dentinal tubules ; third, the dentinal sheaths ; and fourth, the dentinal fibers.

The Matrix.—As previously stated, the matrix is composed of organic and inorganic substances, but the proportionate quantity of the organic and the inorganic constituents is so variable that it is impossible to furnish a definite chemic analysis. The relative quantity of organic and inorganic matter is not only variable in the teeth of different individuals, but is continually changing in the teeth of the same individual, the former being present in larger quantities during youth, and gradually diminishing as age advances. From an examination of perfectly dried dentin, von Bibra furnishes the following approximate analyses:

Organic matter (tooth-cartilage), ... 27.61
Fat, ... 0.40
Calcium phosphate and fluorid, ... 66.72
Calcium carbonate, ... 3.36
Magnesium phosphate, ... 1.08
Other salts, ... 0.83

The organic basis of the matrix appears to be structureless and transparent, and, although closely resembling the matrix of bone, is not identical with it. While the matrix is usually structureless, there are instances in which the presence at one time of connective-tissue fibers is indicated.

The Dentinal Tubules (Figs. 260, 261, 262).—Beginning by a

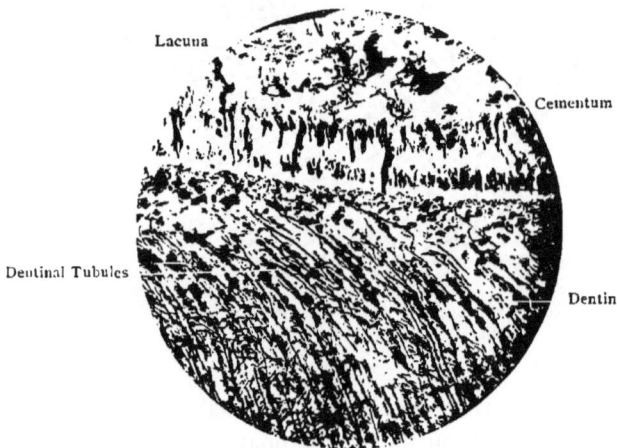

FIG. 260.—DENTIN AND CEMENTUM FROM ROOT OF HUMAN MOLAR.—(*After Geise.*)

free opening about the walls of the pulp-cavity, the dentinal tubules permeate the matrix in all directions. The tubules are generally disposed in a direction perpendicular to the surface, so that in different parts of the tooth they radiate in various directions. Beginning upon the surface of the pulp-cavity, at which point they are of greatest diameter, they pass more or less in a spiral manner toward the surface (Fig. 262), before reaching which they become gradually reduced in size, as a result of the numerous branches which they give off (Fig. 261). The branches given off from the main tubes are quite variable in size, and anastomose with one another or with the branches from other tubules. In the region of the pulp the tubules are so closely associated that but little space is provided for the intertubular substance or matrix; but as the surface is approached, they become more widely separated, and, in consequence, the matrix substance is present in greater abundance. While the general

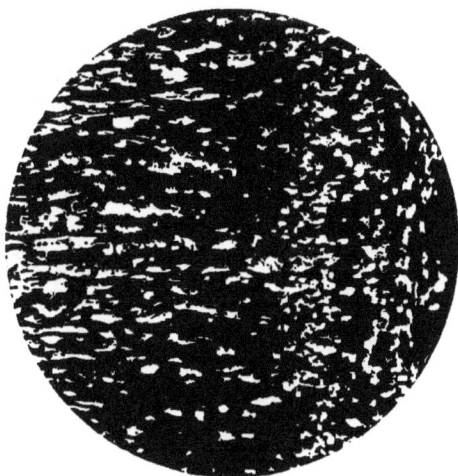

FIG. 261.—LONGITUDINAL SECTION THROUGH ROOT OF HUMAN MOLAR. BRANCHING OF THE DENTINAL TUBULES. × 200.

direction of the tubes is perpendicular, they do not pursue a direct course, but are more or less curved as they pass from within outward. The curvature of the tubuli may be divided into two classes,—long curves and short curves,— usually referred to as the primary and secondary curvatures of the dentinal tubules. The primary curvatures are few in number and are most prominent in the crown, while the secondary curvatures, principally found in the roots, are smaller in size and more numerous. The branches from the main tubes terminate in various ways, either by anastomosis, by gradually fading out into hair-like terminals, or by ending in hooks and loops. In rare instances they are said to enter the substance of the enamel or cementum, but it is doubtful if they do so normally. The branches from a main tube are usually two in number, the latter being almost equal in diameter to the former, and from this first set of branches a number of minute branches are given off almost at right angles. In the crown this latter class of tubules are seldom observed, excepting near the enamel margin, but in the root they are everywhere noticed. Small varicosities

FIG. 262.—TRANSVERSE SECTION THROUGH ROOT OF HUMAN MOLAR, SHOWING THE CURVATURE OF THE DENTIN TUBULES ABOUT THE PULP-CANAL. × 40.

are frequently present, but not in sufficient numbers to produce a striated appearance to the surface of the dentin.

The Dentinal Sheaths.—While the dentinal tubules ramify through the matrix in the form of well-defined channels, the walls of the channels are not formed by the matrix, but by an indestructible substance the exact character of which is not fully understood. The walls of the tubes, or the dentinal sheaths, as they are termed, are believed by some histologists to be calcified, while others, though acknowledging their apparent indestructibility, are doubtful as to the correctness of this theory. Neumann being the first to accurately describe the walls of the tubules, they have become known as "Neumann's sheaths of the dentin tubes." The existence of the dentinal sheaths may best be demonstrated by subjecting the tissue to the action of strong acid for a sufficient time to destroy the intervening matrix, which process usually requires several days. The fibrous mass remaining will be found to contain a collection of tubes, which, however, by careful examination, are found not to be the

FIG. 263.—LONGITUDINAL SECTION THROUGH ROOT OF HUMAN TOOTH, SHOWING SECONDARY CURVATURE OF DENTIN TUBULES. × 40.

dentinal tubules themselves, but the walls of these channels. One authority—Magitot—and, more recently, Sudduth, deny the existence of a wall to the dentin tubes. Tomes, while inclined to the belief that the tubes are provided with definite walls, suggests that they may have been produced artificially during the preparation of the specimen, and that they are only brought into existence by the action of the agents used for this purpose. In conclusion, the same writer adds that that part of the matrix immediately surrounding the fibril differs

in its chemic constituents from those parts containing the body
of the matrix.

The Dentinal Fibers.—Occupying the lumen of each dentin
tube is a soft, elastic fiber, which is continuous with and has its
origin from the odontoblastic cells upon the periphery of the
pulp. The existence of these elongated processes of the odontoblasts of the pulp having first been demonstrated by Tomes,
are otherwise known as Tomes' fibers. By means of these
fibers, which not only fill the lumen of the larger tubes, but the
minute branches as well, the substance of the dentin is both

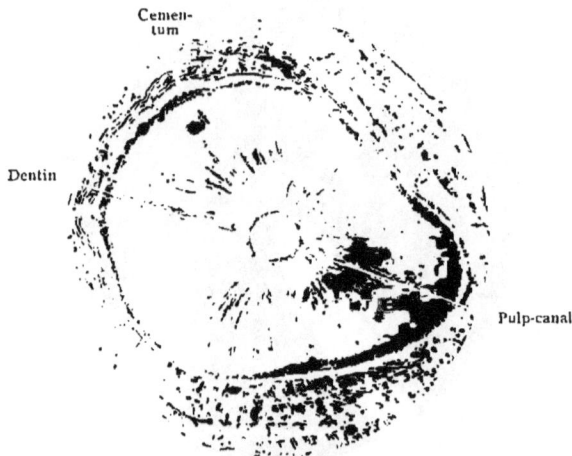

FIG. 264.—TRANSVERSE SECTION THROUGH THE ROOT OF A HUMAN INCISOR, SHOWING
THE DENTIN SURROUNDED BY THE CEMENTUM. × 30.

nourished and rendered extremely sensitive. There is still some
doubt as to the real nature of the fibrils, but, if they are processes from the odontoblasts or cells of the pulp, it would appear
that the substance would be identical with that of the cell-protoplasm. Bodecker claims that they are not round but inclined
to angularity, but Tomes infers that this form has been produced by the action of some reagent. Klein (" Atlas of Histology ") advances the theory that the odontoblasts are active
in the generation of the matrix for the dentin only, and that the

dentinal fibrils are not processes from them, but originate from cells intervening between the odontoblasts and connecting with the dentin tubes. It has never been fully demonstrated that true nerve-fibers enter the dentin along with or in the substance of the dentinal fibril, but, while the evidence is not at present forthcoming, there is but little doubt that the sensitiveness of the dentin is produced by the presence of organized tissue in the tubuli, in the substance of which ramify minute nerve-terminals. The conflicting views of the older writers as

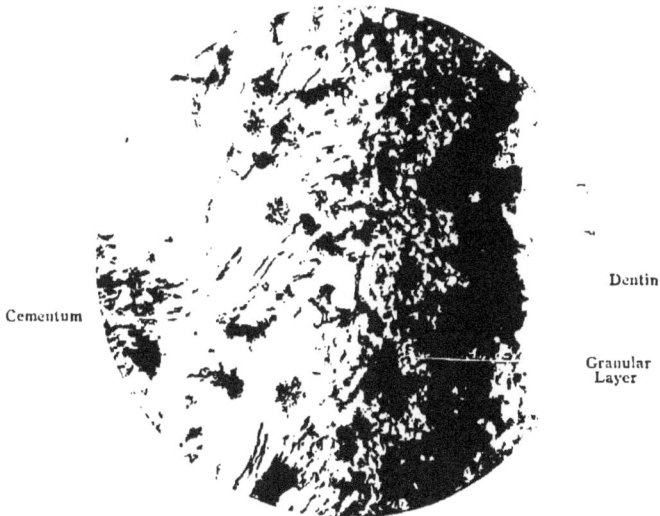

FIG. 265.—TOMES' GRANULAR LAYER. × 40.

to the contents of the dentinal tubuli have been recently commented upon by Dr. J. Leon Williams,* and he has shown that in a certain sense the views of all the investigators previously named are to be given credence. In certain instances he found no fibers given off from the odontoblasts, but from contracted cells lying between the odontoblasts the dentinal fibers were directly observed to proceed. In other instances

* "Items of Interest," vol. xx, No. 5.

the reverse condition would be present, the fibers projecting into the tubules directly from the odontoblasts.

Interglobular Spaces.—In that part of the dentin which immediately underlies the cementum of the root, numerous intercommunicating, irregularly branched spaces are found. These are known as the interglobular spaces (Fig. 265). On account of the granular appearance which this portion of the dentin exhibits under low magnifying power, Tomes has designated it as the "granular layer." The granular layer is also found upon that portion of the dentin which underlies the enamel, but in this region it is far less marked. Many of the dentin tubes have

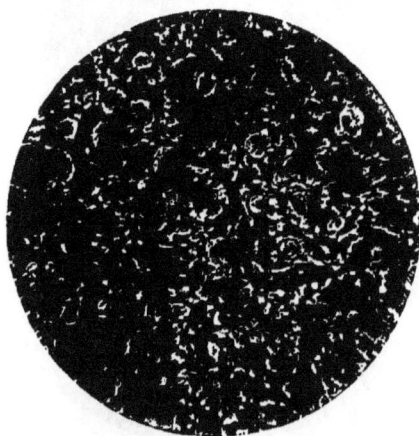

FIG. 266.—SO-CALLED INTERGLOBULAR SPACES IN DRIED SECTION OF DENTIN. X 100.

their endings in these spaces. While the interglobular spaces are most numerous near the peripheral portion of the dentin, they are by no means confined to these parts. They are present in all parts of the dentin, but not so closely associated, and may be observed, when a dried section of dentin is examined, as spaces with irregular outlines and sharp-pointed processes extending in various directions (Fig. 266). The term "interglobular spaces" becomes partly a misnomer when we come to examine the so-called "spaces" more carefully. In normal dentin the "spaces" are filled with a soft, living plasma, having a structural arrangement similar to the general matrix of the dentin, and it is only in a dried specimen that an air-space is found by the shrinking or shriveling of the organic contents. The interglobular spaces forming the granular layer, which are much more numerous, but of smaller size, than those found in the body of the dentin, are also filled with a soft, living plasma, and, according to Bodecker, they communicate, on the one hand, with the dentin fibers in the dentin tubules, and, on the other, with the lacunæ and canaliculi of the cementum. According to Sudduth, the interglobular spaces (so called) are occupied by masses of calcoglobulin which have not become fully calcified.

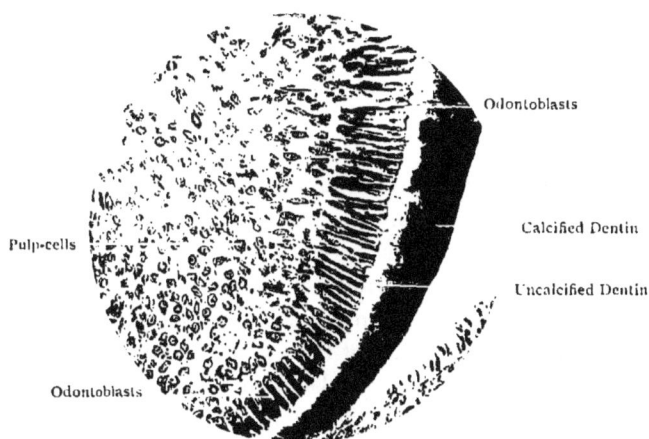

FIG. 267.—PULP AND FORMING DENTIN FROM AN INCISOR TOOTH.—(*After Geist.*)

Dentinification.—The dentin bulb, or papilla from which the dentin is formed, having already been described in part I, the process of calcification will at once be taken up. It will be recalled that calcification of the dentin does not begin until the dentin papilla has developed to the form and size of the dentin of the future tooth-crown. When this has taken place, there is generated upon the surface of the papilla a modified form of connective-tissue cells called odontoblasts (Fig. 267). These cells, which are arranged in a single row upon the exterior of the papilla, vary in form according to their activity. When most active, they are broadest at the extremity directed toward the interior of the papilla. Proceeding from a single odontoblast there may be one or more processes, which are supposed to eventually occupy the tubes of the dentin, as the dentinal fibers. These cells contain an oblong nucleus, which occupies the extremity of the cell most distant from the dentin, but during the period of greatest activity becomes elongated or pointed in the direction of the process. The odontoblastic cells, while actively engaged in the calcifying process, are closely associated or crowded together, but previous to this time there is more or less space between them, which is filled with indifferent tissue. The first layer of dentin being formed upon the surface of the papilla, it will be observed that all additions to its bulk take place from within (the reverse being true of enamel).

As stated elsewhere, calcification of the dentin begins upon the coronal extremities of the crowns, the cutting-edges of the incisors and cuspids, and the summits of the cusps in the cuspidate teeth first receiving their lime-salts. While the odontoblasts undoubtedly superintend the calcifying process, the part taken by these cells appears to be somewhat indefinitely determined. It is generally supposed that the lime-salts are secreted under the superintendency of the odontoblasts. The secretion, however, does not take place around the cells, and in that way completely encapsule them, but around their fibrils. While this is taking place the odontoblasts remain free upon the surface of the pulp, and the fibrils assume their places as the organic dentinal fibrils. As the body of dentin becomes thicker, the odontoblasts are forced to recede, and in so doing the fibers lengthen. The dentinal tubules are, of course, formed in a like manner, the walls of the tube being first calcified from the secretion of lime-salts by the fibrils ; and as the fibrils lengthen by the increasing thickness of dentin and the receding of the odontoblasts, the tubes also lengthen.

Cementum (Fig. 268).—Investing the roots of the teeth is a substance which, both chemically and physically, is closely allied to bone. This external covering is known as the *cementum*, and while generally regarded as being confined to the roots of the teeth, by some it is considered to extend to and completely

invest the crowns during the early part of their existence, in
this latter location being known as the enamel cuticle, or membrane of Nasmyth.

Generally speaking, the cementum begins by a thin margin at
the neck of the tooth or cervical line. It may commence at the
free enamel margin of the crown, or it may slightly overlap this
structure. It is thinnest at the neck of the tooth, and gradually
increases in thickness as the apex of the root is approached.

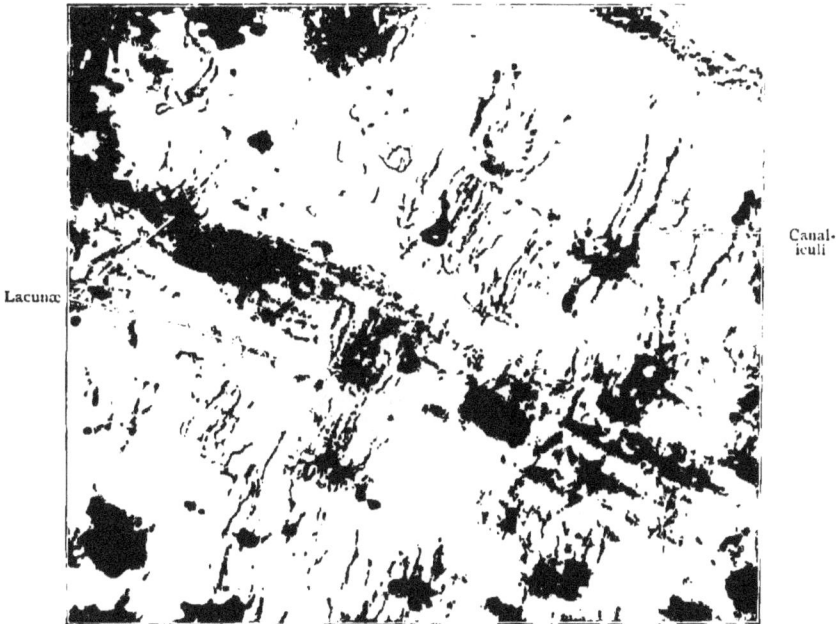

FIG. 208.—CEMENTUM FROM ROOT OF MOLAR. × 200.

In teeth with closely associated roots, the cementum frequently
extends from one root to the other, resulting in a firm, osseous
union. Histologically considered, the structure of cementum,
like ordinary bone, consists of a gelatinous, basal substance,
combined with the salts of lime, and of numerous little hollow
spaces—*lacunœ*. Branching in every direction from the lacunœ
are many minute processes—*canaliculi*.

The Matrix.—The matrix is so nearly identical with that of
bone that it is with difficulty that they can be distinguished. By
decalcification it retains its form and structure, and by the intimate blending of organic and inorganic substances it is provided
with hardness, solidity, and elasticity.   Calcium salts and col-
lagenous fibrils, united by a small amount of cement-substance,
in finer or coarser bundles, compose the ground-substance, or
matrix, of cementum.

Let us first take up the study of this tissue at different periods

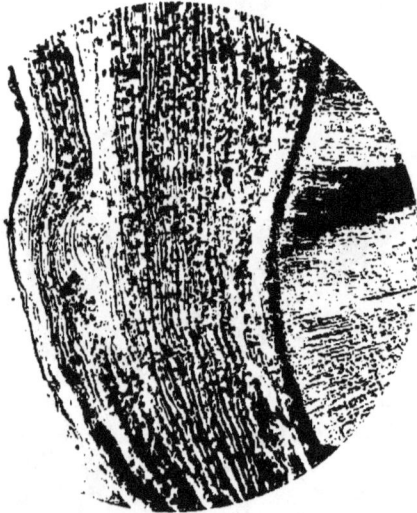

FIG. 269.—LONGITUDINAL SECTION THROUGH ROOT OF HUMAN MOLAR.
INCREMENTAL LINES OF CEMENTUM. × 30.

of its existence, and in this manner learn of its character, its
mode of development, and the changes which take place as its
growth proceeds. The striated markings of the tissue have led
to the belief that there are, during the process of cementification,
periods of activity and periods of rest or little activity. An
examination of the structure under low power (Fig. 269) shows
the incremental lines placed, with more or less regularity, one
beyond the other, and when thus studied adds much to the
strength of the theory of interrupted development.

Figure 270 is prepared from a developing deciduous incisor three months after birth. At this period the developing organ is made up of enamel and dentin alone, the process of cementification not yet being under way. The establishment of the dentinal periphery, which surface is unchangeable, provides a basis for the first layer of cementum generated by the cemento-blasts, which at this period are forming about the inner wall of the tooth-follicle. In close proximity to the surface the inter-globular spaces are observed somewhat widely distributed, and

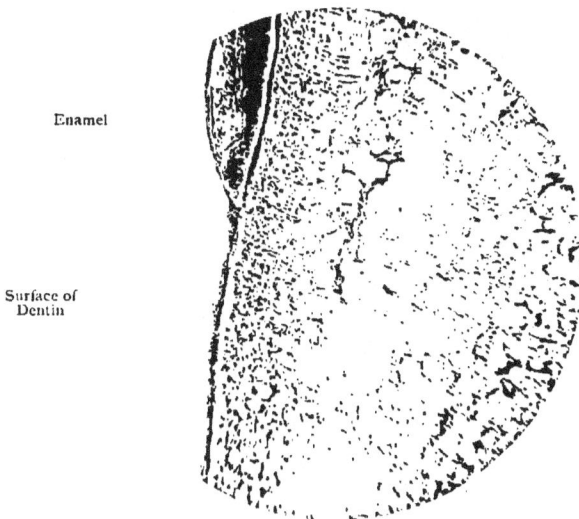

Enamel

Surface of
Dentin

FIG. 270.—SECTION THROUGH DEVELOPING INCISOR, THREE MONTHS AFTER BIRTH.
× 30.

proportionately large in size, resulting in a surface poorly calci-fied and forming a ready attachment for the cemental tissue. Figure 271 shows the process of cementification under way, the section being prepared from a six-month-old tooth. In an ex-amination of the ground-substance of this developing tissue there is an unbroken granular appearance, possessing neither stria-tions, fibers, nor cement-corpuscles. This appearance is one which persists in the oldest or first-formed stratum, and is again noticeable in the outermost or youngest stratum. While the

oldest stratum or strata retain this primary character, this can
not be said of those subsequently laid upon it, for they successively develop in their matrix the partially calcified cells and fibers
from the formative tissue.

Figure 272, taken from a one-year-old tooth, shows a further
advance in the process of cementification. Many of the transverse fibers. of the peridental membrane are observed penetrating the developing tissue, and will, at a later period, by their
partial calcification, become a part of its substance. Already

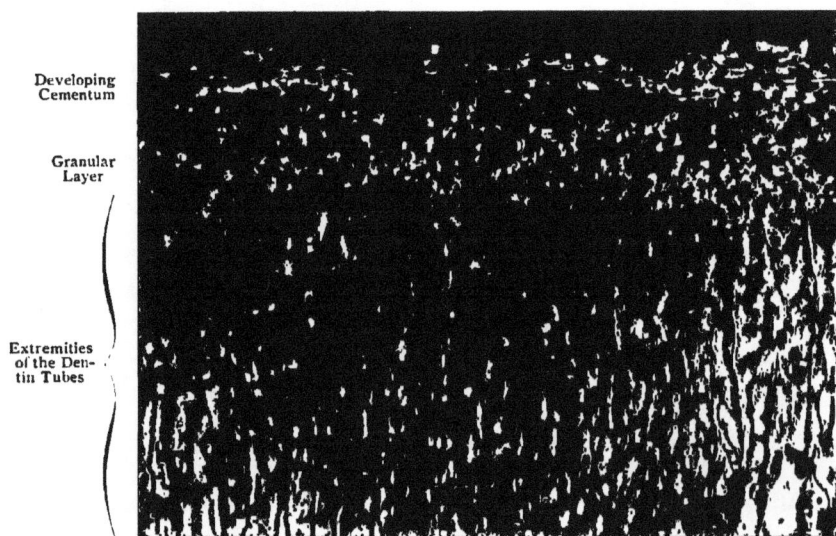

FIG. 271.—DEVELOPING CEMENTUM, FROM SIX-MONTH-OLD TOOTH. ✕ 200.

there has been established an intimate blending of the cemental
tissues with the dentinal tissues through the medium of the
granular layer, and by the further calcification of the latter this
union gradually becomes more thorough. Figure 273 illustrates three distinct zones of developing cementum : the older,
unbroken granular zone at A, now beautifully cemented to
the granular layer; a second or intermediate zone, B, having
encapsuled within its ground-substance many of its formative
cells; and an outer zone, C, but recently laid down, showing

numerous longitudinal, wave-like striations, emblematic of the cementoblastic activity. In this outer zone the minute laminations

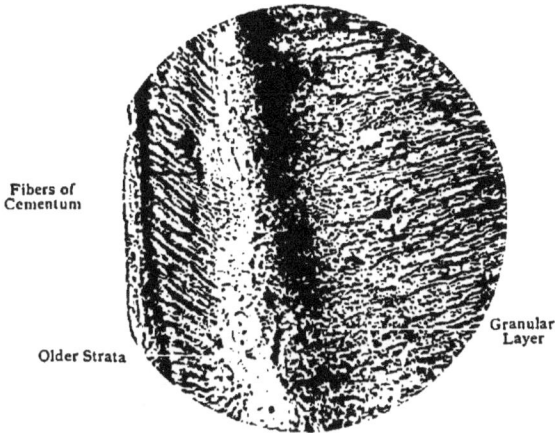

FIG. 272.—SECTION THROUGH ONE-YEAR-OLD TOOTH. ✕ 60.

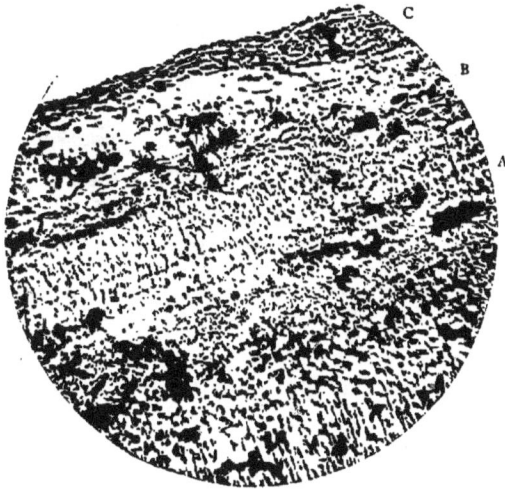

FIG. 273.—DEVELOPING CEMENTUM, FROM TRANSVERSE SECTION OF BICUSPID. ✕ 100.

disappear as the tissue becomes more thoroughly calcified and the matrix gradually partakes of the nature of the older tissue.

The position occupied by the cementum on the root has much to do with its character. In the region of the cervix the cement-corpuscles are few in number, and when present possess extremely short and irregular processes. In the region of the apex the structure is much more complex in character, longitudinal striæ, transverse fibers, cement-corpuscles, and zones of apparently unbroken granular matrix all serving to this end.

To continue the study of this tissue let us examine in detail the lamellæ, the cement-corpuscles, and the cement-fibers.

The Lamellæ.—We are told that the lamellæ are about the

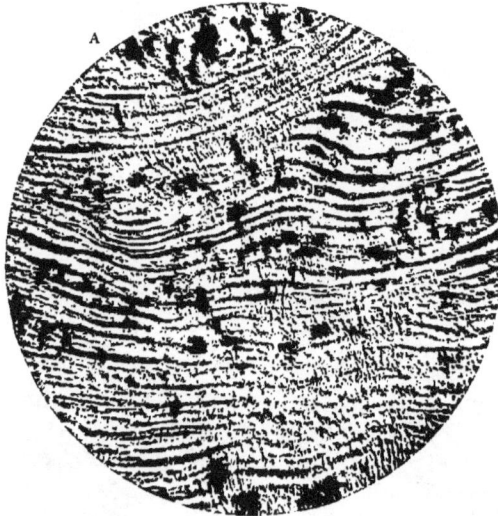

FIG. 274.—TRANSVERSE SECTION FROM ROOT OF BICUSPID, SHOWING VARIATION IN THE DISPOSITION OF THE LAMELLÆ. × 40.

same in number over all parts of the tooth-root, but that they are much thinner at the neck than at the apex. In addition to this they are usually considered as running parallel to, or nearly parallel to, the surface of the dentin. While these statements might, and probably do, describe the disposition of the lamellæ in young cementum, they do not apply with so much certainty to the conditions after the adult period. The lamellæ in the region of the apex are not only of greater width, but are usually greater in number than those occupying the cervix of the same root.

Figure 274 is prepared from a transverse section of an adult bicuspid in the region of the apex, and shows how the disposition of the lamellæ may vary in thin, normal cementum. At A, which represents the granular union of the cementum with the dentin, the incremental lines are observed to follow the surface of the dentin. As the center of the area is approached this regularity is much interfered with, some of the lamellæ being discontinued, others greatly thickened, while the field,

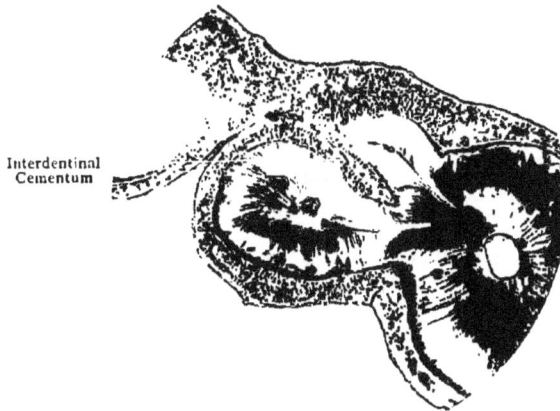

Interdentinal
Cementum

FIG. 275.—TRANSVERSE SECTION THROUGH FUSED ROOTS OF MOLAR TOOTH, SHOWING INTERDENTINAL CEMENTUM. X 30.

taken in its entirety, exhibits anything but regularity in the laying down of the different strata. This same condition may be observed in longitudinal section. While the lamellæ are usually characteristic of the cemental tissues in general, they are seldom found in interdentinal cementum, or that growth which takes place between roots, resulting in their fusion (Fig. 275). This, of course, refers to the tissue as formed between closely associated roots of an individual tooth, and not to that union which sometimes takes place between the roots of different teeth. The interdentinal tissue previously referred to appears to have many characteristics common to itself; thus, the cement-corpuscles are peculiar in form, fibers are few in number, and, as before stated, the lamellæ are not decided.

Cement-corpuscles.—Many of the cementoblasts of the peridental membrane, like the osteoblasts of the periosteum, become encapsuled within the developing tissue, and persist as irregularly shaped spaces, filled with a protoplasmic mass, and are known as cement-corpuscles. These correspond to the lacunæ

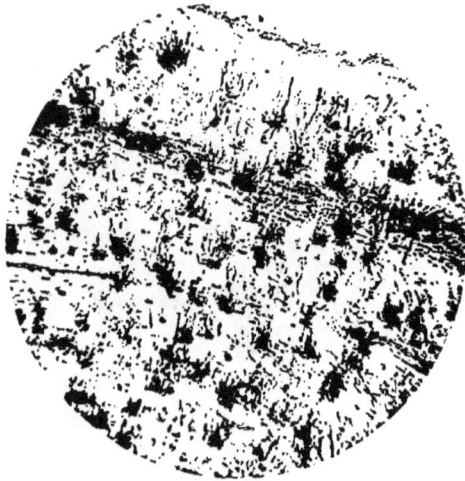

FIG. 276.—CEMENT-CORPUSCLES OF OUTER OR YOUNGER STRATA. × 40.

of bone, but, unlike these, are very variable in size, in form, and in the number and direction of their processes. Figure 268 shows a number of cemental lacunæ and canaliculi. In the majority of instances the body of the corpuscle will be found to be oval or slightly oblong, with its long axis parallel to the surface; but it is by no means uncommon to find them very irregular in outline, with the greatest diameter in the opposite direction. The processes are quite variable in length and irregular in their course, and, while there is a general disposition for them to extend toward the surface, they in many instances radiate in various directions. All of these features are in contradistinction to the lacunæ and canaliculi of bone, which are placed with much more regularity in the osseous matrix, the corpuscles being oblong or cylindric in outline, with their processes about equally distributed in every direction, and uniting directly and positively with the canaliculi of neighboring lacunæ. As previously stated, the cement-corpuscles are very variable in outline, this difference in form appearing to be much influenced by the part of the tooth examined. The younger corpuscles (Fig. 276), or those associated with the outer strata, are usually distinctly outlined and provided with delicate processes, the majority of which are

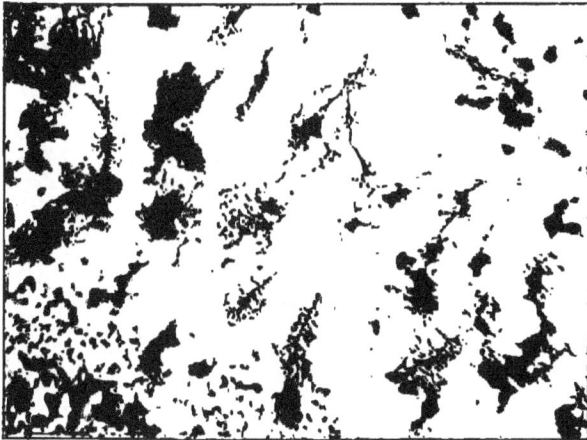

FIG. 277.—CEMENT-CORPUSCLES COMMON TO INTERDENTINAL CEMENTUM. × 100.

directed toward the surface. In the older strata the outlines of the corpuscles are much more irregular, the processes short and extremely clumsy.

The proportionate distribution of the corpuscles to the various parts of the tooth-root is as follows: The innermost or oldest zone and the outermost or youngest zones contain but few; in the intervening strata they are most abundant, especially in the region of the apex, becoming less numerous in passing crown-ward. In interdentinal cementum the corpuscles are somewhat regularly distributed throughout the ground-substance adjacent to the granular layer, but near the center of this confused mass of imperfectly calcified tissue they are seldom present. When the interdentinal space is slight, peculiarly formed corpuscles are often observed (Fig. 277), provided with a long, rod-like, central portion or trunk, from which are given off numerous tree-like branches, the terminals of which are frequently lost in the granular layer upon either side.

Cement-fibers.—In a manner similar to that in which the cementoblasts become encapsuled within the developing cemental tissue forming the cement-corpuscles, many of the fibers

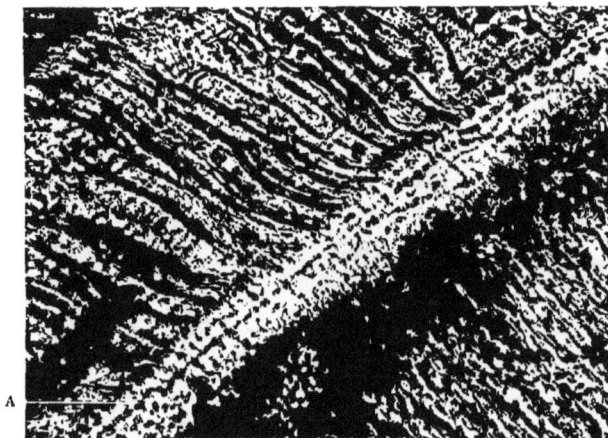

FIG. 278.—TRANSVERSE SECTION THROUGH ROOT OF MOLAR, SHOWING CEMENTAL FIBERS. × 300.

of the peridental membrane undergo a like transformation, and are found in the tissue as more or less imperfectly calcified fibers transversely disposed. By many writers these filamentary, thread-like structures have been compared to the delicate, net-like processes which pass through the concentric lamellæ of bone, serving to hold them together and designated as Sharpey's fibers; but, according to Black, these are the principal fibers of the alveolodental periosteum, and, as already stated, become a part of the cemental tissue during its evolution. In figure 278 the fibers are shown under high power; A represents the

primary or older stratum of the tissue, and it is from the outer
margin of this zone that the fibers first make their appearance,
passing more or less directly in the direction of the surface
until the next incremental line is reached, at which point they
gradually disappear, but recur in the succeeding lamellæ. There
is a marked disposition for the fibers of each concentric lamella
to keep within its borders, or, in other words, to become indi-
vidualized; but in many instances they pass through from one
lamella to another, and occasionally extend unbroken through
the entire thickness of the tissue. It occasionally happens that

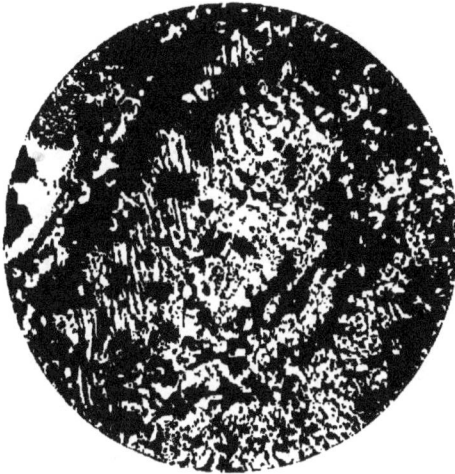

FIG. 279.

the fibers are plentifully distributed to a region comprising three
or four lamellæ, followed by a zone of similar proportions in
which they are entirely absent. The cement-fibers considered,
as the partially calcified residue of the principal fibers of the
peridental membrane, would naturally assume a general direction relative to their manner of distribution before this change
had taken place, and in most instances they are thus disposed.
In figure 279, taken from the center of a long axis of a growing
bicuspid, the disposition of the fibers, which are alone observed
in the second lamella, is slightly crownward. The inclination

for the fibers to be thus disposed is most pronounced in young
cementum, but after middle life, or at a period when the tissue
has greatly increased in thickness, the course of the fibers, even
in the same locality, is greatly at variance.

In figure 272, also from a young tooth, the fibers are shown
springing directly from the peridental membrane, with their
free extremities penetrating this tissue. This illustration is prepared from a transverse section in the cervical region, and the
inclination of the fibers is such as to warrant the belief that
they were some of those whose function it has been to return

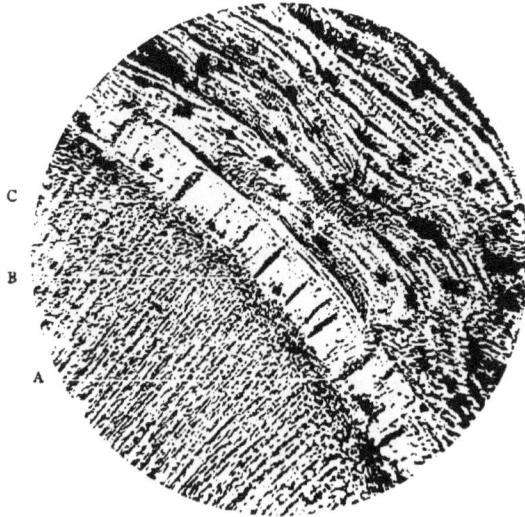

FIG. 280.

the tooth to its normal position when slightly rotated upon itself.
Another class of fibers common to the cement-tissue are those
which appear to be grouped in bundles, springing more or less
regularly, at intervals, from the granular layer and penetrating
the basement layer of the cementum as though serving to tie
this tissue to the periphery of the dentin. In figure 280 a
number of these bundles are shown at A, B, and C. While
the field is but a small proportion of the circumference of the
root, they are observed, under low power, to be distributed in a

like manner to all parts. These circumferential fibers, as they may be called, are also observed in longitudinal section, being distributed with considerable regularity throughout the whole extent of the root. They are also present in the tissue at the earliest period at which its character may be studied, the individual bundles at this time being proportionately larger. These might be, and probably are, considered as prolongations from the dentinal fibers, but it is doubtful if the true fibers of the dentin are ever found penetrating the cementum.

Cementification.—We have seen, in the study of the development of the teeth, that the tooth-generating organs were confined in a closed sac or follicle, and, while the walls of this sac were not directly interested in the calcification of the dentin or enamel, this can not be said of the cementum. Attention has also been directed to the fact that at the time of the eruption of the crown of the tooth a portion of the root only is calcified. As the growth of the root continues, the follicular wall becomes closely adherent to it. Upon the inner face of this vascular membrane a layer of osteoblastic cells (cementoblasts) are generated, and as a result of the calcification of these cells the cementum is formed. It will thus be seen that the process of cementification is but a slightly modified form of subperiosteal bone development. At the beginning of cementum calcification the diameter of the dentin of the root is as great as it will ever be, all additions to its bulk taking place from within. But while the diameter of the dentin is thus fixed, the diameter of the root is increased by the additional layers of cementum as they are deposited upon its surface. As previously stated, a single layer of cementoblasts is first formed in the membrane surrounding the root, these soon becoming inclosed in a spherule of lime. By the time this has taken place another layer makes its appearance, assuming all the characteristics of the first-formed layer. Other layers are formed in turn until the cementum assumes its mature thickness.

The Dental Pulp (Fig. 281).—The tooth-pulp, or formative organ of the dentin, occupies the central or pulp-cavity, and in the fully developed tooth assumes a general outline closely corresponding to the exterior of the organ.* Along with its primary function of generating the dentin, it becomes the medium through which this structure receives its vascular and nervous supply.

Histologically considered, the pulp may be described as a mucus-like, protoplasmic matrix, containing delicate connective-tissue fibers not formed into bundles and numerous nucleated cells, the latter being especially numerous on the periphery of

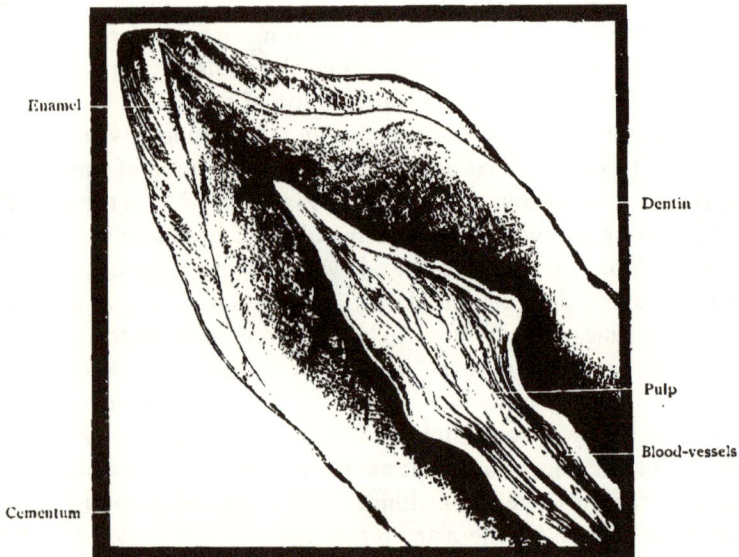

Enamel

Dentin

Pulp

Blood-vessels

Cementum

FIG. 281.— LONGITUDINAL SECTION THROUGH HUMAN CUSPID, SHOWING TOOTH-PULP.—
(*After Geise.*) ✕ 10.

the pulp, or that portion which comes in contact with the dentin. The cells are not closely enough associated to form a complete tissue in themselves, but are found imbedded in a mucoid matrix, with always a definite space between them. In general the cells are elongated or spindle-shaped, with a delicate, hair-like process

---

* The pulp not only occupies the central cavity in the tooth-crown, but the canals of the roots as well; therefore the form of the pulp corresponds to the outline of the pulp-cavity, already described.

attached to either extremity. In the pulp-chamber the cells vary somewhat in outline, in some instances being spheroid, in others appearing as slender filaments, so that the cell proper can scarcely be distinguished from its processes. A third class of cells may be met with, from which three or more filaments are given off. As stated, the distribution of cells varies considerably in different parts of the pulp, this being true not only as regards numbers, but also as to the relations existing between the cells. In the coronal portion of the pulp the position assumed by each individual cell appears to be without

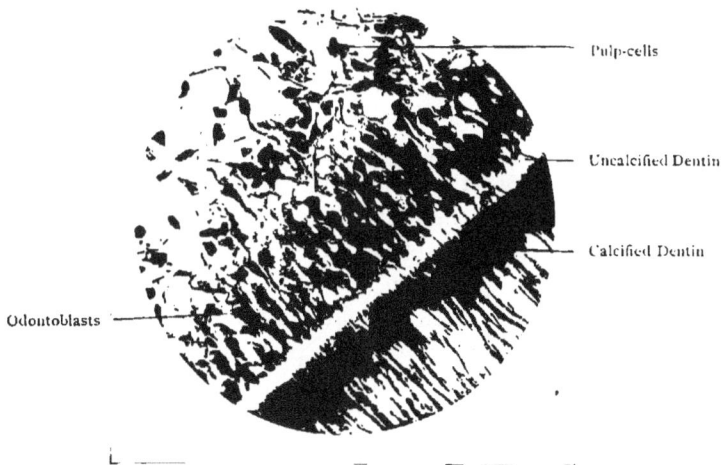

FIG. 282.—SECTION THROUGH PULP AND DENTIN.—(*After Geist.*)

regard to the position of neighboring cells, while in that portion of the pulp occupying the root-canals the cells are arranged parallel with the length of the root. The cells are least in number in the interior of the pulp, but gradually become more plentiful as the periphery is approached.

The Odontoblasts (Fig. 282).—The most active cells of the pulp are those directly on its periphery, in contact with the dentin, and known as the odontoblasts. The odontoblastic layer, otherwise known as the membrana eboris, is composed of a single row of cells, each of which contains, near the extremity most distant from the dentin, a well-defined nucleus. They are large, elongated cells, each furnished with three sets of fibers or processes—the dentinal process, the pulp process, and the lateral processes. The dentinal process or processes—there may be more than one present—communicate with the deeper-lying cells of the pulp, while by means of the lateral processes the cells are brought into communication with neighboring cells. The processes given off in the direction of the dentin, or the dentinal processes, may be one for each cell, in which case they are of considerable size, and are inclined to taper as they enter the substance of the dentin. Again, a single cell may give off a number of smaller processes in this direction. The odonto-blasts vary much in form according to their functional activity. Before the period of dentinification they are spheroid or pyriform, during the period of calcification the dentin extremity becomes somewhat flattened and square, while in advanced years they again return to their primitive, rounded form. Covering the entire surface of the pulp like an epithelium, the odonto-blasts are especially closely associated at the end nearest the dentin, forming an unbroken layer, while the pulpal extremities are inclined to assert their individuality by disassociation.

Blood-vessels of the Pulp.—The pulp is richly supplied with blood-vessels, forming networks extending principally in a direction parallel to the long axis of the tooth, and finally terminate in a capillary plexus closely associated with the odontoblastic layer. The veins of the pulp are ordinarily somewhat larger than the arteries, and form numerous anastomoses. This organ appears to be destitute of lymphatics—at least, none are known to occur in its substance. The blood-vessels of the pulp are provided with a longitudinal layer of thinly distributed muscular fiber, but otherwise the walls of the vessels are noted for their delicacy.

Nerves of the Pulp.—After entering the apical foramen either by one large trunk or by two or more minute ones, the fibers pursue a parallel course, breaking up but little or giving off but few fibers in that portion of the pulp confined to the canal. When the expanded or coronal portion of the pulp is reached, numerous subdivisions occur which are distributed in every direction, and ending in a rich plexus beneath the odontoblastic layer, or membrana eboris. In the body of the pulp the fibers are medullated, but those occupying the periphery are non-medullated and supposed to pass into the dentinal tubes. While this latter hypothesis is in all probability correct, such a distribution of the germinal fibers has never been definitely demonstrated. Two investigators (Ball and Magitot) claim to have partially satisfied themselves in regard to the final distribution of the non-medullated fibers. The former states that he has

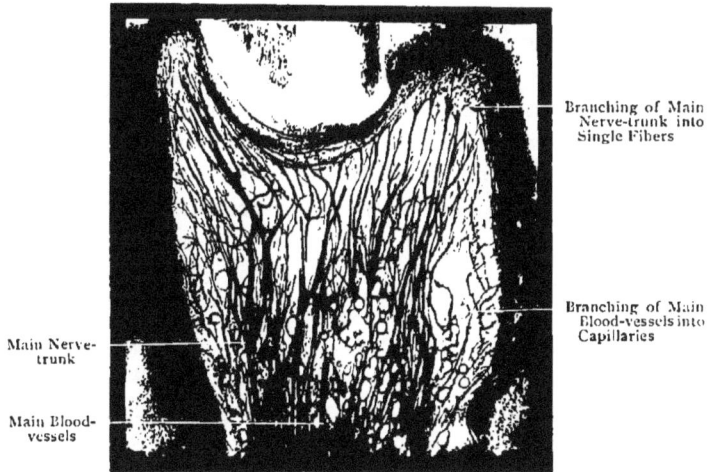

Branching of Main
Nerve-trunk into
Single Fibers

Branching of Main
Blood-vessels into
Capillaries

Main Nerve-
trunk

Main Blood-
vessels

FIG. 283.—DISTRIBUTION OF BLOOD-VESSELS AND NERVES TO THE PULP OF HUMAN MOLAR.—(*After Geist.*) × 20.

traced these fibers into continuity with the larger medullated fibers in the deeper pulp-tissue, and claims to have found them passing through the membrana eboris, beyond which point they assumed a direction parallel to the dentinal tubes. This theory is controverted by Magitot, who claims that the dentinal fibers are, in a measure, themselves prolongations of the nerves, being so constituted through the medium of the branched stellate cells which lie immediately beneath the membrana eboris, and by which the nerves are made continuous.

Nasmyth's Membrane.—Nasmyth's membrane, otherwise known as the enamel cuticle or persistent dentinal capsule, is an exceedingly thin and peculiarly indestructible structure, entirely covering the enamel crown of a tooth. As to the presence of this membrane, which can alone be demonstrated by chemic detachment, there appears no doubt, but in regard to its origin and definite structure much difference of opinion has been expressed. By some writers (Tomes and Magitot) it is maintained that it is continuous with, and similar in structure to, the cementum covering the root, being an extension of the outermost layer in the region of the neck of the tooth; and, in view of the fact that lacunæ are found in its substance, this theory would appear to be correct. On the other hand, it is considered to be a product of the epithelium (Huxley and Kölliker) and in no manner connected with the cementum. In the opinion of the author, it would be difficult to understand how the theory advanced by Tomes could be accepted. During the entire period of saccular development the crown of the tooth is in close relationship to the enamel organ, this structure intervening between the forming enamel and the wall of the tooth-sac, from which the cementum is developed. It would, therefore, appear that this membrane is generated from the epithelium by a change in the character and form of the ameloblastic cells after having completed their function of enamel calcification. Sudduth attributes its formation to a metamorphosis of the ameloblastic layer, the prismatic cells assuming a horizontal direction. The ameloblasts are observed to be prismatic in form up to the point at which the enamel prisms are yet unfinished, but as the surface is approached they are observed to shorten and widen, and near the gum-margin they assume a longitudinal direction instead of being at right angles to the body of the crown. Mrs. Emily Whitman has devoted much time to the study of the development of mammalian teeth, and appears to be of the opinion that the cuticula dentis is the result of a change in the form and character of the enamel cells, this metamorphosis taking place either before or after calcification of the underlying tooth-tissues. Nasmyth's membrane shows many characteristics which differ from those of the body of enamel subjacent to it, serving as an indestructible, highly polished surface-capping to the enamel prisms. The indestructible nature of this membrane by reagents would appear to indicate that in structure it is closely akin to the structure lining the dentinal tubules, the lacunæ, etc.

Alveolodental Periosteum (Fig. 284).—As a general description of this membrane has already been given in part I, it alone remains to treat of its histologic character, which may best be accomplished by first referring to the duties

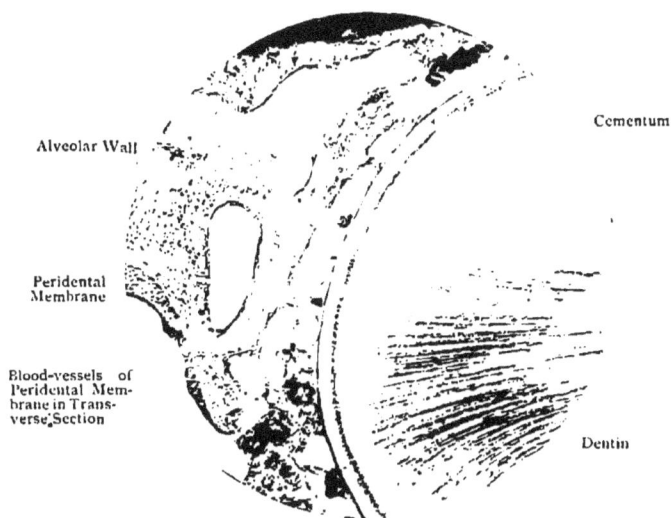

FIG. 284.—TRANSVERSE SECTION THROUGH ROOT OF HUMAN INCISOR AND SURROUNDING ALVEOLAR WALL, WITH PERIDENTAL MEMBRANE INTERVENING. × 40.

which it has to perform. These may be divided into three classes—functional, physical, and sensory. The functional office is accomplished through its cellular elements—the osteoblasts and cementoblasts; the physical office is performed by the fibrous elements, through which the tooth is fixed in its position; and the sensory office through the abundance of nerves, which are richly distributed to all parts of the membrane. We, therefore, find in this structure, besides connective tissue, cells, fibers, nerves, and blood-vessels. The principal cells, as already stated, are the osteoblasts and cementoblasts, but there are also present fibroblasts and osteoblasts.

The osteoblasts, which are instrumental in the building of a portion of the alveolar walls, are found lying against the bone, between the principal fibers. These cells do not appear to be evenly distributed, being numerous and crowded together in some parts, while others will appear to be almost destitute of them. They are most plentiful in the young subject, and seldom present at all in old age. In youth the peridental membrane is thickest, and, as the building of bone occurs on the inner wall of the alveolus, it can only progress as the membrane becomes reduced in thickness. The osteoblasts are polygonal cells, inclining to the oval form, and vary greatly in size, with their longest diameter at right angles with the surface of the forming bone. During the period of the development of the young alveolar wall they are inclined to be crowded together, and are frequently much distorted from pressure upon one another. As age advances this condition becomes less pronounced, and the cells separate into groups.

The Cementoblasts.—Stationed upon the opposite side of the root-membrane, or that in contact with the root of the tooth, are another class of cells—the cementoblasts—or those cells which are concerned in the formation of the matrix and the deposit of the lime-salts from which the cementum is generated. Like the osteoblasts, these cells are found lying between the principal fibers of the root-membrane. These cells differ in form from the osteoblasts, notwithstanding that they have a similar function. Instead of the polygonal form common to the osteoblasts, we find these cells to be more or less flattened, with outlines somewhat irregular. Extending from the body of the cell, which contains a well-defined nucleus, are a number of irregular processes, which penetrate the neighboring fibers or the interfibrous substance. Unlike the osteoblasts, the cementoblasts appear at all times to be evenly distributed over the surface of the cementum, occupying all the space except that taken up by the fibers as they leave the cementum. As to the development of the osteoblasts and cementoblasts, they appear to be carried to the fibrous meshes of the membrane by the blood as leukocytes, or ameboid cells, after which, by differentiation, they become fitted for the development of bone or cementum, and become allied to their respective places, against the surface of one or the other of these structures.

Fibroblasts and Osteoclasts.—Fibroblasts and osteoclasts are also present in the alveolodental periosteum, the former for the purpose of the increase or renewal of the fibrous tissue, the latter being functionally concerned in the removal of a part of the alveolar walls to accommodate the ever-varying position of the teeth, or acting in a similar manner upon the cementum of the root. The osteoclasts, or giant-cells, are generally inclined to the round or oblong form, and usually contain a number of nucleoli. They vary much in size, and are seldom branched or provided with processes. In addition to the four classes of cells already mentioned as being present within the meshes of the fibrous tissue of the root-membrane, there is another class, present, however, during youth only, which appears to be in course of development, and, therefore, without apparent function.

The Fibers of the Alveolodental Periosteum.—The principal fibers of the root-membrane are those which extend from the cementum on one side to the alveolar wall on the other, and firmly fixed at either extremity by penetrating the calcified structures. The fibers are all of the white, or inelastic, connective-tissue variety. It is by means of the connective-tissue fibers that the actual attachment of the membrane both to the bone and to the cementum takes place, the fibers passing directly into the hard tissues, which they traverse for some distance, being here known as Sharpey's fibers.

The arrangement of the fibers is somewhat different over the various parts of the root. In the region of the gingival margin they pass out from the substance of the cementum, retaining their solid form or dividing into fasciculi of finer fibers. In general the bulk of fibers lie parallel with one another, deviating only to give place to blood-vessels and nerves. There is some variation in the distribution of the fibers about the different gingival surfaces. Upon the labial and lingual surfaces they pass out directly into the fibrous tissue of the gum, and soon become lost in this tissue. On the mesial and distal surfaces the fibers passing the lower margin of the alveolar wall join the fibers of the neighboring tooth. This disposition for the fibers to bend toward the adjacent tooth is first observed at the various angles of the gingival margin. All about the free border of the gum the fibers from the peridental membrane assist in forming this tissue, which is covered by a dense epithelial coating of moderate thickness, surrounded or surmounted by the peridental fibers. As the border of the alveolar wall is approached, the fibers are observed to pass under the proper tissues of the gum, and unite with the outer periosteal layer overlying the outer alveolar wall. The fibers immediately within the alveolus are slightly inclined in an apical direction, while those occupying the central portion of the membrane, or that midway between the apex and the gingiva, pass nearly straight across from the cementum to the bone. It is in this locality that the largest and strongest fibers are found. As the apex of the root is approached, the inclination of the fibers is crownward from the cementum to the alveolar wall. In this situation the single fibers are inclined to break up into fasciculi. Immediately surrounding the apex of the root the fibers are irregular during youth, but are disposed more regularly or fan-like in older subjects.

While this account briefly furnishes a description of the distribution of the fibers in various locations, and is in most instances correct, they occasionally vary from this arrangement. While in most respects the fibers of this membrane closely resemble the corresponding fibers of attached periosteum, they possess some peculiarities. It might be supposed that the fibers passing out from the cementum would in some way differ from those springing from the alveolar wall, but, with the exception of being somewhat less in size, they are otherwise of the same character.

Interfibrous Elements.—Besides the various forms of cells, blood-vessels, and nerves, there is present in the peridental membrane an interfibrous tissue. This tissue is principally

•

composed of the fibroblasts belonging to the principal fibers, and other fibroblasts accompanied by delicate fibers which appear to be independently distributed. This interfibrous tissue, which is thus seen to be ordinary fibrous connective tissue, appears to pervade the entire membrane wherever sufficient space is found to permit of its presence. In some parts of the membrane this tissue appears to be more plentiful than the principal fibers themselves. The interfibrous tissue also forms an investment for the blood-vessels and nerves in addition to the tissues properly belonging to their walls.

INDEX.

421

www.ingramcontent.com/pod-product-compliance
Lightning Source LLC
Chambersburg PA
CBHW021348210326
41599CB00011B/791